SMART BREEDING

Molecular Interventions and Advancements for Crop Improvement

SMART BREEDING

Molecular Interventions and Advancements for Crop Improvement

Edited by

Kailash Chandra

Subhash Chand

Ravi Prakash Saini

Rohit Sharma

AAP | APPLE ACADEMIC PRESS

First edition published 2024

Apple Academic Press Inc.
1265 Goldenrod Circle, NE,
Palm Bay, FL 32905 USA

760 Laurentian Drive, Unit 19,
Burlington, ON L7N 0A4, CANADA

CRC Press
2385 NW Executive Center Drive,
Suite 320, Boca Raton FL 33431

4 Park Square, Milton Park,
Abingdon, Oxon, OX14 4RN UK

Library and Archives Canada Cataloguing in Publication

Title: Smart breeding : molecular interventions and advancements for crop improvement / edited by Kailash Chandra, Subhash Chand, Ravi Prakash Saini, Rohit Sharma.
Names: Chandra, Kailash, editor. | Chand, Subhash, editor. | Saini, Ravi Prakash, editor. | Sharma, Rohit (Crop scientist), editor.
Description: Includes bibliographical references and index.
Identifiers: Canadiana (print) 20230441718 | Canadiana (ebook) 20230441777 | ISBN 9781774913314 (hardcover) | ISBN 9781774913321 (paperback) | ISBN 9781003361862 (ebook)
Subjects: LCSH: Plant breeding. | LCSH: Crop improvement.
Classification: LCC SB123 .S63 2023 | DDC 631.53—dc23

Library of Congress Cataloging-in-Publication Data

CIP data on file with US Library of Congress

ISBN: 978-1-77491-331-4 (hbk)
ISBN: 978-1-77491-332-1 (pbk)
ISBN: 978-1-00336-186-2 (ebk)

About the Editors

Kailash Chandra, PhD

Assistant Professor, Plant Breeding and Genetics, Sri Karan Narendra Agriculture University, Jobner, Jaipur, Rajasthan, India

Kailash Chandra, PhD, is currently working as an Assistant Professor in the Department of Plant Breeding and Genetics at Sri Karan Narendra Agriculture University, Jobner, Rajasthan, India. His research mainly focuses on pulse breeding, barley breeding, biotic stress, and molecular breeding. He has published more than 25 research papers in various national and international journals. He has also published book chapters with various reputed publishers, including Elsevier, Springer, Nova, IntechOpen, and CRC Press. He is also serving as a life member of various scientific societies, such as the Indian Society of Genetics and Plant Breeding, Indian Society of Plant Breeders, and the Indian Science Congress Association. Dr. Chandra is the recipient of a Young Scientist Award from the Agricultural Technology Development Society, Uttar Pradesh, India. He has obtained his BSc (Ag.) degree from the University of Agricultural Sciences, Dharwad, Karnataka (ICAR-National Talent Scholarship) and MSc (Ag.) from the University of Agricultural Sciences, Bangalore, Karnataka (ICAR-Junior Research Fellowship). He earned his PhD degree from Banaras Hindu University, Varanasi, Uttar Pradesh (ICAR- Senior Research Fellowship), India. He has also qualified ICAR-NET, CSIR-UGC NET and JRF in life science.

Subhash Chand, PhD
Scientist (Genetics and Plant Breeding),
AICRP on Forage Crops and Utilization
ICAR-Indian Grassland and Fodder Research Institute,
Jhansi, India

Subhash Chand, PhD, has completed his PhD from Division of Genetics, ICAR-Indian Agricultural Research Institute, New Delhi and presently posted at ICAR-Indian Grassland and Fodder Research Institute, Jhansi, India. He has five years of experience in fodder crops, including fodder oat, fodder maize, berseem, lucerne, guinea grass, and BN hybrids. He has published more than fifteen research papers in international and national journals, including *Frontiers in Plant Sciences, Journal of Plant Growth Regulation, Grass and Forage Science, Crop and Pasture Science, Sustainability, Plant Breeding, Indian Journal of Genetics and Plant Breeding,* and others. He has also published several book chapters in books published by Elsevier, Springer, IntechOpen, Nova, CRC Press, Apple Academic Press, and others. He has edited five books on fodder crops published by AICRP on Forage Crops and Utilization, including Database of Forage Crop Varieties: 2020 and Technological Advances in Forage Crop Protection. His specialization areas are genetics and plant breeding, quantitative genetics, disease resistance breeding, and molecular breeding. He was also awarded ICAR-NTS, ICAR-JRF and ICAR-SRF during his academic period. Dr. Chand was graduated from UAS-Dharwad (Karnataka) and earned his postgraduate degree at GBPUA&T-Pantnagar (Uttarakhand), India.

Ravi Prakash Saini, PhD
Scientist (Agricultural Biotechnology),
ICAR-Indian Grassland and Fodder Research Institute,
Jhansi, India

Ravi Prakash Saini, PhD, is posted at the ICAR-Indian Grassland and Fodder Research Institute, Jhansi, as a scientist (agricultural biotechnology), where he has been involved in various research programs on forage and fodder crops. He has published research papers in reputed peer-reviewed international and national journals, such as *Plos One, Journal of Plant Biochemistry and Biotechnology, and Archives of Phytopathology and Plant Protection.* He has published book chapters in books published by CRC Press (Taylor and Francis Group), Nova, and others. His specialization is in tissue culture, molecular cloning, and basic molecular techniques like polymerase chain reaction, northern and southern hybridization, etc. Dr. Saini earned his PhD in Molecular Biology and Biotechnology from the ICAR-Indian Agricultural Research Institute, New Delhi, India, and his PhD on biotic stress tolerance, in which he used host-delivered RNA interference (HD-RNAi) technology as a proof of concept and developed transgenic tobacco lines against *H. armigera*. He graduated from UAS-Dharwad (Karnataka) and earned his postgraduate degree at from ICAR-IARI, New Delhi, India. He was awarded NTS, ICAR-JRF, and CSIR fellowships during his academic period.

Rohit Sharma, PhD
Department of Plants and Crops, Faculty of Bioscience Engineering, Ghent University, Belgium

Rohit Sharma, PhD, is affiliated with the Department of Plants and Crops, Faculty of Bioscience Engineering, Ghent University, Belgium. His research mainly focuses on RNAi in crop protection, and he also has a great interest in various molecular techniques in crop development, including molecular markers and genome editing. He has published multiple research papers in reputed international journals and book chapters, such as with Elsevier, Springer, Nova, IntechOpen, CRC Press, and others. He completed his PhD at Ghent University, Ghent, Belgium. He received his BSc (Ag.) degree from the University of Agricultural Sciences, Dharwad, Karnataka (ICAR-National Talent Scholarship), and MSc in Crop Science from the University of Hohenheim, Stuttgart, Germany.

Contents

Contributors

Sylvester N. T. T. Addy
Legumes and Oil Seed Improvement Department, Council for Scientific and Industrial Research-Crops Research Institute (CSIR-CRI), Fumesua-Kumasi, Ghana

John K. Ahiakpa
Organization of African Academic Doctors (OAAD), Off Kamiti Road, Nairobi, Kenya; Research Desk Consulting Limited, Kwabenya-Accra, Ghana

Aparna
Department of Genetics and Plant Breeding, GBPUAT, Pantnagar, Uttarakhand, India; Departments of Agriculture, Jagan Nath University, Jaipur, Rajasthan, India

Bhabendra Baisakh
Department of Plant Breeding and Genetics, College of Agriculture, Odisha University of Agriculture and Technology, Bhubaneswar, Odisha, India

Javaid A. Bhat
International Genome Center, Jiangsu University, Zhenjiang, China

Subhash Chand
AICRP on Forage Crops and Utilization, ICAR–Indian Grassland and Fodder Research Institute, Jhansi, Uttar Pradesh, India

Rakesh Chaudhary
Department of Genetics and Plant Breeding, Rani Lakshmi Bai Central Agricultural University, Jhansi, Uttar Pradesh, India

Charupriya Chauhan
Department of Genetics and Plant Breeding, GBPUAT, Pantnagar, Uttarakhand, India

Divya Chauhan
Banasthali Vidyapeeth, Jaipur, Rajasthan, India

Sharda Choudhary
Senior Scientist, ICAR–NRC on Seed Spices, Ajmer, Rajasthan, India

Tapas Ranjan Das
Indian Agricultural Research Institute, Regional Station, Pusa, Samastipur, Bihar, India

Manasi Dash
Department of Plant Breeding and Genetics, Odisha University of Agriculture and Technology, Bhubaneswar, Odisha, India

Indra Deo
Department of Genetics and Plant Breeding, GBPUAT, Pantnagar, Uttarakhand, India

Lakshya Goyal
Department of Plant Breeding and Genetics, Punjab Agricultural University, Ludhiana, Punjab, India

Indu
Division of Crop Improvement, ICAR–Indian Grassland and Fodder Research Institute, Jhansi, Uttar Pradesh, India

Anjali Joshi
Forest Genetics and Tree Improvement Division, Arid Forest Research Institute, Jodhpur, Rajasthan, India

Mandakini Kabi
Department of Plant Breeding and Genetics, Faculty of Agriculture, Sri Sri University, Cuttack, Odisha, India

Benjamin Karikari
Department of Agricultural Biotechnology, Faculty of Agriculture, Food and Consumer Sciences, University for Development Studies, Nyankpala Campus, Tamale, Ghana.

Mahipal Singh Kesawat
Department of Plant Breeding and Genetics, Faculty of Agriculture, Sri Sri University, Cuttack, Odisha, India

Sanjeev Kumar
Division of Crop Improvement, ICAR–Indian Institute of Sugarcane Research, Lucknow, Uttar Pradesh, India

Varsha Kumari
Department of Plant Breeding and Genetics, SKN Agriculture University, Jobner, Rajasthan, India

Priyanka Kumawat
Department of Agronomy, SKN Agriculture University, Jobner, Rajasthan, India,

Amar Kant Kushwaha
Division of Crop Improvement and Biotechnology, ICAR–Central Institute of Subtropical Horticulture, Lucknow, Uttar Pradesh, India

Dalpat Lal
Departments of Agriculture, Jagan Nath University, Jaipur, Rajasthan, India

D. B. K. V. Mani
Department of Plant Breeding and Genetics, College of Agriculture, Odisha University of Agriculture and Technology, Bhubaneswar, Odisha, India

Abinash Mishra
Department of Plant Breeding and Genetics, College of Agriculture, Odisha University of Agriculture and Technology, Bhubaneswar, Odisha, India

Ankita Mishra
Department of Plant Breeding and Genetics, College of Agriculture, Odisha University of Agriculture and Technology, Bhubaneswar, Odisha, India

Rajdeep Mundiyara
Department of Plant Breeding and Genetics, Agriculture University, Jodhpur, Rajasthan, India

Sunil Kumar Nair
Department of Genetics and Plant Breeding, College of Agriculture, IGKV, Raipur, Chhattisgarh, India

Gyanisha Nayak
Department of Genetics and Plant Breeding, Indira Gandhi Krishi Vishwavidyalaya, Raipur, Chhattisgarh, India

Himanshu Pandey
P.G. Department of Agriculture Khalsa College Amritsar, Punjab, India

Geeta Prasad
ICAR–National Bureau of Plant Genetic Resources, New Delhi, India

Mohammad Taqi Rabbani
Department of Agronomy and Plant Breeding, Faculty of Agriculture, Balkh University, Mazar I Sharif, Balkh, Afghanistan

Shyam Singh Rajput
Department of Plant Breeding and Genetics, SKN Agriculture University, Jobner, Rajasthan, India

Reena Rani
Division of Plant Improvement and Pest Management, ICAR–Central Arid Zone Research Institute, Jodhpur, Rajasthan, India

K. T. Ravikiran
Regional Research Station, ICAR–Central Soil Salinity Research Institute Lucknow, Uttar Pradesh, India

Kajal Samantara
Institute of Technology, University of Tartu, Tartu, Estonia

S. Mukesh Sankar
Division of Genetics, ICAR–Indian Agricultural Research Institute, New Delhi, India

Archana Sanyal
Scientist, Division of Plant Improvement and Pest Management, ICAR–Central Arid Zone Research Institute, Jodhpur, Rajasthan, India

Mahesh Sharma
Department of Plant Breeding and Genetics, SKN Agriculture University, Jobner, Rajasthan, India

Radheshyam Sharma
Biotechnology Center, Jawaharlal Nehru Krishi Vishwa Vidyalaya, Jabalpur, Madhya Pradesh, India

Aalok Shiv
Division of Crop Improvement, ICAR–Indian Institute of Sugarcane Research, Lucknow, Uttar Pradesh, India

Agnija Sibadatta
Department of Plant Breeding and Genetics, Odisha University of Agriculture and Technology, Bhubaneswar, Odisha

Bhuri Singh
Department of Plant Breeding and Genetics, Agriculture University, Kota, Rajasthan, India

Devendra Singh
Institute of Biosciences and Technology, Shri Ramswaroop Memorial University Barabanki, Uttar Pradesh, India

Gurvinder Singh
Department of Plant Pathology, Dr. Y.S. Parmar University of Horticulture and Forestry, Nauni, Solan, Himachal Pradesh, India

Suman Singh
Departments of Agriculture, Jagan Nath University, Jaipur, Rajasthan, India

Sukhdeep Singh Sivia
Department of Genetics and Plant Breeding, Chaudhary Charan Singh Haryana Agricultural University (CCSHAU), Hisar, Haryana, India

Harsha Somashekar
Plant Cytogenetics Laboratory, National Institute of Genetics (NIG), The Graduate University for Advanced Studies (SOKENDAI), Mishima, Shizuoka Prefecture, Japan

S. L. Soumya
Protection of Plant Varieties and Farmers' Rights Authority (PPVFRA), Ministry of Agriculture and Farmers Welfare, Government of India, New Delhi, India

Sunayana
Plant Breeder, RRS, Abohar, Punjab Agricultural University, Ludhiana, Punjab, India

J. S. Suvadra
Department of Plant Breeding and Genetics, Odisha University of Agriculture and Technology, Bhubaneswar, Odisha, India

S. K. Tripathy
Department of Agricultural Biotechnology, College of Agriculture, Odisha University of Agriculture and Technology, Bhubaneswar, India

Atul Kumar Upadhyay
Department of Biotechnology, Thapar Institute of Engineering and Technology, Patiala, Punjab, India

Abbreviations

4CL	4-coumarate ligase
5-MTHF	5-Methylenetetrahydrofolate
AB	ascochyta blight
ABA	abscisic acid
ABI	applied biosystems
AC	amylose content
AD	activation domain
ADF	acidic detergent fiber
ADL	acidic detergent lignin
ADP	Andean diversity panel
AFLP	amplified fragment length polymorphisms
AGI	above-ground index
AI	artificial intelligence
ALS	angular leaf spot
AM	association mapping
AMDROUT	Asia maize drought tolerance
ANOVA	analysis of variance
ANR	anthocyanidin reductase
AP-PC	arbitrarily primed PCR
AsA	ascorbic acid
Avr	avirulence
BAC	bacterial artificial chromosome
BAP	6-benzylaminopurine
BC	backcross
BCMV	bean common mosaic virus
BI	background-independent
BIL	backcross inbred line
BLB	bacterial leaf blight
BSA	bulk segregant analysis
BSR-Seq	bulked segregant RNA-Seq
BV	breeding value
BWA	burrows-wheeler aligner
Ca	calcium
CAPS	cleaved amplified polymorphic sequences

Cas9	CRISPR-associated protein 9
CBB	common bacterial blight
CC	cellulose content
CDC	crop development center
CDDP	conserved DNA-derived polymorphism
CENH3	centromeric histone 3
CGS	candidate gene sequencing
CIM	composite interval mapping
CLCuKoV	cotton leaf curl Kokhran virus
CLS	cercospora leaf spot
cM	centiMorgan
CMLM	compression MLM
CO	crossing over
CoRAP	conserved region amplification polymorphism
CRISPR	clustered regularly interspaced short palindromic repeats
CRISPRi	CRISPR-interference
CRoPS	complexity reduction of polymorphic sequences
CRSA	crop root system architecture
CSL	chromosome substitution line
CSSLs	chromosome segment substitution lines
CWR	crop wild relatives
DA	drought avoidance
DALP	direct amplification of length polymorphisms
DArT	diversity array technology
ddRAD-Seq	double digest restriction-associated DNA sequencing
DE	drought escape
DEGs	differentially expressed genes
DGN	DNA-guided nuclease
DHs	doubled haploids
DIDS	4,4′-diisothiocyano-2,2′-tilbenedisulfonic acid
DIRT	digital imaging of root traits
DMC	dry matter content
DMC1	disrupted meiotic cDNA 1
DNA	deoxyribonucleic acid
DR	drought recovery
DSBs	double-strand breaks
DT	drought tolerance
DTFF	days to first flowering
ED	enzymatic discoloration

EMS	ethyl methane sulfonate
EPSPS	endolpyruvylshikimate-3-phosphate synthase
E-QTLs	epistatic QTLs
EST	expressed sequence tag
EβF	(E)-β-farnesene
FAO	Food and Agriculture Organization
FE	fiber elongation
Fe	iron
FL	fiber length
FM	fiber micronaire
FMs	functional markers
FP	full-ripe period
FS	fiber strength
FU	fiber uniformity
FW	fusarium wilt
GAB	genomics-assisted breeding
GAPIT	genome association and prediction integrated tool
GBS	genotyping by sequencing
GE	genome editing
GEBVs	genomic estimated breeding values
GEI	genotype-by-environment interaction
GLM	general linear model
GM	genetic modification
GPC	grain protein content
gRNAs	guide RNAs
GR-RSC	genome reduction on restriction site conservation
GS	genetic selection
GT	gelatinization temperature
GTM	gene-targeted markers
GW	grain weight
GWAS	genome wide association studies
HC	hemicellulose content
HDR	homology directed repair
HG	heterodera glycines
HIFs	heterozygous inbred families
HMW-GS	high molecular weight glutenin subunits
HNs	homing meganucleases
HP	half-ripe period
HR	homologous recombination

hsp	heat-shock proteins
Hsp70	heat shock protein 70
HT	high-throughput
HTP	high-throughput phenotyping
ICARDA	International Center for Agricultural Research in the Dry Areas
ICRISAT	International Crops Research Institute for the Semi-Arid Tropics
InDels	insertions or deletions
iPBS	inter-primer binding site
IRAP	inter-retrotransposon amplification polymorphism
ISAP	inter-SINE amplified polymorphism
ISJ	intron splice junction
iSNAP	inter small RNA polymorphism
ISSR	inter-sample sequence repeats
IT	intron-targeting
ITP	intron-targeting polymorphism
LD	linkage disequilibrium
LED	light-emitting diode
LGs	linkage groups
LINEs	long interspersed nuclear elements
LOD	logarithm of odds
LTRs	long terminal repeats
LYC	lycopene content
MAB	marker-assisted breeding
MABC	marker-assisted backcrossing
MAF	minimum allele frequency
MAGIC	multiparent advanced generation intercross
MAGP	marker-assisted gene pyramiding
MAI	marker-assisted introgression
MARB	marker-assisted reverse breeding
MARS	marker-assisted recurrent selection
MAS	marker-assisted selection
Mbp	megabase pair
MDP	Mesoamerican diversity panel
Mg	magnesium
MGS	mapped genome space
MIM	multiple interval mapping
MiRNA FMs	microRNA functional markers

miRNAs	microRNAs
MLM	mixed linear model
MLO	mildew-resistant locus
Mn	manganese
MNs	meganucleases
Mo	molybdenum
MPSS	massively parallel signature sequencing
mQTLs	metabolite QTLs
MS	mass spectrometry
MT	million tons
MTA	marker-trait association
NAM	nested association mapping
NBS	nucleotide binding site
nc-RNAs	non-coding RNAs
NDF	neutral detergent fiber
NGM	next-generation mapping
NGS	next-generation sequencing
NHEJ	non-homologous end joining
NIKS	needle in the k-stack
NIL	near-isogenic line
NIR	near-infrared imaging
NMR	nuclear magnetic resonance
ORFs	open reading frames
PAAP	promoter anchored amplified polymorphism
PAAP-RAPD	promoter anchored amplified polymorphism based on RAPD
PAM	protospacer adjacent motif
PCR	polymerase chain reaction
pegRNA	prime editing guide RNA
PFR	Plant and Food Research Limited
PGRC	Plant Gene Resources of Canada
PGRs	plant genetic resources
PIC	polymorphism information content
PIP	potential intron polymorphism
PLUG	PCR-based landmark unique gene
PM	powdery mildew
PPO	polyphenol oxidase
PPRs	pathogen and pattern resistance receptors
PRI	photochemical reflectance index

PRR	pythium root rot
PS	phenotypic selection
PSbMV	pea seed-borne mosaic virus
PSF	phosphorus fluoride sulfide
PUE	phosphorus use efficiency
PUFA	polyunsaturated fatty acid
QTL	quantitative trait loci
RAPD	random amplified polymorphic DNA
RBMs	RNA-based markers
REMAP	retrotransposon microsatellite amplification polymorphism
RFLP	restriction fragment length polymorphism
RGAP	resistance-gene analog polymorphism
RGB	red-green-blue
RGENs	RNA-guided endonucleases
RGMs	resistance-gene-based markers
RGN	RNA-guided nuclease
RIL	recombinant inbred line
RISC	RNA-induced silencing complex
RMRB	RNAi-mediated reverse breeding
RNAi	RNA-interference
RNA-Seq	RNA sequencing
ROAM	random-open-parent association mapping
ROS	reactive oxygen metabolism
RPG	recurrent parent genome
RRL	reduced representation library
RT	reverse transcriptase
RVDs	repeat variable di-residues
RYMV	rice yellow mottle virus resistance
SA	sodium azide
SAGE	serial analysis of gene expression
SAHN	sequential agglomerative hierarchical non-overlapping
SBS	sequencing by synthesis
SCN	soybean cyst nematode
SE	stably expressed
sgRNA	single guide RNA
sgRNA: Cas9	single guide RNA and Cas9 complex
SHOREmap	SHOrt REad map
SIM	simple interval mapping
SINEs	short interspersed nuclear elements

siRNA	short-interfering RNAs
SLAF-seq	specific-locus amplified fragment sequencing
SMA	single-marker analysis
SMD	sterility mosaic disease
SMS	single molecule sequencing
SNF	symbiotic nitrogen fixation
SNP	single nucleotide polymorphism
SPAD	soil-plant analysis development
SPC	seed protein concentration
SPC	soluble protein content
SRAP	sequence-related amplified polymorphism
SS	starch synthase
SSAP	sequence-specific amplified polymorphism
SSC	soluble solid content
SSCP	single strand conformation polymorphisms
SSD	single-seed decent
SSNs	site-specific nucleases
SSR	simple sequence repeats
SSRs	site-specific recombinases
STMS	sequence-tagged microsatellite site
SY	seed yield
SYMRK	targeting symbiosis receptor-like kinase
TA	titratable acidity
TALENs	transcription activator-like effector nucleases
TBP	tubulin-based polymorphism
TE	transposable element
TFMs	targeted fingerprinting markers
THF	tetrahydrofolate
TILLING	targeting induced local lesions in genomes
TPE	target population of environments
tracrRNA	trans-activating CRISPR RNA
TRAP	targeted region amplified polymorphism
TSC	tuber starch content
TuMV	turnip mosaic virus
UAV	unmanned aerial vehicle
UN	United Nations
USDA	United States Department of Agriculture
VIGS	virus-induced gene silencing
VSC	visual score

WES	whole-exome sequencing
WGRS	whole-genome re-sequencing
WGS	whole genome scanning
WGS	whole genome sequencing
ZFNs	zinc finger nucleases

Foreword - D. K. Yadava

Plant breeding has played a substantial role in meeting the food and nutritional requirements of both economically developing and developed nations. The world's population is increasing day by day and is expected to reach nine billion by the next 25 years. With the unprecedented human population growth, global warming and climate change are extremely affecting global food and nutritional security. Therefore, there is a dire need to breed climate-resilient cultivars in agriculturally important crops and make them more tolerant to prevailing climatic conditions without any yield penalty.

The present edited book *Smart Breeding: Molecular Interventions and Advancements for Crop Improvement,* is well written with the objective to provide comprehensive information about the advancements made in the field of molecular breeding. The content of each chapter is presented in very lucid and easy-to-understand language along with relevant examples.

The contributions of all four editors are commendable. They have done a comprehensive exercise in collecting the content of each and every chapter and in interpreting every topic in a very simple way. The contribution led by the editors of the book also deserves special appreciation for the excellent compilation of all the topics related to modern molecular breeding and molecular biology. I am very sure that this book will be very useful for students, researchers as well as teachers of various disciplines.

I once again congratulate all the authors for their efforts and hard work.

—D. K. Yadava
Assistant Director General (Seed),
Indian Council of Agricultural Research,
Krishi Bhavan, New Delhi, India

Foreword - J. S. Sandhu

Plant breeding has made a tremendous contribution to modifying wild-type genotypes for better performance and increased quality to meet the changing human needs and climatic conditions. In the past, conventional plant breeding led by dedicated plant breeders could meet the global food demand and consumer preferences. Besides policy interventions, increased crop productivity and biofortification could be the best ways to alleviate malnutrition problems in developing nations, mainly in Asian and African countries.

As per the FAO, the global population is expected to reach 9.15 billion by the end of 2050, posing a severe threat to the global food system. In addition, biotic (insects and diseases) and abiotic (drought, heat, cold, salinity, alkalinity, etc.) stresses, land degradation, and urbanization thwart crop productivity and affect global nutritional security under unprecedented climate change and global warming. Therefore, plant breeders must develop genotypes with high yield potential, tolerance to prevailing stresses, and which are nutritionally enriched end products. Thus, conventional breeding approaches only would not be sufficient to achieve the target; therefore, using contemporary cutting-edge biotechnological interventions with previous ones could accelerate breeding cycles per year, provide more target-specific alterations, and have a more precise insertion of target gene's with minimal time and in a highly predictable and accurate manner.

The present edited book *Smart Breeding: Molecular Interventions and Advancements for Crop Improvement*, is designed to provide up-to-date information about molecular markers and their applications, allele mining, QTL mapping approaches, MutMap, and the use of marker-assisted selection for biotic and abiotic stress tolerance in agriculturally important crops. The chapters also highlight the pre-breeding and reverse-breeding approaches and their utility to broaden the pre-existing genetic base of germplasm for better use. Contemporary cutting-edge biotechnological interventions such as genome editing (GE) tools (TALENs and CRISPR/Cas9) and next-generation sequencing (NGS) approaches are also included, and their role in crop improvement has been described elaborately.

I hope this book will be helpful to students, teachers, plant breeders, and research scholars due to its comprehensive subject coverage and in-depth analysis of recently available literature.

The contribution led by Dr. Kailash Chandra from our university is much appreciable, and the other editors, Dr. Subhash Chand (ICAR-IGFRI, Jhansi), Dr. Ravi PrakashSaini (ICAR-IGFRI, Jhansi) and Dr. Rohit Sharma (Ghent University, Belgium), deserve compliments for this excellent book on recent advancements in plant breeding. I am not only confident but also sure that readers from different disciplines will find this book useful for their academic and research purposes.

I once again congratulate all the editors for this particular publication and best wishes for their future endeavors.

—Prof. J. S. Sandhu

Vice-Chancellor,

Sri Karan Narendra Agriculture University, Jobner,

District, Jaipur, Rajasthan, India

Foreword - B. D. Singh

Plant breeding aims to modify the genotypes of plant species to enhance their usefulness to humans. This application of genetic principles has made a substantial contribution to global food and nutritional needs despite many human-accelerated challenges, such as genetic erosion, climate change, global warming, urbanization, land degradation, and other adverse factors. Traditional plant breeding came into the limelight on the global scene when it enabled the '*Green Revolution*' (the 1960s to 1980s) by developing short-duration, fertilizer-responsive, and high-yielding varieties, particularly in wheat and rice. However, the incessantly increasing human population necessitates that we continue to increase global crop production despite the increasingly adverse effects of climate change, especially increasing temperatures and declining precipitation. Conventional breeding methods alone would not be capable of achieving the required increases in food production, and they must be supplemented with the recent advances in biotechnology, genomics, and bioinformatics to reinforce the toolkit of plant breeders for creating novel, useful variation by identification, isolation, and direct manipulation of the target genes involved in crop performance.

The present book has been designed to provide adequate information on the molecular technologies applicable to plant breeding. The chapters included in *Smart Breeding: Molecular Interventions and Advancements for Crop Improvement* cover the recent advances in molecular markers and their applications; the utility of MAS for the improvement of various traits under different stresses, QTL mapping, and their application to improve quantitative traits, and allele mining. The book also highlights pre-breeding, reverse breeding, genome editing (GE), and NGS approaches and tailoring different economic attributes to improve crop productivity in various crops. This book is intended for graduate students, and researchers involved in crop improvement.

I am delighted that my student, Kailash Chandra, an alumnus of Banaras Hindu University, is the lead editor of the present book and has been the key contributor to the conceptualization and development of the present volume. I would like to commend the other editors and appreciate the efforts of all the authors in creating their well-prepared chapters. I hope this book will find ready acceptance as an excellent source material among graduate students and research workers.

—B. D. Singh

Ex-Rector, Banaras Hindu University,
Varanasi, Uttar Pradesh, India

Preface

Recent studies have estimated that the global food demand would be increased by 35% to 56% from 2010 to 2050; however, the expected population at risk would be changed from –91% to +8% during the same period. Besides the unprecedented growth of the human population, other factors such as climate change, increased rate of urbanization, and land degradation are putting pressure on the available natural resources and, ultimately, on the global food system. Therefore, to meet burgeoning global food demand, we must incorporate advanced technologies with conventional breeding approaches to develop or improve crop varieties for various biotic (insect, disease, etc.) and abiotic (heat, drought, cold, salinity, etc.) stresses. Climate-smart and biofortified crop varieties are the basic need of the hour to address the current challenges, and biotechnological approaches can align with existing breeding tools to meet the current requirement.

The chapters included in *Smart Breeding: Molecular Interventions and Advancements for Crop Improvement* provide an enlightening scientific view by nationally and internationally leading scientists on the latest advances in plant science, including recent advances of molecular markers, reducing breeding cycles employing marker-assisted selection, QTL mapping, allele mining, genome editing tools (ZFNs, TALENs, and CRISPR/Cas9), reverse breeding, pre-breeding, and next-generation sequencing approaches tailoring different economic traits to improve crop productivity in various crops for the benefit of humankind.

We are confident that this book will improve the scientific knowledge and understanding of under- and postgraduate students and researchers on cutting-edge technologies. Each chapter has carefully been written scientifically by keeping the end-users need.

Last but not least, we thank the editors and publishing managers of *Apple Academic Press* and contributing authors for their extensive help, support, and patience throughout the process of drafting and publishing this book.

—*Editors*

Introduction

This book *Smart Breeding: Molecular Interventions and Advancements for Crop Improvement*, is a compilation of 14 chapters. Each chapter is focused on novel breeding and biotechnological approaches for crop improvement. The subject area of each chapter has been designed, to be lucid and straightforward, encompassing comprehensive information on each topic. The book will surely help academicians, researchers, and students build strong concepts about molecular breeding and recent advances in plant breeding such as pre-breeding, reverse breeding, allele mining, next-generation sequencing (NGS), etc.

The authors have discussed molecular markers, genesis, and their significant applications for crop improvement in Chapter 1. Molecular markers are classified either into hybridization, PCR-based, or sequencing-based markers and play a vital role in gene isolation, genome sequencing, and understanding evolutionary relationships among related organisms. In Chapter 2, the authors seek your attention towards QTL mapping, a modern breeding approach for mapping quantitative traits. In plants, most yield-related traits are quantitative and governed by several genes (major and minor). In Chapter 3, the authors have explained various breakthrough technologies developed for the improvement of pulses. Pulse molecular breeding and genomics have shown dramatic progress in the last 5–10 years. Advances in 'omics' approaches like genomics, transcriptomics, phenomics, proteomics, metabolomics, and ionomics have enriched understanding of the complex traits in pulses. In Chapters 4 and 5, the authors have narrated the most recent and versatile approaches for precise gene alterations in the genome of organisms by using genome editing (GE). The authors have provided comprehensive information about the application of genome editing tools in various crops for improving desirable traits by citing several examples. In the past decade, genome editing has emerged as a powerful technology for improving desirable traits in organisms.

In Chapter 6, the authors have explained one of the challenges that the entire world currently faces, i.e., climate change. They have discussed the necessity and development of climate-resilient crops that can withstand hostility or adversities of climate without substantially compromising yield. In Chapter 7, the authors discuss reverse breeding, a novel plant breeding technique that allows breeders to develop parental lines directly from the heterozygous plant within a shorter time frame. Reverse breeding produces perfectly complementing homozygous parental lines employing engineered meiosis. In Chapter 8, the authors explain various pre-breeding approaches for biotic and abiotic stress management in crop plants. Pre-breeding allows us to explore an organism's gene pool resources (wild relatives, ancestors, and landraces) and guides us to move desirable genes from such resources into the present-day cultivars. In Chapter 9, the authors discuss the design and importance of MutMap and its variant in modern breeding. MutMap is one of the novel gene mapping approaches that allows rapid identification of causal nucleotide changes of mutants by whole-genome resequencing. Chapter 10, focusses on the allele mining approach, a way to find existing variation in an organism's gene pool by scanning the entire gene (from the promoter region to its transcription termination site, including introns).

Furthermore, in Chapter 11, the authors discuss the most important application of molecular markers in breeding, i.e., marker-assisted selection, where molecular markers have played a significant role in accelerating breeding programs—several examples of marker-assisted selection are quoted in rice and maize crop improvement. In continuation, Chapter 12 also discusses the role of marker-assisted selection in improving qualitative traits in various agriculturally important crops. Marker-assisted selection provides an easy way to search for the desirable genotype in a mapping population and reduces the breeding cycle with more precision and accuracy. In Chapter 13, the authors mainly discuss the recent breeding techniques available for pulse crop improvement. Chapter 14 explains the role of next-generation sequencing (NGS) platforms in crop improvement. The authors summarize the technical know-how of each sequencing platform and its merits and demerits.

We are sure that this book will enhance readers' knowledge about the recent developments in breeding and biotechnological approaches concerning crop improvement.

CHAPTER 1

Insights into Molecular Markers and Applications in the 21st Century

GYANISHA NAYAK,[1] AGNIJA SIBADATTA,[2] J. S. SUVADRA,[2] and MANASI DASH[2]

[1]*Department of Genetics and Plant Breeding, Indira Gandhi Krishi Vishwavidyalaya, Raipur, Chhattisgarh, India*

[2]*Department of Plant Breeding and Genetics, Odisha University of Agriculture and Technology, Bhubaneswar, Odisha, India*

ABSTRACT

The identification and study of genetic diversity in plants can facilitate an improved understanding of the molecular basis of a variety of biological events. Because of the sheer impossibility of sequencing the entire plant kingdom, molecular markers and their association with the phenotypes provide the necessary landmarks for determining genetic variation and heritability. Different categories of molecular markers have been developed, such as RFLP, RAPD, SSR, AFLP, etc., and are in use for the ecological, taxonomical, phylogenetic, evolutionary, and genetic studies pertaining to plant sciences that have outfitted the crop improvement. These techniques have been well adopted, along with their advantages and limitations. Over the past few years, advanced techniques of a new class have emerged from the amalgamation of existing basic techniques to amalgamate the advantages that increase its sensitivity and resolution toward detection of genetic discontinuity and uniqueness. For the construction of conserved DNA and gene family-based makers, length polymorphism of exon-intron structures in various extensively dispersed and regular gene families or plant genes

Smart Breeding: Molecular Interventions and Advancements for Crop Improvement. Kailash Chandra, Subhash Chand, Ravi Prakash Saini, and Rohit Sharma (Eds.)

is used. Transposable elements (TEs) that are abundantly scattered in plant genomes could be utilized to create fingerprints as well (transposable element-based markers). Different gene-targeting techniques can also be used to construct markers using RNA information (RNA-based markers). Another form of marker involves the use of customary features of resistance genes of the plant to detect genetic variation or produce bands associated with a specific phenotype (resistance-gene-based markers). Even though resistance genes might be under specialized evolutionary pressure, conserved DNA-based techniques could have limited genome coverage and may be unable to uncover genetic variability. Particularly when fingerprints are obtained from the expressed or transcribed regions of the genome, such approaches are potent enough to generate phenotypically connected functional markers (FMs) (targeted fingerprinting markers) despite the drawbacks. Hence this chapter deals with a detailed account of advanced molecular markers and their applicability in plant breeding.

1.1 INTRODUCTION

In recent years, we have seen incredible progress in the use of molecular techniques to plant breeding programs, such as genetic engineering and marker-associated breeding. Molecular markers are molecules that can be used to distinguish between various strains of the same species or between distinct species. It is important to understand how a marker is connected to the desired phenotype in plant breeding, although this is irrelevant to genetic diversity or phylogenetic investigations. Attributes such as independent, polymorphic, and stable, able to provide sufficient resolution quickly at a cheap cost is what makes an ideal marker. Many other features may be essential depending on the nature of the study. Markers can be inherited in one of two ways, one being dominant/recessive and the other being co-dominant. A marker is considered to be co-dominant when there can be a separation between the genetic patterns of homozygotes from that of heterozygotes. Co-dominant markers are usually more informative than dominant markers (Boopathi, 2012). Different molecular markers are classified as DNA-hybridization-based [restriction fragment length polymorphism (RFLP) and diversity array technology (DArT)], polymerase chain reaction (PCR) based [random amplified polymorphic DNA (RAPD), amplified fragment length polymorphism (AFLP), inter-sample sequence repeats (ISSR)] and sequence-based markers [single nucleotide polymorphism (SNP)]. RFLP markers are the first generation of DNA markers and co-dominant in nature. Differences between individuals in terms of the size of fragments generated by restriction enzyme digestion of genomic DNA and detected by hybridization with probes are termed as RFLP. An RFLP marker is resulted when a single restriction enzyme creates fragments of varying lengths from the same stretch of genomic DNA from different individuals, strains, or related species. They are a type of Southern-Blotting-based markers. With the advent of PCR technology, hybridization-based techniques have lost their prominence, resulting in an extensively used classification of molecular markers into PCR and non-PCR based. RAPD is a dominant marker system based on PCR. When DNAs are subjected to PCR, utilizing a single oligonucleotide (typically 10 bases long) and random primer, it identifies a genomic region that is amplified in some but not in other individuals of a population. This technique produces high degrees of polymorphism while being simple and easy to use. However, RAPDs are more susceptible to experimental conditions than RFLPs, making consistency and reproducibility more difficult. Alternative PCR-based markers such as AFLP, SSR, and others have been developed as a result. AFLPs are dominant markers and have some characteristics with both RFLP and RAPD as they use restriction enzyme-digested genomic DNA as a template for PCR amplification. When compared to RAPDs, they provide more information and are more reproducible. SSRs are tandem repeats of small nucleotide motifs of around 2–6 bases in length, such as (GT)n, (AAT)n, and (GATA)n. They are characterized by their co-dominant nature, reproducibility, locus-specificity, hyper-variability, and random genome-wide distribution in most cases. SSR markers were once the most used PCR-based markers, but SNP markers are progressively replacing them. A single nucleotide base difference between two DNA sequences at the same genomic region of different individuals, strains, or lines of a species is called SNP. Since a single location in a given strand can have A, T, G, or C, each indicating one SNP allele, an SNP locus can have four alleles. SNPs are co-dominant markers that are commonly related to genes and present in the simplest/ultimate form of polymorphism, making them particularly appealing as genetic markers in genetic research and breeding. When these methods are combined with NGS (next-generation sequencing), have produced ultrahigh-throughput, cheaper assays for various novel marker technologies.

The major advantage of random DNA markers is not requiring any prior sequence information from the organism being studied. The majority of dominant markers are produced at random along the entire length genome, sampling various loci. These procedures are technically straightforward, inexpensive, and produce a number of markers for each sample. However, they have several drawbacks, including drifting of fragments of similar size arising from independent loci across diverse analyzed samples, heteroduplex formation, nested priming, collision of the equisized fragments in one-line, artifactual segregation distortions caused by off-target/erroneous scoring of loci, poor gel resolution or hidden co-dominance. Functional markers (FMs) pose for a replacement to random DNA markers for overcoming these issues. FMs are formed when polymorphic sites within the genes produce phenotypic trait variation (Andersen & Lübberstedt, 2003). FMs are directly associated with the allele of the trait of interest, unlike random DNA markers (Varshney et al., 2005). As a result, marker-assisted breeding (MAB) could be efficiently carried out using FMs than random DNA markers. It's crucial to distinguish between FMs and gene-targeted markers (GTM) because not all GTMs are engaged in variation of phenotypic attributes and hence may not become functional. GTMs are an effective tool to evaluate gene functionality, and if they are related with qualities of interest, they can be utilized directly in breeding programs (Zhen et al., 2018). Many such advanced marker systems have recently been developed (Figure 1.1).

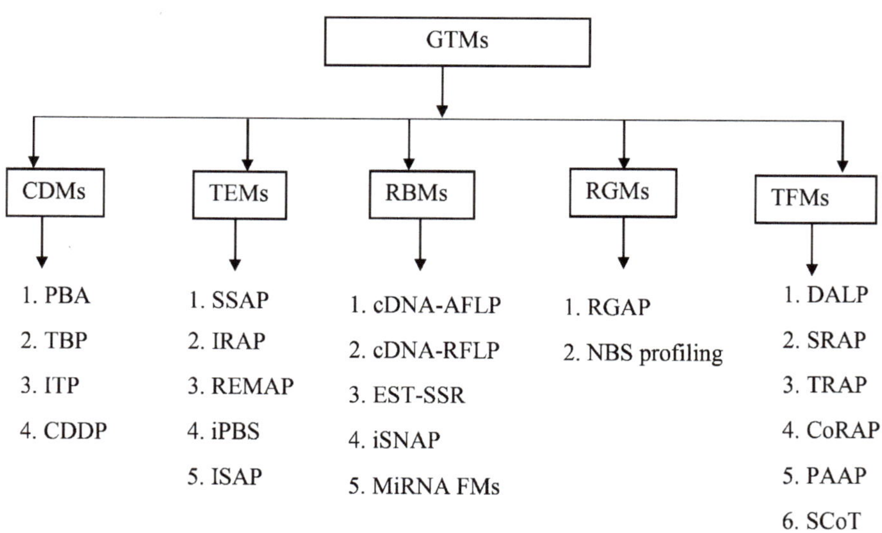

FIGURE 1.1 Gene-tagged marker systems and its classification.

1.2 CONSERVED DNA MARKERS AND GENE FAMILY-BASED MARKERS (CDMS)

This group of markers can be thought to be derived from the GTMs that use length polymorphisms in exon-intron structures in a variety of frequently distributed and common gene families or plant genes. When functionality and resolution given by the fast evolution of SSRs or slowly evolving DNA sequences is an issue, CDMs can be used. Multi-locus markers, varied in length and functionally relevant to a specific trait, are created using these techniques from arbitrarily allocated members of a targeted gene (family). To merge the benefits of multiloci profile development along with its functionality, there is a need to design conserved DNA-based primers without any previous knowledge of the entire genome. Plant genomes contain a large number of gene families that can be targeted using approaches like those described in subsections.

1.2.1 CYTOCHROME P450-BASED ANALOGS (PBA)

Cytochromes P450 (CYPs) are a monooxygenase superfamily of enzymes that contain heme as a cofactor. Animals, plants, fungi, protists, bacteria, and archaea, as well as viruses, all have CYP enzymes. These proteins play a role in the manufacture of defense chemicals, fatty acids, and hormones in plants. Polymorphisms of gene analogs of the mammalian cytochrome P450 monooxygenase was examined and concluded as an efficient tool in studying functional genomics in plants (Yamanaka et al., 2003). The sequence variation of P450 gene analogs in plants was found beneficial for investigations both at the functional and genome-wide scales (Somerville & Somerville, 1999). According to the genomic annotation of *Arabidopsis*, nearly 0.9% of the 29,000 genes in the Arabidopsis genome (26 pseudogenes and 272 genes) are potential Cyt P450 genes. This implies that these genes are quite diversified, making them suitable for use in diversity assessments. Putri et al. (2020) used P450-based Analogs (PBA) to study genetic diversity in turmeric accessions.

1.2.2 TUBULIN-BASED POLYMORPHISM (TBP)

Tubulin can refer to either the tubulin protein superfamily of globular proteins or one of its constituent proteins in molecular biology. Microtubules, a significant component of the eukaryotic cytoskeleton, are formed when α- and β-tubulins polymerize. Microtubules play a major role in a various biological processes, including mitosis. Tubulin-binding medications destroy malignant cells by interfering with the dynamics of microtubules, which are necessary for DNA segregation and thus cell division. A new molecular marker-based method, TBP viz., tubulin-based polymorphism is discovered by Bardini et al. (2004), which inspects the presence of intron-specific DNA polymorphisms of the plant-tubulin gene family. The tubulin proteins' multifunctional and crucial role is displayed in the conserved segments of their primary amino acid sequence. Primers that amplify the first intron of various β-tubulin isotypes, discovering specific fingerprints, can be used to study the ubiquitous aspect of this gene family. The procedure is quick, easy, and reliable, and it doesn't require any prior knowledge of the plant genome in question. TBP's ability to differentiate between the accessions and species has been successfully demonstrated in oilseed rape, coffee, and lotus. TBP was able to detect certain genetic variations in all of these cases when used in combination with a rather simplified and easily recognizable pattern of DNA amplification. Breviario et al. (2007) developed a further variation of the technique, which they called combinatorial TBP (cTBP). Rabokon et al. (2017) used this h-TBP (horse-TBP) and TBP to differentiate between the accessions of a wild relative of wheat, i.e., *Aegilops biuncialis* Vis., and suggested that both the techniques uncover DNA polymorphism level sufficiently high. After a successful classification of various species and their wild relatives within the Papilionoideae sub-family, Braglia et al. (2020) inferred that TBP can be used as a versatile tool for plant genotyping.

1.2.3 INTRON-TARGETING POLYMORPHISM (ITP)

Introns are non-coding portions of an RNA transcript or the DNA that encodes it that are removed before translation through splicing. Intron polymorphism have long been thought to be a source of variation as they show the slow rate of sequence evolution, which is thought to occur under minimal restrictions and in accordance with the neutral theory of sequence evolution. Intron length polymorphism has recently been discovered to be a practical and trustworthy source of information also due to its good interspecies transferability, according to recent reports. Because introns directly signify the variation occurring inside genes, they could be used to create genetic maps (Han et al., 2006). Intron indels (insertion-deletions) are becoming increasingly useful genetic markers for a variety of plant species (Vali et al., 2008). The basic approach as explained by Weining & Langridge, (1991) is termed as intron-targeting (IT) that takes into account the intron splice junction (ISJ) primers. In this preliminary study was carried out on the intron length polymorphism of α-amylase gene family. The absence of large plant genomic datasets, however, has slowed down the development of primers and full exploitation of this method. This technique has been referred to under many names separately, including exon-primed intron-crossing (EPIC) PCR (Palumbi & Baker, 1994), intron-flanking primers (Wei et al., 2005), conserved-intron scanning primers (CISP; (Feltus et al., 2006)), potential intron polymorphism (PIP; (Yang et al., 2007)) and PCR-based landmark unique gene (PLUG) markers (Ishikawa et al., 2007). There are slight differences to be stated between these techniques. The key to successful cross-species amplification capability and transferability of IT markers is conservation of intron-exon junctions and gene structures across related genomes in various taxa. Wang et al. (2017) constructed a total of 359 IT markers for tracing the 4VS chromatin in the wheat background and achieved a remarkable success rate and specificity (64.62%). Thus, ITP appears a promising candidate in developing chromosome-specific markers in any other crops also.

1.2.4 CONSERVED DNA-DERIVED POLYMORPHISM (CDDP)

Collard & Mackill (2009) devised a unique approach for producing plant DNA markers. This approach uses data mining as a tool to find short, conserved amino acid sequences in proteins followed by the development of PCR primers based on the DNA sequence. This approach enhances reproducibility by using a single longer primer amplification with a high annealing temperature. For applications such as QTL mapping, this method could be employed in conjunction with or instead of other technically simple dominant marker methods (Collard & Mackill, 2009). CDDP has a higher reproducibility than random DNA markers. The advantage of CDDP over random DNA markers is high reproducibility. Assessment of genetic diversity using the CDDP markers has been reported in *Solanum dulcamara* (Poczai et al., 2011), Chrysanthemum (Li et al., 2013), Chickpea (Hajibarat et al., 2015), wheat (Hamidi et al., 2014) and *Pistacia vera* (Aouadi et al., 2019) (Table 1.1).

TABLE 1.1 Merits and Demerits of the CDMs

Markers	Merits	Demerits
PBA	• Abundantly present • High reproducibility • High polymorphism • Prior sequence information is not required • Locus specific	• Occurrence of homoplasy (high) • Heteroduplexes not reported, but may occur
TBP	• High reproducibility • High polymorphism • No prior sequence information is required • Locus specific	• Occurrence of homoplasy (low) • Heteroduplexes not reported, but may occur
ITP	• High reproducibility • Locus specific	• Less abundance • Prior sequence information is required • Occurrence of homoplasy (low) • Heteroduplex formation occurs
CCDP	• High reproducibility	• Prior sequence information is required • Occurrence of homoplasy (high) • Heteroduplexes may occur

Source: Adapted from Poczai et al., 2013.

1.3 TRANSPOSABLE ELEMENT-BASED MARKERS (TEMS)

A transposable element (TE) is a DNA sequence that is mobile within a genome, tending to mutations and change of genetic identity and genome size of the cell. Duplication of the same genetic material occurs frequently as a result of transposition. Barbara McClintock discovered this in 1950 and was awarded the Nobel Prize in 1983. According to its ability of transposition, TEs are classified into one of two classes: (i) copy and paste (Class I TEs); and (ii) cut and paste (Class II TEs). Class I TEs are copied in two phases: At first, they are transcribed from DNA to RNA, and then there is reverse transcription of RNA to DNA. Later, this cloned DNA then transpose itself at different locations in the genome. The reverse transcriptase (RT), which is usually encoded by the TE itself, catalyzes the reverse transcription phase. There is no RNA intermediate in the cut-and-paste transposition mechanism of class II TEs. Retrotransposons with long terminal repeats (LTRs), Retro-posons with long interspersed nuclear elements (LINEs, LINE-1s, or L1s), and Retroposons with short interspersed nuclear elements (SINEs) are the three primary classes of retrotransposons. Class-I elements, retrotransposons, have some specific benefit due to its 'copy-and-paste' mechanism. Hence, these are most commonly used for the development of markers.

1.3.1 SEQUENCE SPECIFIC AMPLIFIED POLYMORPHISM (SSAP)

Sequence-specific amplified polymorphism (SSAP) was first described by Waugh et al. in 1997. The SSAP method might be thought of as a variant of anchored PCR or rather a refinement of AFLP marker. In SSAP, the genomic DNA is digested with two restriction enzymes, one that makes infrequent cuts while another one cuts frequently (generally PstI and MseI). Post digestion, the restricted DNA fragments are ligated to short double-stranded adaptors having known sequences. Then, using the adapter-homologous primers, a pre-selective PCR amplification is performed. The pre-amplification process is used in order to reduce the genomic complexity and improve reproducibility. This is followed by the use of a retrotransposon-specific primer in combination with a rare or a frequent site adaptor primer to conduct a selective amplification. Primers commonly anneal to the LTR regions of retrotransposons or to the element's internal components. The distance between the transposon inserted region and the adjoining restriction cut region determines the size of the fragments, with changes in insertion sites between genome's distinct banding pattern could be readily observed (Syed & Flavell, 2006).

1.3.2 INTER-RETROTRANSPOSON AMPLIFICATION POLYMORPHISM (IRAP)

The IRAP and retrotransposon microsatellite amplification polymorphism (REMAP) marker systems were described by Kalendar et al. (1999). The procedure required for both differ from SSAP as they do not require restriction enzyme digestion or ligation, and the products could be resolved using standard agarose gel electrophoresis without the requirement of a sequencing instrument. The IRAP method depicts the use of amplified DNA sequence between two retroelements, called the retrotransposon insertional polymorphisms. The amplification of DNA is carried out using one or two primers pointing outwards from an LTR. IRAP can be identified either with a single primer that matches the 5′ or 3′ end of the LTR but is orientated away from the LTR or two or more primers. As for the method involving two primers, it could come from the same retrotransposon element family or even separate ones. The PCR products, and followed by the fingerprint patterns, result from the amplification of numerous target sites inside the genome. The intricacy of the pattern obtained will be affected by the retrotransposon copy number, which reflects the genome size, as well as by their pattern of its insertion and also by the size of the retrotransposon families chosen for analysis.

1.3.3 RETROTRANSPOSON MICROSATELLITE AMPLIFICATION POLYMORPHISM (REMAP)

The REMAP approach is identical to IRAP, except that one of the two primers matches an SSR motif at the 3′ end with one or more non-SSR anchor nucleotides. Plant and animal genomes contain microsatellites in the form of (NNNN)n, (NNN)n, or (NN)n. Primers constructed for the unique sequences flanking microsatellites are generally used to detect differences in the number of SSR units in a microsatellite. Otherwise, the part of the genome existing adjacent to microsatellites may be amplified by ISSR (Zietkiewicz et al., 1994), in a way identical to IRAP. Anchor nucleotides are attached at the 3′ end of the SSR primer in REMAP to prevent primer slippage within the SSR, which would result in a "stutter" pattern in the fingerprint, as well as to prevent detection of recurrence within the SSR. REMAP uses the same type of primer to that of ISSR and IRAP. As it appears that SSR primers in REMAP should also generate ISSR markers and LTR primers should also give IRAP markers, this is not often the case in practice. This is most likely owing to a combination of factors, such as genome shape and PCR competition. Because they provide consistent and repeatable banding profiles, REMAP, and IRAP have been applied both separately and in combination to examine genetic divergence in various plant taxa (Poczai et al., 2013).

1.3.4 INTER-PRIMER BINDING SITE (IPBS) POLYMORPHISM

The requirement of sequence data to create element specific primers is a major shortcoming of all retrotransposon-based molecular marker approach. The basic requirement is the sequence of an LTR end, which could be obtained either from a database or by cloning and sequencing of the genomic DNA flanking the conserved retrotransposon regions.

Even with its rapid nature of retrotransposon isolation with conserved TE primers via the PCR has been developed, the requirement of a huge number of clones to acquire very few acceptable primer sequences complicates the process. There are no conserved motifs in the LTRs, which would allow them to be amplified directly by PCR. Kalendar et al. (2010) proposed the inter-primer binding site (iPBS) that has helped overcome this problem by exploiting retrotransposons' PBS sites, which are shared by LTR transposons and includes 18 nucleotides complementary sequence limited to some specific of tRNAs. Primers of variable length (12–18 bp) are used to anneal to these sites. Retrotransposons must have opposite directionality and be close enough to each other in order to amplify the intergenic regions with iPBS. Because these amplicons contain the LTR motif, the technique can also be used to isolate retrotransposons and scan the genome. Using the conserved parts of PBS sequences for direct spotting of polymorphism among individuals, polymorphism within transcription profiles, fast cloning of LTR sites from genomic DNA, and database search for LTR retrotransposons, the iPBS method has resulted in the establishment of a practically universal and extremely proficient mechanism.

1.3.5 INTER-SINE AMPLIFIED POLYMORPHISM (ISAP)

This system is built on a class of dispersed repetitive DNA sequences entitled as SINEs. These are the retrotransposons that are not LTRs and are amplified by reverse transcription. The retrotransposition is accomplished using an RNA intermediary, and the copy thus obtained is integrated at a different location in the genome, while the source copy is maintained. Since they are nonautonomous retroelements, they rely on proteins encoded by similar LINEs (long interspersed nuclear elements) for their function (Ohshima & Okada, 2005). The "copy and paste" technique leads SINEs to evolve and diversify, resulting in different families with copy numbers tending to a few thousand. SINEs are short chromosomal segments of about 100–500 nucleotides that are extensively dispersed. Multiple SINE families exist in a species at the same time, and the majority of copies are found in euchromatic areas, often near or within genes (Wenke et al., 2011). SINEs have been found in flowering plants in a number of studies. They were discovered in angiosperms of all levels, gymnosperms, monocots, and dicots. Ever since the development of SINE-Finder, a bioinformatic tool, it is now possible to identify SINEs in any genome quickly (Wenke et al., 2011). The ISAP marker system was created by Seibt et al. (2012) using the potato (*Solanum tuberosum*) as the model plant. ISAP markers were employed by Sormin et al. (2021) to genotype *Cucumis melo* accessions and examine their transferability to *Coleus spp.*

TE markers can be applied in plant breeding in the following ways. In wheat, barley, pigeon pea, common bean, sweet potato, azuki bean, strawberry, flax, potato, field pea, and other plants, retrotransposons were used for cultivar differentiation along with genetic diversity analysis (Bhat et al., 2020). SINE, a non-LTR-Retrotransposon, was used to genotype plants quickly and reliably (Wenke et al., 2015). Genetic variability and diversity analyze were performed using DNA transposons in rice (Park et al., 2003), Chenopodiaceae subfamily Betoideae (Grzebelus et al., 2011), wheat (Yaakov et al., 2012), maize (Zerjal et al., 2012) and so on. Because of their genome-wide occurrence and high level of polymorphism, both class I and class II TE markers are currently utilized in linkage and association mapping (AM). Retrotransposon-based molecular markers were used to construct linkage maps in Japanese gentian, lentil, rye, sunflower, sweet potato and other polyploid species because they showed an exceedingly high percentage of simplex markers (90%) when compared to other molecular markers of polyploid species. Other uses of TE markers accounts for analyzing somatic polymorphism in *Vitis vinifera*, elucidating species discrimination and geographic organization in Madagascan coffee species, mutation breeding in sorghum, and evaluating the outcrossing rate of wild barley populations and barley landraces (Bhat et al., 2020) (Table 1.2).

1.4 RNA-BASED MARKERS (RBMS)

Considering morphological markers then isoenzyme markers to random DNA markers to now FMs and GTMs centering the polymorphic locations within genes, mapping markers has significantly evolved. Even though the use of these recently emerging marker techniques is not yet as vast as that of the previously well-established methods, such as SSR markers, the number of studies utilizing these advanced methods in plant science is still showing a growing trend. Gupta & Rustgi (2004), for the first time reviewed the molecular markers originating from the expressed or transcribed regions of genomes.

TABLE 1.2 Merits and Demerits of the TEMs

Markers	Merits	Demerits
SSAP	• Abundantly present • High reproducibility • High polymorphism • Locus specific • No heteroduplexes	• Prior sequence information is required • Occurrence of homoplasy (low)
IARP/REMAP	• Abundantly present • Locus specific • No heteroduplexes	• Prior sequence information is required • Occurrence of homoplasy (medium)
iBPS	• Abundantly present • High reproducibility • High polymorphism • No prior sequence information is required • Locus specific • No heteroduplexes	• Occurrence of homoplasy (low)
ISAP	• Abundantly present • High reproducibility • Locus specific • No occurrence of homoplasy • No heteroduplexes	• Prior sequence information is required

Source: Adapted from Poczai et al., 2013.

1.4.1 CDNA-AFLP

Bachem et al. (1996) invented this marker system which is comprising of four basic steps for creating fingerprints. The first step is to extract RNA from plant tissues, further utilized to make cDNA. Restriction enzyme of one or more kind is employed, with cDNA being the main template, is followed in the same way as that of AFLP marker. A genome-wide transcriptome map was created using cDNA-AFLP for Arabidopsis (Brugmans et al., 2002) and Potato (Ritter et al., 2008). It's also been used to identify alterations in gene expression in *Triticum aestivum* (Kashkush et al., 2002), *Angelica sinensis* (Oliv.) Diels (Guang et al., 2011) and cotton (Claverie et al., 2012). Xiao et al. (2016) used cDNA-AFLP to identify transcriptionally regulated genes during mismatched interactions between non-heading Chinese cabbage and

H. parasitica. Archangi et al. (2019) used cDNA-AFLP markers studied correlation of plant height with several seed yield (SY)-related traits in 15 cumin (*Cuminum cyminum*) accessions for irrigated and drought stress conditions.

1.4.2 CDNA-RFLP

cDNA clones to be utilized directly as RFLP probes was demonstrated by Bryan et al. (1999) in his experiments. These markers could be transformed into particular PCR markers, which can then be employed in gene tagging or diagnostics. The cDNA clones could be utilized for RFLP analysis with no further screening. Such an approach has been successfully applied to a variety of plant species, i.e., sunflower (Gentzbittel et al., 1999), wheat (Akhunov et al., 2003) and oil palm (Singh et al., 2008).

1.4.3 EST-SSR

Expressed sequence tag (EST) is a small sub-sequence of a cDNA sequence. ESTs could be utilized in detection of gene transcripts also they can be useful for gene identification and sequence determination. ESTs are read in a one direction and serve as a strong foundation for assessment of genetic diversity and analysis of gene expression analysis. Generating 5′ and 3′ ESTs, the expressed genes can be sequenced in both directions after being converted to the cDNA. The latter are more frequently found in untranslated regions, whereas 5′ ESTs are linked to protein coding. In terms of amplification and detection, EST-SSRs are identical to typical genomic (gSSR) microsatellites. Because EST-SSRs are created from the transcribed segment of the genome, the fundamental variation is in primer creation and primer position. *In-silico* approaches are used to extract them directly from sequence data. When compared to genomic SSR markers, EST generated gSSRs have more chances to be identified within the functional sequences and thus provide more information. When compared to gSSRs, its most essential property is that they are easier to transmit between distantly related species. Such markers have been found to be useful in the analysis of rice (Cho et al., 2000), alpine lady-fern (Woodhead et al., 2003), and the genus *Medicago* (Eujayl et al., 2004). Lu et al. (2020) identified and established a new set of polymorphic EST-SSR markers in *Sophora japonica* and evaluated its genetic diversity. Transcriptome sequencing was done in *Populus wulianensis* (Wu et al., 2020) and *Schisandra chinensis* (Shi et al., 2020) with the help of EST-SSR markers which can be further used in genetic research.

1.4.4 INTER SMALL RNA POLYMORPHISM (ISNAP)

Endogenous small nc-RNAs (non-coding RNAs) usually consist of 20–24 nucleotides, are found widely across the genome and portrays essential regulatory roles usually in eukaryotes. Gui et al. (2011) employed a new type of PCR-based marker termed inter small RNA polymorphism (iSNAP), which used multi-mapped small RNA sequences (iSNAP). iSNAP primers were built using multi-mapped short RNA and its two end-conserved flanking sequences. This technique can be used for genome mapping (e.g., rice) and fingerprinting even in species with minimal genetic diversity (e.g., tobacco), and so gives a new molecular marker technology opportunity (Gui et al., 2011).

1.4.5 MICRORNA FUNCTIONAL MARKERS (MIRNA FMS)

MicroRNAs (miRNAs) are a new class of short RNAs with 18–25 nucleotides (Ambros, 2004) that plays important role in nutrient deficit (Sunkar et al., 2007), plant growth and development (Glazinska et al., 2009), and other abiotic stress (Jian et al., 2010). miRNAs that silence gene expression at the post-transcriptional level by either cleaving or inhibiting translation of a target mRNA via complementation between the miRNA and target mRNA (Meyers et al., 2008) could be phylogenetically conserved. MicroRNA discovery, detection, and analysis have increased as a result of recent advancements in next-generation or advanced sequencing technology (Kozomara and Griffiths-Jones, 2011). With the growing number of miRNA sequences, it is now feasible to create miRNA-based markers, a new sort of marker that's simple to use and detect. Htwe et al. (2015) created a miR1511-InDel marker that was utilized in analyzing genetic variation in a huge sample of *Glycine soja* and *Glycine max*, concluding that miRNA1511-InDel can be used as a new functional marker type for genetic diversity research, gene mapping, and molecular breeding. MiRNA functional markers (FMs) have been successfully developed in other major crops like tobacco, rice, and soybean (Htwe et al., 2015) (Table 1.3).

TABLE 1.3 Merits and Demerits of the RBMs

Markers	Merits	Demerits
cDNA-AFLP	• Abundantly present • High reproducibility • High polymorphism • Prior sequence information is not required • No heteroduplexes	• Occurrence of homoplasy (medium)
cDNA-RFLP	• Abundantly present • High reproducibility • Prior sequence information is not required • Locus specific • No heteroduplexes	• Occurrence of homoplasy (low)
EST-SSR	• High reproducibility • Locus specific	• Prior sequence information is required • Occurrence of homoplasy (medium) • Heteroduplex formation occurs
iSNAP	• Abundantly present • High reproducibility • High polymorphism • Locus specific • Homoplasy not reported	• Prior sequence information is required • Heteroduplexes may occur
MiRNA FMs	• High reproducibility • High polymorphism • Locus specific	• Occurrence of homoplasy (low) • Prior sequence information is required

Source: Adapted from Poczai et al., 2013.

1.5 RESISTANCE-GENE-BASED MARKERS (RGMS)

As per the gene-for-gene hypothesis given by Flor (1946), "resistance-gene markers utilize specific features of genes involved in plant defense mechanisms that makes them a unique group within GTM." To defend against diseases, plants have developed active and passive defensive systems. Adaptive and innate immune responses both are active systems. Adaptive immunity is mostly utilized to combat viruses and is based on the RNAi response. R proteins (resistance proteins) and PPRs (pathogen and pattern resistance receptors) are used in innate immunity, which allows the plant to protect itself against a wide range of pathogens (Jones & Dangl, 2006).

PPRs recognize molecular patterns associated with microbes or pathogens that are conserved across pathogens of a given class (Wolpert et al., 2002). R proteins then recognize avirulence (Avr) factors that aren't shared by all infections. Induced R protein signaling results in the generation of reactive oxygen species and induces a hypersensitive response also called programmed cell death, which is known to kill the cells under consideration (Gebhardt & Valkonen, 2001). According to current study, cell death does not directly prevent the disease from spreading; rather, an unknown mechanism blocks the pathogen's migration in the surrounding surviving tissue (Tian & Valkonen, 2012). As every R gene is responding to a distinct pathogenic Avr gene, R-protein mediated borne immunity is also known as gene-to-gene resistance (Gebhardt & Valkonen, 2001).

1.5.1 RESISTANCE-GENE ANALOG POLYMORPHISM (RGAP)

According to the findings of Leister et al. (1996), employing R-like gene specific primers, precise PCR markers related to the R-genes could be generated rapidly. In numerous areas of research, RGAP has been proved to be feasible. It has been utilized in several research to develop R genes linked molecular markers that provide resistance to pathogens, such as and Strip rust resistance in wheat (Yan et al., 2003) and FBH resistance in wheat (Mazaheri et al., 2008). It's also been beneficial for delineating R gene domains (viz., NBS and LRR domains, e.g., disease resistance genes in bean, Mutlu et al., 2006) for analyzing genetic variability, e.g., in rice (Ren et al., 2013) and mapping of resistance genes, e.g., all-stage stripe rust resistance gene in *Triticum aestivum-Haynaldia villosa* (Lu et al., 2013) and stripe rust resistance gene in wheat (Zheng et al., 2014).

1.5.2 NUCLEOTIDE-BINDING SITE (NBS) PROFILING

Nucleotide binding site (NBS) profiling is a modern molecular marker approach devised by Van der Linden et al. in 2004. This method addresses resistance genes and their analogs specifically. Plants have a large number of resistance (R) genes with a NBS that is spread across every chromosome (Monosi et al., 2004). Primers centering various conserved motifs within the NBS domain (P loop, GLPL motif and kinase-2 motif) might be utilized as a point of initiation. As a result, this technique generates gene-specific markers, whereas other marker techniques, such as AFLP generate markers at random throughout the genome. NBS profiling produces a repeatable polymorphic multi-locus banding pattern, which has been used to successfully identify and map RGAs in apple, potato, and lettuce (Syed et al., 2006). NBS profiling was used to evaluate genetic diversity in durum wheat accessions (Mantovani et al., 2007) and *Gossypium raimondii* (Wei et al., 2013). Wang et al. (2008) used NBS profiling to carry out phylogenetic studies and discrimination in a part of the section Petota of the genus *Solanum*. SolariX is a repository of DNA sequence tags from the NBS domain of disease resistance genes of potato, *Solanum tuberosum* Group Tuberosum. Prakash et al. (2020) explained the use of SolariX database to detect polymorphisms associated with NBS-LRR R gene variants that give particular disease resistance and to create genetic markers for selection (Table 1.4).

TABLE 1.4 Merits and Demerits of the RGMs

Markers	Merits	Demerits
RGAP	• Abundantly present • Prior sequence information is not required • Locus specific • Heteroduplexes not reported	• Occurrence of homoplasy (low)
NBS-profiling	• Abundantly present • High reproducibility • High polymorphism • Prior sequence information is not required • Locus specific • No heteroduplexes	• Occurrence of homoplasy (low)

Source: Adapted from Poczai et al., 2013.

1.6 TARGETED FINGERPRINTING MARKERS (TFMS)

A new family of markers termed, targeted fingerprinting markers (TFMs), has been designed to make use of growing knowledge of genomic components (TFMs). As per definition these multi-locus markers are created in a semi-random and targeted manner at multiple parts of the genome, and apparently matching the polymorphism in the gene or any gene-related region, regardless of its role. That implies the marker systems above-mentioned are (gene)-targeted markers that may or may not produce variation in phenotypic trait fingerprints. TFM markers work by integrating the best aspects of multiple fundamental approaches while also integrating methodological changes to improve resolution and sensitivity when detecting genetic differentiation and uniqueness.

1.6.1 DIRECT AMPLIFICATION OF LENGTH POLYMORPHISMS (DALP)

Desmarais and coworkers developed the DALP technique in the year 1998. This method produces genomic fingerprinting and allows DNA polymorphism sequencing in nearly all species using an arbitrarily primed PCR (AP-PCR). Each oligonucleotide pair was intended to create a unique multiple-banding pattern, while the complete set of resulting fragments may be detected by directly using the same two universal M13 sequencing primers. It brings together the benefits of a high-resolution fingerprinting approach with the ability to characterize polymorphisms. The application of family members as templates in the multi-locus detection step permits the direct detection of allele transmission, along with rapid marker mapping and selection of loci linked to certain traits or disorders. This method can be used to spot the microsatellite DNA loci and micro-deletions/insertions in population genetics investigations, but it could be even used in other fields of biology, including genome mapping to define polymorphic sequence tagged sites that can be immediately located on a genetic map (Desmarais et al., 1998). The method's key benefit is that it may produce banding patterns with a less number of primers using simple combinations and just altering a single primer between experiments. According to studies using DALPs, findings may be acquired consistently and quickly for a variety of applications, including identifying new monolocus co-dominant markers (Perrot-Minnot et al., 2000), investigation of genetic mapping (Langar et al., 2003) and population diversity (Ma et al., 2008).

1.6.2 SEQUENCE-RELATED AMPLIFIED POLYMORPHISM (SRAP)

Sequence-related amplified polymorphism (SRAP) was developed by Li & Quiros (2001) to discover a molecular approach in genetic variation within the open reading frames (ORFs) of plants genomes and related organisms. This technique proceeds by amplification with two primers. The primers are relatively long (17–18) nucleotides having a core sequence of 13 to 14 base length, while the first 10–11 base commencing at the 5′ end are the sequences of no particular significance (called filler sequences). CCGG sequence follows as the forward primer whereas AATT as the reverse primer. At the 3′ terminus, three selective nucleotides follow the core. The filler sequences have forward and reverse primers that ought to be distinct from one another and could be 10–11 bases long. These markers were used to successfully tag the glucosinolate desaturation gene BoGLS-ALK by Li & Quiros (2001). SRAPs has been readily detected in other crops such as rice, potato, apple, citrus, and celery, lettuce, garlic, Chinese cabbage, rapeseed (Li & Quiros, 2001). SARPs can be successfully employed to study genetic diversity in *Limonium sinense* (De & Daizhen, 2015) and cumin (Bhatt et al., 2017). Apart from genetic diversity analysis, SRAPs holds the capacity to enhance the current status of molecular tools for studies related to diversity through mapping and tagging of genes, sex determination and hybrid detection (Aneja et al., 2012).

1.6.3 TARGETED REGION AMPLIFIED POLYMORPHISM (TRAP)

The advent of large-scale DNA sequencing technology has resulted in a massive load of sequence data from different plant species. Hu & Vick (2003) developed targeted region amplified polymorphism (TRAP) by using bioinformatics techniques and information sourcing from the EST database to construct polymorphic markers revolving around the targeted gene sequences candidate. To construct markers, this technique employs two 18-nucleotide primers. The fixed primer is constructed from the database's targeted EST sequence; the arbitrary primer, on the other hand, is an arbitrary sequence having an AT- or GC-rich core to anneal with an intron or exon, respectively. The production of universal primers restrictions the use of this approach to species having known ESTs or needs to create new sequence information for the development of primer. Despite such drawback, it has been extensively accepted for various grounds in several plant species, e.g., for genotyping in lettuce (Hu et al., 2005), assessing genetic diversity in sugarcane (Alwala et al., 2006), *Lentinula edodes* (Xiao et al., 2010) and sweet sorghum (Khidr et al., 2020).

1.6.4 CONSERVED REGION AMPLIFICATION POLYMORPHISM (CORAP)

Establishment of a novel PCR-based molecular marker technique, based on the techniques of SRAP and TRAP which is called conserved region amplification polymorphism (CoRAP) was reported by Wang et al. (2009) for genotyping of *Salvia miltiorrhiza*. Like TRAP, it takes into account two different primers; one being fixed and another being arbitrary primer. The fixed primer is derived from GenBank through the target EST sequences information (has specific binding sites at the exon of the target sequence), whereas, the core sequence (CACGC) of the latter is a conserved segment usually found in introns. This new genotyping approach is quick effective, and can be replicated (Wang et al., 2009).

1.6.5 *PROMOTER ANCHORED AMPLIFIED POLYMORPHISM (PAAP)*

Genes' transcription begins at a region of DNA sequence called the promoter. They are an essential element for expression vectors as they facilitate binding of the RNA polymerase to template DNA. mRNA transcribed from the DNA is further translated into a functional protein by action of RNA polymerase. As a result, the promoter region determines when and where your gene of interest is expressed in the organism. "The structure of promoter sequences varies considerably based on the gene, and numerous short motifs acting as recognition sites for proteins involved in transcription initiation" (Lemon & Tjian, 2000). This property of promoters allows them to be tagged with degenerate primers to make length polymorphisms that may be detected by electrophoresis. Pang et al. (2009) made an to develop molecular marker systems by designing several short nucleotide primers to search for polymorphisms associated with high fiber yield and quality in cotton and called them as promoter anchored amplified polymorphism based on RAPD (PAAP-RAPD). Mokate et al. (2017) used this PAAP-RAPD markers to study genetic diversity in soybean.

1.6.6 *START CODON TARGETED (SCOT) POLYMORPHISM*

A novel technique for the development of plant DNA markers based on the short, conserved sites flanking the ATG start codon in plant genes was developed by Collard and Mackill in 2009. This method employs one primer PCR with a higher annealing temperature. In SCoTs utility of primer pairs was suggested by Gorji et al. (2011). "SCoT markers can be easily replicated, while its reproducibility doesn't primarily dependent on the primer length and annealing temperature" (Gorji et al., 2011). This method can be used in isolation or in combination with other approaches for applications such as genetic studies, bulk segregant analysis (BSA), and QTL mapping, performed in laboratories employing agarose gel electrophoresis (Collard & Mackill, 2009b). Applicability of this marker in diversity analysis has been seen in many crops, e.g., mango (Luo et al., 2010), *Boehmeria nivea* (Satya et al., 2015), *Triticum turgidum* var. *durum* (Etminan et al., 2016), *Maytenus emarginata* (Jatan et al., 2018) and Manilkara (Vanijajiva, 2020) (Table 1.5).

TABLE 1.5 . Merits and Demerits of the TFMs

Markers	Merits	Demerits
DALP	• Abundantly present • High polymorphism • Prior sequence information is not required	• Occurrence of homoplasy (medium) • Heteroduplex formation occurs
SRAP	• Abundantly present • High reproducibility • High polymorphism • Prior sequence information is not required • Locus specific	• Occurrence of homoplasy (medium) • Heteroduplex formation may occur
TRAP	• High reproducibility • High polymorphism • Locus specific	• Prior sequence information is required • Occurrence of homoplasy (medium) • Heteroduplex formation may occurs
CoRAP	• Homoplasy not reported	• Prior sequence information is required • Heteroduplex formation occurs
PAAP	• Abundantly present • No prior sequence information is required	• Low reproducibility • High homoplasy • Heteroduplex formation occurs
SCoT	• Abundantly present • Prior sequence information is not required	• High homoplasy • Heteroduplex formation may occur

Source: Adapted from Poczai et al., 2013.

1.7 CONCLUSION

Although the well-established methods, such as random DNA markers, are widely used, unlike the novel marker approaches in plant breeding, the number of studies on these newly discovered markers are increasing due to the fact that they have the ability to provide supplementary information sources. Such novel markers with better efficiency have been developed more for important crops like rice, potato, maize, etc., as compared to the minor crops, for which far less study has been conducted. There still exists unavailability of markers in alternative scientific domains like molecular biology and phylogenetics, in which target organisms have less economic value thus not having any previous knowledge of sequence or accessibility of genome information for constructing the primer. The main hindrance in case of certain new discoveries is the lack of preliminary genetic information, which is tedious and hefty laboratory work. The cost of producing GTM will decrease with the reducing cost of DNA sequencing through advanced and highly effective technologies. The increasing research in these new markers is indicative of the fact that they could be valuable for a variety of applications. Moreover, these methods seem to have more accuracy than the DNA markers which make use of some unknown genetic reorganizations. Majority of strategies presented here are likely to produce better organized datasets that might be employed individually or in collaboration with sequencing level characters in sectors of plant sciences where they are not yet applied.

KEYWORDS

- **amplified fragment length polymorphism**
- **diversity array technology**
- **functional markers**
- **genetic diversity**
- **molecular breeding**
- **molecular markers**
- **polymerase chain reaction**

REFERENCES

Akhunov, E. D., Goodyear, A. W., Geng, S., Qi, L. L., Echalier, B., Gill, B. S., Miftahudin, et al., (2003). The organization and rate of evolution of wheat genomes are correlated with recombination rates along chromosome arms. *Genome Research, 13*, 753–763.

Alwala, S., Suman, A., Arro, J. A., Veremis, J. C., & Kimbeng, C. A., (2006). Target region amplification polymorphism (TRAP) for assessing genetic diversity in sugarcane germplasm collections. *Crop Science, 46*, 448–455.

Ambros, V., (2004). The functions of animal microRNAs. *Nature, 431*(7006), 350–355.

Andersen, J. R., & Lübberstedt, T., (2003). Functional markers in plants. *Trends in Plant Science, 8*, 554–560.

Aneja, B., Yadav, N. R., Chawla, V., & Yadav, R. C., (2012). Sequence-related amplified polymorphism (SRAP) molecular marker system and its applications in crop improvement. *Molecular Breeding, 30*, 1635–1648.

Aouadi, M., Guenni, K., Abdallah, D., Louati, M., Chatti, K., Baraket, G., & Salhi, H. A., (2019). Conserved DNA-derived polymorphism: New markers for genetic diversity analysis of Tunisian *Pistacia vera* L. *Physiology and Molecular Biology of Plants, 25*(5), 1211–1223.

Archangi, A., Heidari, B., & Mohammadi, N. G., (2019). Association between seed yield-related traits and cDNA-AFLP markers in cumin (*Cuminum cyminum*) under drought and irrigation regimes. *Industrial Crops and Products, 133*, 276–283.

Bachem, C. W., Van-Der, H. R. S., Bruijn, S. M., Vreugdenhil, D., Zabeau, M., & Visser, R. G., (1996). Visualization of differential gene expression using a novel method of RNA fingerprinting based on AFLP: Analysis of gene expression during potato tuber development. *Plant Journal, 9*, 745–753.

Bardini, M., Lee, D., Donini, P., & Mariani, A., (2004). Tubulin-based polymorphism (TBP): A new tool, based on functionally relevant sequences, to assess genetic diversity in plant species. *Genome, 291*, 281–291.

Bhat, R. S., Shirasawa, K., Monden, Y., Yamashita, H., & Tahara, M., (2020). Developing transposable element marker system for molecular breeding. *Legume Genomics Methods and Protocols* (Vol. 2107, pp. 233–251). Springer Nature.

Bhatt, J., Kumar, S., Patel, S., & Solanki, R., (2017). Sequence-related amplified polymorphism (SRAP) markers based genetic diversity analysis of cumin genotypes. *Annals of Agrarian Science, 15*(4), 434–438.

Boopathi, N. M., (2012). *Genetic Mapping and Marker Assisted Selection: Basics, Practice and Benefits* (p. 60). Springer Science & Business Media.

Braglia, L., Gavazzi, F., Laura, M., Silvia, G., Peter, N., & Breviario, D., (2020). On the applicability of the tubulin-based polymorphism (TBP) genotyping method: A comprehensive guide illustrated through the application on different genetic resources in the legume family. *Plant Methods, 16*, 86.

Breviario, D., Baird, W. V., Sangoi, S., Hilu, K., Blumetti, P., & Gianì, S., (2007). High polymorphism and resolution in targeted fingerprinting with combined β-tubulin introns. *Molecular Breeding, 20*, 249–259.

Brugmans, B., Del-Carmen, A. F., Bachem, C. W. B., Van, O. H., Van-Eck, H. J., & Visser, R. G. F., (2002). A novel method for the construction of genome wide transcriptome maps. *Plant Journal, 31*, 211–222.

Bryan, G. J., Stephenson, P., Collins, A., Kirby, J., Smith, J. B., & Gale, M. D., (1999). Low levels of DNA sequence variation among adapted genotypes of hexaploid wheat. *Theoretical and Applied Genetics, 99*, 192–198.

Cho, Y. G., Ishii, T., Trmnykh, S., Chen, X., Lipovich, L., McCouch, S. R., Park, W. D., et al., (2000). Diversity of microsatellites derived from genomic libraries and GenBank sequences in rice (*Oryza sativa* L.). *Theoretical of Applied Genetics, 100*, 713–722.

Claverie, M., Souquet, M., Jean, J., Forestier-Chiron, N., Lepitre, V., Pré, M., Jacobs, J., et al., (2012). cDNA-AFLP-based genetical genomics in cotton fibers. *Theoretical and Applied Genetics., 124*, 665–668.

Collard, B. C. Y., & Mackill, D. J., (2009a). Conserved DNA-derived polymorphism (CDDP): A simple and novel method for generating DNA markers in plants. *Plant Molecular Biology Reporter, 27*, 558–562.

Collard, B. C. Y., & Mackill, D. J., (2009b). Start codon targeted (SCoT) Polymorphism: A simple, novel DNA marker technique for generating gene-targeted markers in plants. *Plant Molecular Biology Reporter, 27*, 86.

Desmarais, E., Lanneluc, I., & Lagnel, J., (1998). Direct amplification of length polymorphisms (DALP), or how to get and characterize new genetic markers in many species. *Nucleic Acids Research, 26*(6), 1458–1465.

Etminan, A., Pour-Aboughadareh, Mohammadi, R., AhmadiRad, R., Noori, A., Mahdavian, Z., & Moradi, Z., (2016). Applicability of start codon targeted (SCoT) and inter-simple sequence repeat (ISSR) markers for genetic diversity analysis in durum wheat genotypes. *Biotechnology and Biotechnological Equipment, 30*(6), 1075–1081.

Eujayl, I., Sledge, M. K., Wang, L., May, G. D., Chekhovskiy, K., Zwonitzer, J. C., & Mian, M. A. R., (2004). *Medicago truncatula* EST-SSRs reveal cross-species genetic markers for *Medicago* spp. *Theoretical and Applied Genetics, 108*, 414–422.

Feltus, F. A., Singh, H. P., Lohithaswa, H. C., Schulze, S. R., Silva, T. D., & Paterson, A. H., (2006). A comparative genomics strategy for targeted discovery of single nucleotide polymorphisms and conserved-noncoding sequences in orphan crops. *Plant Physiology, 140*, 1183–1191.

Ge, D., & Daizhen, Z., (2015). Application of sequence-related amplified polymorphism to genetic diversity analysis in *Limonium senense*. *Journal of Genetics, 94*, e35–e38.

Gebhardt, C., & Valkonen, J. P., (2001). Organization of genes controlling disease resistance in the potato genome. *Annual Review on Phytopathology, 39*, 79–102.

Gentzbittel, L., Mestries, E., Mouzeyar, S., Mazeyrat, F., Badaoui, S., Vear, F., Tourvieille De, L. D., & Nicolas, P., (1999). A composite map of expressed sequences and phenotypic traits of the sunflower (*Helianthus annuus* L.) genome. *Theoretical and Applied Genetics, 99*, 218–234.

Glazinska, P., Zienkiewicz, A., Wojciechowski, W., & Kopcewicz, J., (2009). The putative miR172 target gene in APETALA2-like is involved in the photoperiodic flower induction of *Ipomoea nil*. *Journal of Plant Physiology, 166*(16), 1801–1813.

Gorji, A. M., Poczai, P., Polgar, Z., & Taller, J., (2011). Efficiency of arbitrarily amplified dominant markers (SCOT, ISSR and RAPD) for diagnostic fingerprinting in tetraploid potato. *American Journal of Potato Research, 88*, 226–237.

Grzebelus, D., Stawujak, K., Mitoraj, J., & Szklarczyk, M., (2011). Dynamics of Vulmar/Vul-MITE group of transposable elements in *Chenopodiaceae* subfamily betoideae. *Genetica, 139*(9), 1209–1216.

Guang, Y., Jinao, D., Hui, Y., Bingsheng, S., & Ziqing, H., (2011). *cDNA-AFLP* analysis of gene expression differences between the flower bud and sprout-shoot apical meristem of *Angelica sinensis (Oliv.)* Diels. *Genetics and Molecular Biology, 34*(2), 274–279.

Gui, Y., Yan, G., Bo, S., Tong, Z., Wang, Y., Xiao, B., Lu, X., et al., (2011). iSNAP: A small RNA-based molecular marker technique. *Plant Breeding, 130*, 515–520.

Gupta, P. K., & Rustgi, S., (2004). Molecular markers from the transcribed/expressed region of the genome in higher plants. *Functional and Integration Genomics, 4*, 139–162.

Hajibarat, Z., Saidi, A., Hajibarat, Z., & Talebi, R., (2015). Characterization of genetic diversity in chickpea using SSR markers, start codon targeted polymorphism (SCoT) and conserved DNA-Derived polymorphism (CDDP). *Physiology and Molecular Biology of Plants: An International Journal of Functional Plant Biology, 21*(3), 365–373.

Hamidi, H., Talebi, R., & Keshavarzi, F., (2014). Comparative efficiency of functional gene-based markers, start codon targeted polymorphism (SCoT) and conserved DNA-derived polymorphism (CDDP) with ISSR markers for diagnostic fingerprinting in wheat (*Triticum aestivum L.*). *Cereal Research Communications, 42*(4), 558–567.

Han, Z., Wang, C., Song, X., Guo, W., Gou, J., Li, C., Chen, X., & Zhang, T., (2006). Characteristics, development and mapping of *Gossypium hirsutum* derived EST-SSRs in allotetraploid cotton. *Theoretical and Applied Genetics, 112*, 430–439.

Htwe, N. M. P. S., Luo, Z. Q., Jin, L. G., Nadon, B., Wang, K. J., & Qiu, L. J., (2015). Functional marker development of miR1511-InDel and allelic diversity within the genus glycine. *BMC. Genomics, 16*, 467.

Hu, J., & Vick, B. A., (2003). Target region amplification polymorphism: A novel marker technique for plant genotyping. *Plant Molecular Biology Report, 21*, 289–294.

Hu, J., Ochoa, O. E., Truco, M. J., & Vick, B. A., (2005). Application of the TRAP technique to lettuce (*Lactuca sativa L.*) genotyping. *Euphytica, 144*, 225–235.

Ishikawa, G., Yonemaru, J., Saito, M., & Nakamura, T., (2007). PCR-based landmark unique gene (PLUG) markers effectively assign homoeologous wheat genes to A, B and D genomes. *BMC Genomics, 8*, 135.

Jian, X., Zhang, L., Li, G., Zhang, L., Wang, X., Cao, X., Fang, X., & Chen, F., (2010). Identification of novel stress-regulated microRNAs from *Oryza sativa* L. *Genomics, 95*(1), 47–55.

Jones, J. D. G., & Dangl, J. L., (2006). The plant immune system. *Nature, 444*, 323–329.

Kalendar, R., Antonius, K., Smýkal, P., & Schulman, A. H., (2010). iPBS: A universal method for DNA fingerprinting and retrotransposon isolation. *Theoretical and Applied Genetics, 12*, 11419–11430.

Kalendar, R., Grob, T., Regina, M., Suoniemi, A., & Schulman, A., (1999). IRAP and REMAP: Two new retrotransposon-based DNA fingerprinting techniques. *Theoretical of Applied Genetics, 98*, 704–711.

Kashkush, K., Feldman, M., & Levy, A. A., (2002). Gene loss, silencing and activation in a newly synthesized wheat allotetraploid. *Genetics, 160*, 1651–1659.

Khidr, Y. A., Mekuriaw, S. A., Hegazy, A. E., & Amer, E., (2020). Suitability of target region amplified polymorphism (TRAP) markers to discern genetic variability in sweet sorghum. *Journal of Genet Engineering & Biotechnology, 18*, 59.

Kozomara, A., & Griffiths-Jones, S., (2011). miRBase: Integrating microRNA annotation and deep-sequencing data. *Nucleic Acids Research, 39*, 152–157.

Langar, K., Lorieux, M., Desmarais, E., Griveau, Y., Gentzbittel, L., & Bervillé, A., (2003). Combined mapping of DALP and AFLP markers in cultivated sunflower using F9 recombinant inbred lines. *Theoretical and Applied Genetics, 106*, 1068–1074.

Leister, D., Ballvora, A., Salamini, F., & Gebhardt, C., (1996). A PCR-based approach for isolating pathogen resistance genes from potato with potential for wide application in plants. *Nature Genetics, 14*, 421–429.

Lemon, B., & Tjian, R., (2000). Orchestrated response: A symphony of transcription factors for gene control. *Genes and Development, 4*, 2551–2569.

Li, G., & Quiros, C., (2001). Sequence-related amplified polymorphism (SRAP), a new marker system based on a simple PCR reaction: Its application to mapping and gene tagging in *Brassica. Theoretical and Applied Genetics, 103*, 455–461.

Li, T., Guo, J., Li, Y., Ning, H., Sun, X., & Zheng, C., (2013). Genetic diversity assessment of chrysanthemum germplasm using conserved DNA-derived polymorphism markers. *Scientia Horticulturae, 162,* 271–277.

Lu, H., Dong-Fang, M. A., Mao-Lin, H. U., Miao-Miao, H. E., Yan, L. U., & Jin-Xue, J., (2013). Genetic analysis and molecular mapping of an all-stage stripe rust resistance gene in *Triticum aestivum-Haynaldia villosa* translocation line V3. *Journal of Integrative Agriculture, 12*(12), 2197–2208.

Lu, Y., Zhang, C., Li, X., Liang, Y., Wang, Y., & Li, W., (2020). Development of EST-SSR markers and their application in the analysis of the genetic diversity of *Sophora japonica* Linn. *Trees, 34*, 1147–1156.

Luo, C., He, X. H., Chen, H., Ou, S. J., & Gao, M. P., (2010). Analysis of diversity and relationships among mango cultivars using start codon targeted (SCoT) markers. *Biochemical Systematics and Ecology, 38*, 1074–1086.

Ma, Y. S., Yu, H., Li, Y. Y., Yan, H., & Cheng, X., (2008). A study of genetic structure of *Stephania yunnanensis* (*Menispermaceae*) by DALP. *Biochemical Genetics, 46*, 227–240.

Mantovani, P., Van, D. L. C. G., Maccaferri, M., Sanguineti, M. C., & Tuberosa, R., (2007). Nucleotide-binding site (NBS) profiling of genetic diversity in durum wheat. *Genome, 49*, 1473–1480.

Mazaheri, M., Naghavi, M. R., Ghaffari, M. R., Pirseyedi, S. M., Ghareyazie, B., Abdemishani, S., & Mardi, M., (2008). Resistance gene analogue polymorphism (RGAP) markers co-localize with the major QTL of fusarium head blight (FHB) resistance, Qfhs.ndsu-3BS in wheat. *Iranian Journal of Biotechnology, 6*(2), 119–123.

Meyers, B. C., Axtell, M. J., Bartel, B., Bartel, D. P., Baulcombe, D., Bowman, J. L., Cao, X., et al., (2008). Criteria for annotation of plant microRNAs. *Plant Cell, 20*(12), 3186–3890.

Mokate, Y. S., Chimote, V. P., Deshmukh, M. P., Gawai, K., Kale, A. A., & Jadhav, A. S., (2017). Use of promoter-anchored RAPD analysis for divergence assessment in soybean as against conventionally used ISSR assay. *Applied Biological Research 19*(1), 1–9.

Monosi, B., Wisser, R. J., Pennill, L., & Hulbert, S. H., (2004). Full genome analysis of resistance gene homologues in rice. *Theoretical and Applied Genetics, 109*, 1434–1447.

Mutlu, N., Miklas, P., & Coyne, D., (2006). Resistance gene analog polymorphism (RGAP) markers co-localize with disease resistance genes and QTL in common bean. *Molecular Breeding, 17,* 127–135.

Ohshima, K., & Okada, N., (2005). SINEs and LINEs: Symbionts of eukaryotic genomes with a common tail. *Cytogenetics Genome Research, 110*, 475–490.

Palumbi, S., & Baker, C., (1994). Contrasting population structure from nuclear intron sequences and mtDNA of humpback whales. *Molecular Biology and Evolution, 11*, 426–435.

Pang, M., Percy, R. G., Hughs, E., & Zhang, J., (2009). Promoter anchored amplified polymorphism based on random amplified polymorphic DNA (PAAP-RAPD) in cotton. *Euphytica, 167*, 281–291.

Park, K. C., Lee, J. K., Kim, N. H., Shin, Y. B., Lee, J. H., & Kim, N. S., (2003). Genetic variation in *Oryza* species detected by MITE-AFLP. *Genes and Genetics System, 78*(3), 235–243.

Perrot-Minnot, M. J., Lagnel, J., Migeon, A., & Navajas, M., (2000). Tracking paternal genes with DALP markers in a pseudo arrhenotokous reproductive system: Biparental transmission but haplodiploid-like inheritance in the mite *Neoseiulus californicus*. *Heredity, 84*, 702–709.

Poczai, P., Varga, I., Bell, N. E., & Hyvönen, J., (2011). Genetic diversity assessment of bittersweet (*Solanum dulcamara, Solanaceae*) germplasm using conserved DNA-derived polymorphism and intron-targeting markers. *Annals of Applied Biology, 159*, 141–153.

Poczai, P., Varga, I., Laos, M., Cseh, A., Bell, N., Valkonen, J. P., & Hyvönen, J., (2013). Advances in plant gene-targeted and functional markers: A review. *Plant Methods, 9*, 6.

Prakash, C., Trognitz, F. C., Venhuizen, P., Von, H. A., & Trognitz, B., (2020). A compendium of genome-wide sequence reads from NBS (nucleotide binding site) domains of resistance genes in the common potato. *Science Report, 10*, 11392.

Putri, T. K., Putri, A. A., Noladhi, W., Tarkus, S., Vergel, C., & Agung, K., (2020). Genetic diversity of 64 turmeric accessions from Indonesia based on p450-based analogue (PBA) marker. *Buletin Penelitian Tanaman Rempah dan Obat., 31*(2), 123–134.

Rabokon, A., Demkovych, A., Sozinov, A., Kozub, N., Sozinov, I., Pirko, Y., & Blume, Y., (2017). Intron length polymorphism of β-tubulin genes of *Aegilops biuncialis Vis. Cell Biology International, 43*(9), 1031–1039.

Ren, J., Yu, Y., Gao, F., Zeng, L., Lu, X., Wu, X., Yan, W., & Ren, G., (2013). Application of resistance gene analog markers to analyses of genetic structure and diversity in rice. *Genome, 56*(7), 377–387.

Ritter, E., Ruiz-de-Galarreta, J. I., Van-Eck, H. J., & Sánchez, I., (2008). Construction of a potato transcriptome map based on the cDNA–AFLP technique. *Theoretical and Applied Genetics, 116*, 1003–1013.

Satya, P., Karan, M., Jana, S., Mitra, S., Sharma, A., Karmakar, P. G., & Ray, D. P., (2015). Start codon targeted (SCoT) polymorphism reveals genetic diversity in wild and domesticated populations of ramie *(Boehmeria nivea L. Gaudich.)*, a premium textile fiber producing species. *Meta Gene, 3*, 62–70.

Seibt, K. M., Wenke, T., Wollrab, C., Junghans, H., Muders, K., Dehmer, K. J., Diekmann, K., & Schmidt, T., (2012). Development and application of SINE-based markers for genotyping of potato varieties. *Theoretical and Applied Genetics, 125*, 185–196.

Shekhawat, J. K., Rai, M. K., Shekhawat, N. S., & Kataria, V., (2018). Start codon targeted (SCoT) polymorphism for evaluation of genetic diversity of wild population of *Maytenus emarginata. Industrial Crops and Products, 122*, 202–208.

Shi, G., Li, S., Wang, Z., Sun, D., Zhang, S., Guo, J., & Ai, J., (2020). EST-SSR markers development and application in an important medicinal plant, *Schisandra chinensis* (*Schisandraceae*). *Biotechnology & Biotechnological Equipment, 34*(1), 1114–1121.

Singh, R., Panandam, J., Rahman, R., Suan-Choo, Cheah., & Tan, S., (2008). Identification of cDNA-RFLP markers and their use for molecular mapping in oil palm *(Elaeis guineensis)*. *Asia Pacific Journal of Molecular Biology and Biotechnology, 16*, 53–56.

Somerville, C., & Somerville, S., (1999). Plant functional genomics. *Science, 285*, 380–383.

Sormin, S., Purwantoro, A., Setiawan, A. B., & Teo, C. H., (2021). Application of inter-SINE amplified polymorphism (ISAP) markers for genotyping of *Cucumis melo* accessions and its transferability in *Coleus* spp. *Biodiversitas, 22*(5), 2918–2929.

Sunkar, R., Chinnusamy, V., Zhu, J., & Zhu, J. K., (2007). Small RNAs as big players in plant abiotic stress responses and nutrient deprivation. *Trends in Plant Science, 12*(7), 301–309.

Syed, N. H., & Flavell, A. J., (2006). Sequence-specific amplification polymorphisms (SSAPs): A multi-locus approach for analyzing transposon insertions. *Nature Protocol., 1*, 2746–2752.

Tian, Y. P., & Valkonen, J. P., (2012). Mapping of the a virulence determinant of potato virus Y strain O corresponding to the gene Ny for hypersensitive resistance in potato. *Molecular Plant Microbe Interact*. doi: http://dx.doi.org/10.1094/MPMI-09-12-0219-R.

Vali, U., Brandstrom, M., Johansson, M., & Ellegren, H., (2008). Insertion-deletion polymorphisms (indels) as genetic markers in natural populations. *BMC Genetics, 9*, 8.

Vanijajiva, O., (2020). Start codon targeted (SCoT) polymorphism reveals genetic diversity of Manilkara in Thailand. *Biodiversitas., 21*(2), 666–673.

Varshney, R. K., Graner, A., & Sorrells, M. E., (2005). Genomics-assisted breeding for crop improvement. *Trends in Plant Science, 10*, 621–630.

Wang, H., Dai, K., Xiao, J., Yuan, C., Zhao, R., Doležel, J., Wu, Y., et al., (2017). Development of intron targeting (IT) markers specific for chromosome arm 4VS of *Haynaldia villosa* by chromosome sorting and next-generation sequencing. *BMC Genomics, 18*, 167.

Wang, M., Van, D. B. R., Van, D. L. G., & Vosman, B., (2008). The utility of NBS profiling for plant systematics: A first study in tuber-bearing solanum species. *Plant System Evolution, 276*, 137.

Wang, Q., Zhang, B., & Lu, Q., (2009). Conserved region amplification polymorphism (CoRAP), a novel marker technique for plant genotyping in *Salvia miltiorrhiza. Plant Molecular Biology Report, 27*, 139.

Waugh, R., McLean, K., Flavell, A. J., Pearce, S. R., Kumar, A., Thomas, B. B. T., & Powell, W., (1997). Genetic distribution of BARE-1-like retrotransposable elements in the barley genome revealed by sequence-specific amplification polymorphisms (S-SAP). *Molecular Genetics and Genomics, 253*, 687–694.

Wei, H., Fu, Y., & Arora, R., (2005). Intron-flanking EST-PCR markers: From genetic marker development to gene structure analysis in *Rhododendron. Theoretical and Applied Genetics, 111*, 1347–1356.

Wei, H., Wei, L., Sun, X., Zhu, S., & Zhu, J., (2013). Systematic analysis and comparison of nucleotide-binding site disease resistance genes in a diploid cotton *Gossypium raimondii. Plos One, 8*(8), e68435.

Weining, S., & Langridge, P., (1991). Identification and mapping of polymorphisms in cereals based on the polymerase chain reaction. *Theoretical and Applied Genetics, 82*, 209–216.

Wenke, T., Döbel, T., Sörensen, T. R., Junghans, H., Weisshaar, B., & Schmid, T., (2011). Targeted identification of short interspersed nuclear element families shows their widespread existence and extreme heterogeneity in plant genomes. *Plant Cell, 23*, 3117–3128.

Wenke, T., Seibt, K. M., Dobel, T., Muders, K., & Schmidt, T., (2015). Inter-SINE amplified polymorphism (ISAP) for rapid and robust plant genotyping. In: *Plant Genotyping* (pp. 183–192). Springer.

Wolpert, T. J., Dunkle, L. D., & Ciuffetti, L. M., (2002). Host-selective toxins and avirulence determinants: What's in a name? *Annual Review of Phytopathology, 40*, 251–285.

Woodhead, M., Russell, J., Squirrell, J., Hollingsworth, P. M., Cardle, L., Ramsay, L., et al., (2003). Development of EST-SSRs from the alpine lady-fern, *Athyrium distentifolium. Molecular Ecology Notes, 3*, 287–290.

Wu, Q., Zang, F., Xie, X., Ma, Y., Zheng, Y., & Zang, D., (2020). Full-length transcriptome sequencing analysis and development of EST-SSR markers for the endangered species *Populus wulianensis. Scientific Reports, 10*, 16249.

Xiao, D., Liu, S. T., Wei, Y. P., Zohu, D. Y., Hou, X. L., Li, Y., & Hu, C. M., (2016). cDNA-AFLP analysis reveals differential gene expression in incompatible interaction between infected non-heading Chinese cabbage and *Hyaloperonospora parasitica. Horticultural Research, 3*, 16034.

Xiao, Y., Liu, W., Lu, Y. Y., Gong, W. B., & Bian, Y. B., (2010). Applying target region amplification polymorphism markers for analyzing genetic diversity of Lentinula edodes in China. *Journal of Basic Microbiology, 50*(5), 475–483.

Yaakov, B., Ceylan, E., Domb, K., & Kashkush, K., (2012). Marker utility of miniature inverted repeat transposable elements for wheat biodiversity and evolution. *Theoretical and Applied Genetics, 124*(7), 1365–1373.

Yamanaka, S., Suzuki, E., Tanaka, M., Takeda, Y., Watanabe, J. A., & Watanabe, K. N., (2003). Assessment of cytochrome P450 sequences offers a useful tool for determining genetic diversity in higher plant species. *Theoretical and Applied Genetics, 108*, 1–9.

Yan, G. P., Chen, X., Line, R. F., & Wellings, C., (2003). Resistance gene-analog polymorphism markers co-segregating with the YR5 gene for resistance to wheat stripe rust. *Theoretical and Applied Genetics, 106*, 636–643.

Yang, L., Jin, G., Zhao, X., Zheng, Y., Xu, Z., & Wu, W., (2007). PIP: A database of potential intron polymorphism markers. *Bioinformatics, 23*, 2174–2177.

Zerjal, T., Rousselet, A., Mhiri, C., Combes, V., Madur, D., Grandbastien, M. A., Charcosset, A., & Tenaillon, M. I., (2012). Maize genetic diversity and association mapping using transposable element insertion polymorphisms. *Theoretical and Applied Genetics, 124*(8), 1521–1537.

Zhen, Q., Fang, T., Peng, Q., Liao, L., Zhao, L., Owiti, A., & Han, Y., (2018). Developing gene-tagged molecular markers for evaluation of genetic association of apple SWEET genes with fruit sugar accumulation. *Horticulture Research, 5*, 14.

Zheng, J., Yan, Z., Zhao, L., Li, S., Zhang, Z., Garry, R., Yang, W., & Pu, Z., (2014). Molecular mapping of a stripe rust resistance gene in wheat line C15. *Journal of Genetics, 93*, 443–450.

Zietkiewicz, E., Rafalski, A., & Labuda, D., (1994). Genome fingerprinting by simple sequence repeat (SSR)-anchored polymerase chain reaction amplification. *Genomics, 20*, 176–183.

CHAPTER 2

QTL Mapping: Principle, Approaches, and Applications in Crop Improvement

MANASI DASH and ABINASH MISHRA

Department of Plant Breeding and Genetics, College of Agriculture, OUAT, Bhubaneswar, Odisha, India

ABSTRACT

Agriculturally important traits are mostly quantitative traits under polygenic control. The genomic regions associated with a principal quantitative trait are referred to as quantitative trait loci or QTLs. QTL analysis is employed to find these genomic regions with the help of mapping populations, molecular markers, and robust statistical software. Hence, the development of gene-specific genetic or molecular markers followed by screening of the mapping population with the identified polymorphic markers, constructing linkage map, and QTL identification using the statistical program are the key steps in QTL mapping. Initially, molecular markers are tested for polymorphism, which may be codominant or dominant markers. Codominant markers are visualized as differences in band size whereas dominant markers as either presence or absence of bands. Among various DNA marker types, SSR and SNPs are now being commonly employed in making linkage map and QTL analysis. For QTL analysis, a suitable biparental mapping population is essential which can be generated in various ways. Some commonly used biparental mapping populations include F_2-population, F_2-derived F_3-population ($F_{2:3}$), doubled haploids (DHs), recombinant inbred lines (RILs), and near isogenic lines (NILs). In order to generate higher recombination events compared to that in biparental mating designs, MAGIC populations are

Smart Breeding: Molecular Interventions and Advancements for Crop Improvement. Kailash Chandra, Subhash Chand, Ravi Prakash Saini, and Rohit Sharma (Eds.)

also being frequently used in crop improvement strategies. In this, the large numbers of inbred parents are inter-crossed in a specific pattern to pyramid the genes of all the parents into a single line. Once markers and populations are ready, the next step is the methodology for QTL detection, involving various statistical programs based on single-marker analysis (SMA), simple interval mapping (SIM) or composite interval mapping (CIM). The most important use of QTL mapping is the marker-assisted selection to overcome the laborious conventional selection procedures and to improve selection efficiency.

2.1 INTRODUCTION

Many traits such as economic yield, quality, and stress resistance, are regulated by several genes exerting a small additive effect on the concerning traits. Such complex characters are called polygenic or quantitative traits, and the genes are accordingly described as polygenes or minor genes. The regions within a genome harboring minor genes associated with a principal quantitative trait are designated as QTLs. The term QTL was coined by Gelderman in 1975. Analysis of quantitative traits is quite complicated, but a major headway happened in the 1980s due to considerable progress in the use of DNA markers. These markers aided in identifying loci associated with the quantitative traits. The principal use of these molecular markers in plant breeding has mostly been confined to the construction of linkage maps for genetically divergent crop species (Collard et al., 2005). The QTLs are characterized by Mendelian inheritance, have small additive effects, and are controlled by several genes that are influenced variably by the environment. They may be dominant or codominant. Using QTLs, linkage maps are employed to identify chromosomal regions harboring important genes, regulating simple traits and quantitative traits. The linkage map construction and QTL analysis that helps to locate regions within the genomic background associated with the quantitative traits is best described as mapping of loci associated with quantitative traits (QTL mapping) (also called 'genome mapping') (McCouch & Doerge, 1995; Mohan et al., 1997; Paterson et al., 1997). DNA markers that are tightly flanked to agronomically important genes referred to as 'molecular or gene tagging,' are used for marker-assisted selection (MAS) in crop improvement programs (Ribaut & Hoisington, 1998). Hence, generation of crop-specific genetic or molecular markers followed by screening the mapping population with the identified polymorphic DNA markers, constructing association maps and QTL identification using the statistical program are the basic steps in QTL mapping (Meena et al., 2017; Goel et al., 2019). Molecular markers have been extensively used in crop improvement program in varying crop species, including rice (Lu et al., 1997; Xiao et al., 1998; Hittalmani et al., 2003), wheat (Kumar et al., 2006; Li et al., 2017; Ramaya et al., 2017), maize (Guo et al., 2008), chickpea (Rakshit et al., 2003; Churchill et al., 2004; Radhika et al., 2007; Iruela et al., 2009), common bean (Adam-Blondon et al., 1994; Schneider et al., 2001), mung bean (Humphry et al., 2003), pea (Dirlewanger et al., 1994), oilseed rape (Fan et al., 2010; Wu et al., 2013; Yang et al., 2012), groundnut (Luo et al., 1997), soybean (Kulkarni et al., 2017) and tomato (De Vicente & Tanksley, 1993; Alpet et al., 1995; Grandilo & Tanksley, 1996; Kinkade et al., 2003).

2.2 GENETIC MARKERS

Genetic markers are important tools of crop breeding and crop biotechnology. Primarily, genetic markers are allelic variants of the gene or genetic loci that are transmitted to subsequent generations. Simply speaking, a genetic marker, particularly a molecular marker, which is usually a stretch of DNA sequence, sometimes a gene with a known location on a chromosomal thread, can be effectively employed without any ambiguity for identifying individuals or species. Genetic markers with close proximity to certain desirable genes are used in MAS with the adoption of molecular tagging methods (Collard et al., 2005; Kumawat et al., 2020).

Generally, genetic markers may be of the following types:

1. **Phenotypic Markers:** Also called 'classical' or 'visible' markers, these are usually morphological characters (Khan & Spoor, 2001; Douaihy et al., 2012). Primitive crop improvement activities were mostly confined to the selection of easily observable phenotypic traits, including pubescence color, seed coat color, inflorescence color, stem length, presence of colored rinds, etc. Liu & co-workers (1991) introgressed *Rht 10* (reduced height) in the genetic background of *Taigu* nuclear *male-sterile* gene line using backcross (BC) programs. They generated a tight linkage between the dwarfing gene *Rht 10* and the male sterile gene *Ta1*. The dwarfism was subsequently used as a principal criterion for the selection of male sterile lines (Liu, 1991).

2. **Cytological Markers:** These confer polymorphism in terms of chromosome banding patterns and structural features w.r.t. chromosome length, centromere position, etc., in the chromosomes (Gill et al., 1991; Endo, 1996). To address the profuse impediments in the use of morphological markers in crop improvement program, significant efforts were made in the use of cytological markers in the field of plant breeding. The banding pattern or karyotype exhibited by means of different forms, including color, width, and length of bands, were used as 'landmarks' not only for chromosome characterization but also for physical mapping and linkage group identification.
3. **Biochemical Markers:** These include isozymes and protein markers (Joshi & Nguyen, 1997; Joshi et al., 1997). Isozymes being the structural or allelic variant of the enzymes, exhibit differential electrophoretic mobility due to differences in their molecular weights; and hence, they can be effectively used in genetic diversity analysis (Xu, 2010).
4. **Molecular Markers:** Also called DNA markers, these are principal marker types due to their abundant nature, evolving from mutations such as point mutations, insertions, deletions, or due to replication error of repeat sequences in DNA. Molecular markers that indicate the differences between individuals are called polymorphic markers, and the markers, unable to differentiate individuals at DNA level, are referred to as monomorphic markers (Paterson et al., 1997, 1998; Hittalmani et al., 2003; Kinkade & Foolad, 2003; Collard et al., 2005; Wang et al., 2019). The markers may be codominant type or dominant type. Codominant markers specify the differences in band size of DNA, whereas in the case of dominant markers, the bands are either present or absent (Figure 2.1). Hence, codominant markers can discriminate between homozygote plants from the heterozygotes, while dominant markers are unable to do so. DNA markers are broadly classified into following classes based on the technique used for their detection: (i) first-generation/hybridization-based/low through-put: RFLP, VNTR, DArT, etc.; (ii) second-generation/polymerase chain reaction (PCR)-based/medium through-put: RAPD, ISSR, SSR, etc.; and (iii) third generation/DNA sequence-based/high throughput: SNPs (Gupta et al., 1997; Jones et al., 1997; Joshi et al., 1997; Collard et al., 2005).

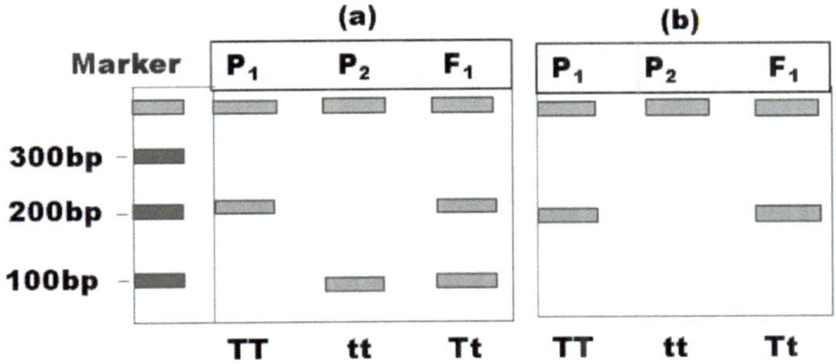

FIGURE 2.1 Banding pattern exhibited by (a) codominant; and (b) dominant markers.

2.3 LINKAGE MAPS

It is a schematic representation of a 'road map,' portraying various molecular or DNA markers in a precise direction and order in a chromosome along with the actual distances between them (Figure 2.2). The distance is assessed from the recombination frequencies between pairs of markers (Paterson et al., 1997, 1998; Collard et al., 2005). A Linkage map is predominantly used in discovering chromosomal positions harboring gene of interest and QTLs, linked with the traits under study. Therefore, such types of maps are designated as QTL maps (or genetic maps). Markers with a recombination frequency of more or equal to 50% are generally located distantly either on the same or on different chromosomes and hence described as 'unlinked.' For the construction of linkage map, the requisites are: (i) mapping population derived from the desirable polymorphic parents; (ii) phenotyping of the population; (iii) genotyping the entire population along with the parents; and (iv) application of software programs for calculating the genetic distances.

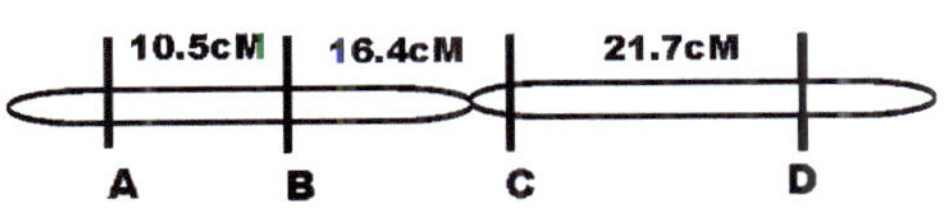

FIGURE 2.2 A hypothetical linkage map constructed using four different genes A, B, C, and D.

2.4 MAPPING POPULATIONS

A plant population derived from hybridization between judiciously selected parental lines and used for locating the position of genetic markers on chromosomes is commonly called a mapping population. The parental lines should be polymorphic for the trait to be studied. Moreover, the trait should be significantly heritable. Hence, it is always desirable to choose parents having extremes phenotypes and confirm the traits heritability. Thus, two or more genetically diverse parents are crossed to develop a mapping population followed by handling the progeny plants in a predefined method (Figure 2.3). The developed mapping population can be used for estimating the 'genetic' distances between genes or pairs of loci, mapping the genes to specific locations in the chromosomes, or for finding DNA markers that are linked to the loci or gene of interest. As the use of a suitable mapping population is very significant in the success of QTL mapping, the procedure to develop some of the commonly used mapping populations and their features are discussed in subsections.

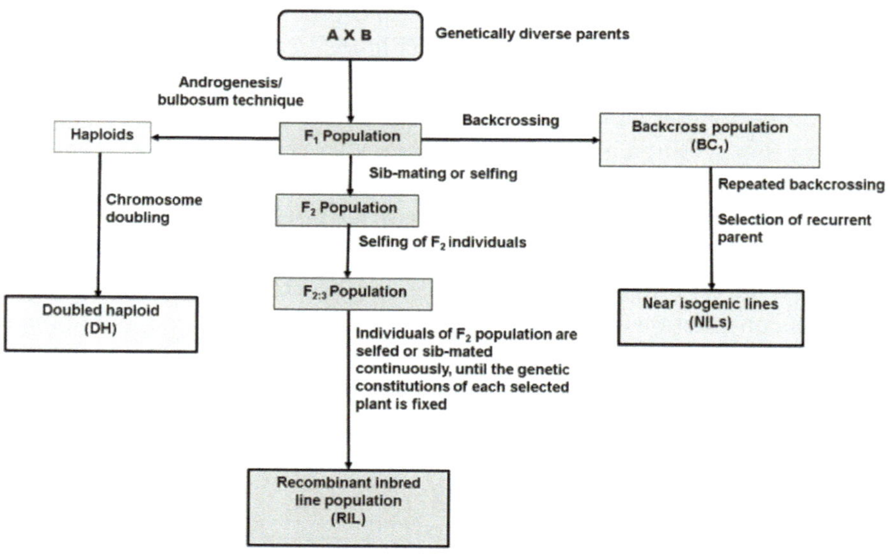

FIGURE 2.3 The methods to develop different types of biparental mapping population.
Source: Modified from: Singh & Singh, (2015).

2.4.1 F_2 POPULATIONS

The F_1 individuals from a cross between two genetically diverse parents are selfed or sib-mated to produce this mapping population (Meksem & Kahl, 2005; Meena et al., 2017). The F_1 individuals are usually heterozygous for all the loci. Hence, each of the F_2 individuals is likely to have obtained a distinct combination of linkage blocks (parts of a chromosome) from the two genetically diverse parents. This distinctiveness is the basis for identifying linkage between loci pairs. For each dominant marker, the expected segregation ratio is 3:1, and for codominant markers, the ratio is 1:2:1 (Table 2.1) (Collard et al., 2005; Meksem & Kahl, 2005). It requires only two generations to develop an F_2 population. However, the population cannot be easily maintained as the F_3 plants are heterozygous due to meiotic recombination events during selfing, but this problem can be greatly addressed by maintaining the F_2 plants as clones (Meksem & Kahl, 2005). A maize F_2 population that was derived from a cross involving a maize landrace and an elite inbred was subjected to QTL analysis. Around 29 putative QTLs affecting 12 principal traits were detected, and a major QTL for ear height (*EH4*) was fine mapped into a 3.23 Mb region on chromosome 8 (Winter & Kahl, 1995). Similar research was carried out involving 170 F_2 cotton individuals. Furthermore, a high-density linkage map harboring 3978 SNP markers and 47 QTLs associating with six traits of early maturity indices was reported (Li et al., 2017).

TABLE 2.1 Anticipated Segregation Ratio for Molecular Markers

Type of Mapping Population	**Segregation Ratio**	
	Codominant Marker	**Dominant Marker**
(a) F_2	1:2:1	3:1
(b) Backcross with the parent having the dominant allele	1:1	1:0
(c) Backcross with the parent having the recessive allele	1:1	1:1
(d) Doubled haploid and RILs	1:1	1:1
(e) NILs	1:1	1:1

2.4.2 F_2 DERIVED F_3 POPULATION

To generate this mapping population, the F_2 individuals are selfed for one generation. Then the seeds are harvested from each individual F_2 plants separately so that the mapping population contains a representative of each F_2 plant. This forms one $F_{2:3}$ population.

The DNA of individual F_2 plant is extracted or a DNA bulk is obtained from a minimum of 20 plants from each $F_{2:3}$ progeny. This bulked DNA is assumed to denote the genotype of the parental F_2 plant (Meksem & Kahl, 2005). Thus, the F_2 genotypes can be reconstituted. Such a population helps in mapping recessive genes. A QTL mapping experiment was done involving $F_{2:3}$ maize individuals produced from the cross Chang 7–2/Zheng 58. A genetic map involving 180 polymorphic SSR markers was constructed, and five QTLs controlling test weight were identified (Ding et al., 2010; Ding, 2011).

2.4.3 BACKCROSS (BC) POPULATION

The F_1 plants are crossed back with either of the two parents to generate a BC mapping population. Genetic analysis is possible if in the BC generations there is detectable phenotypic segregation for the trait of interest. This population has the advantage of easily identifying QTLs that are less epistatic and have linkage drag, especially when one of the parents is wild type. However, this method is not useful for outcrossing or highly heterozygous parents. To reduce the number as well as the size of the donor fragments, repeated crossing back to the recurrent parent is carried out, and as a result, advanced BC lines are developed (Meksem & Kahl, 2005; Meena et al., 2017). From an interspecific BIL population in cotton, derived from a cross between CRI36 (*G. hirsutum*) and Hai 7124 (*G. barbadence*), a novel gene (GhPIN3) for plant height has been identified (Ma et al., 2018). The gene has high expression in apical tissues.

2.4.4 DOUBLED HAPLOIDS (DHS)

These lines comprise of two identical sets of chromosomes and are completely homozygous. They are usually produced by doubling the haploid chromosomes of F_1 plants (Meksem & Kahl, 2005). Haploids can be produced by various methods like elimination of chromosomes from one parent of a cross by wide hybridization, androgenesis, and the *in vitro* culture of unfertilized ovaries and ovules. Parthenogenesis can also be employed (development of an embryo by pseudogamy, semigamy or apogamy) to produce haploid plants. In another approach, haploid-inducing lines, i.e., inducer-based approach, are used to produce haploids (Meksem & Kahl, 2005). DHs are very useful as the time required to obtain fixed inbred is significantly reduced thus, enabling the evaluation of genetically fixed hybrid components at an early stage of the plant selection process. This method has the disadvantage of creating few recombination between linked markers. This technique was applied in constructing a genetic map of a maize cultivar by using 240 doubled haploid lines and 964 SNPs (Shi et al., 2018).

2.4.5 RECOMBINANT INBRED LINES (RILS)

These lines are the most often used mapping population for QTL analysis (Li et al., 2017; Meena et al., 2017). RILs are obtained from continuous selfing or sib mating of an F_2 population until homozygosity is achieved. During selfing, one seed from each line (similar to SSD line) serves as a source for developing the next generation. In the progeny of such lines, no further segregation occurs, as the recombination can no longer change the genetic architecture. RILs can be used as a permanent mapping population that can be reproduced indefinitely. The RILs are thus true breeding homozygous fixed lines that can be used in replicated experiments across different environments or locations or years. As recombination's are more frequent in RILs than in F_2 populations, better resolution is achieved in genetic mapping. The main disadvantage of such a population is that it takes 5–6 years for their development, and also, it is not possible in some species where selfing is a limiting factor. QTL mapping in two soybean RIL populations, namely of ZH (Zhonghuang 24/Huaxia 3) and GB (Gizaol/Brazil 13), has been done for leaf type controlling traits (Wang et al., 2019).

2.4.6 NEAR-ISOGENIC LINES (NILS)

Lines analogous to the anomaly of a single chromosome fragment comprising one to a few loci are called nearly isogenic lines. The development of NILs necessitates many generations of backcrossing supported by MAS. The end product of repeated backcrossing is a backcross inbred line (BIL) which has an almost identical genome as that of the recurrent parent except for the target trait or locus. One or more generations of selfing then produces the final BILs after the final BC. Two more rounds of self-pollination are done at the final stage of the backcrossing program to fix the genomic segments of the donor as well as for visualization of traits under the control of recessive genes (Meksem & Kahl, 2005). The NILs have the advantage that the genetic background remains constant; therefore, these are very precise and statistically strong. Such populations can be very well used in the validation of pilot experiments. However, they are useful only for specific QTLs.

2.4.7 INTROGRESSION LINES

To estimate the effect of small segments of chromosomes, a multitude of introgression lines can be generated, each having a different portion of genomic DNA (Meksem & Kahl, 2005; Perpina et al., 2016). Such collection is often referred to as Exotic Libraries. Such lines can be developed by six generations of recurrent backcrossing assisted with marker-aided selection. This is followed by one or more generations of self-pollination to make the plants homozygous for the introgressed DNA fragments. A genetic library comprising introgression lines was developed from the Japanese cultivar Ginsen Makuwa (var. *makuwa*) into the genetic background of the French Charentais-type cultivar Vedrantasis (var. *cantalupensis*). Phenotyping of the BC generation and the final set of 27 ILs allowed the identification of three stable QTLs for flowering and fruit quality (Perpina et al., 2016).

2.4.8 MULTIPARENT MAPPING POPULATION

The above mapping populations are examples of biparental mapping populations. The occurrence of few recombination events during the development of such biparental mapping populations allows localization of QTL up to only 10–20 cM intervals. So, to overcome these limitations of the biparental mapping population, the development of a multiparent mapping population was proposed. Several multiparent cross populations like 'nested association mapping' (NAM), 'Multi-parent Advanced Generation Inter-Crosses' (MAGIC), and 'random-open-parent association mapping' (ROAM) using different mating designs have been developed (Xiao et al., 1998; Yu & Buckler, 2006). Of these NAM and MAGIC are two commonly employed multiparent mapping populations (Figures 2.4(a) and (b)). The NAM population was initially used in maize to dissect the genetic architecture of some complex traits like flowering duration, leaf architecture, etc. (Yu & Buckler, 2006; Yu et al., 2008). Also, attempts have been made to develop NAM populations in rice, barley, wheat, sorghum, soybean, etc. (Hu et al., 2018).

In multiparent mating designs (also called MAGIC or funnel breeding design), several homozygous/pureline/inbred parents are mated numerous times in a pre-defined pattern to accumulate the genetic background of all the parents into a single plant (Figure 2.4(b)). This method was proposed by Mackay & Powell, (2007) in plants where all possible crosses in half diallel fashion were done and followed by continuous selfing of the F_1 to

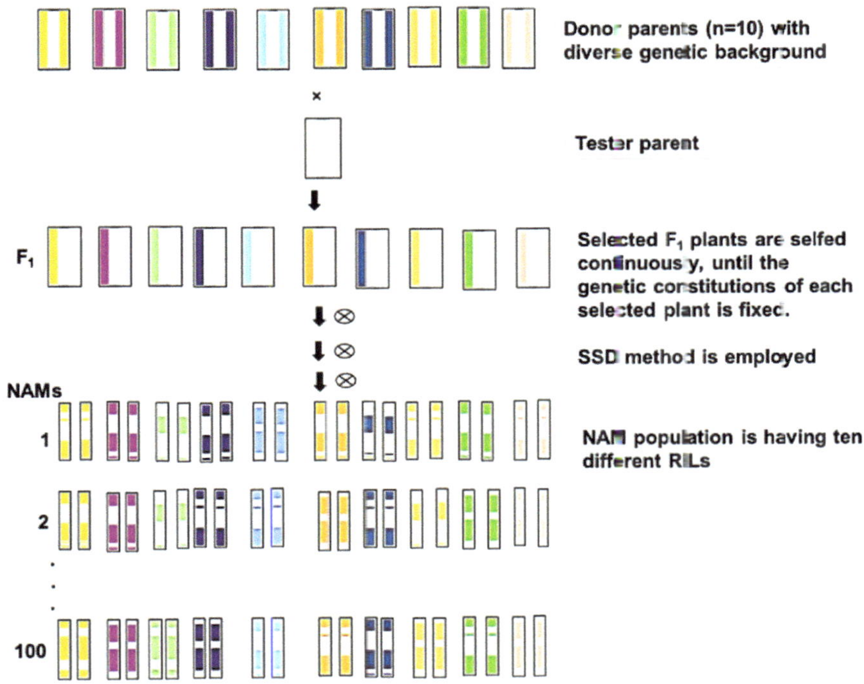

FIGURE 2.4(A) Nested association mapping (NAM) population design.

Source: Modified from: Yu & Buckler (2006).

develop RILs. Usually, three or more 'RILs' (also called MAGIC lines) can be developed from each cross. Such 'MAGIC' lines obtained from all the developed crosses are grouped to create the MAGIC population. It leads to highly diverse lines, each with a unique mosaic of parental alleles (Figure 2.4(b)). The use of a large number of parents and recombination events of a MAGIC population has a clear advantage over a biparental population, while in both the designs, information on the pedigree and genetic structure is available (Churchill et al., 2004; Cavanagh et al., 2008). Now, these methods have the opportunity to be used for fine mapping in crops. MAGIC populations (Table 2.2) have been developed in many crops like rice, tomato, sorghum, barley, and faba bean (Stadlmeier et al., 2018). One considerable benefit of using multi-parent population is the inclusion of a large proportion of genetic variability present within the selected parental lines. Especially in self-pollinated species, a large number of such lines can be created easily and in a cost-effective way. MAGIC populations are also an ideal

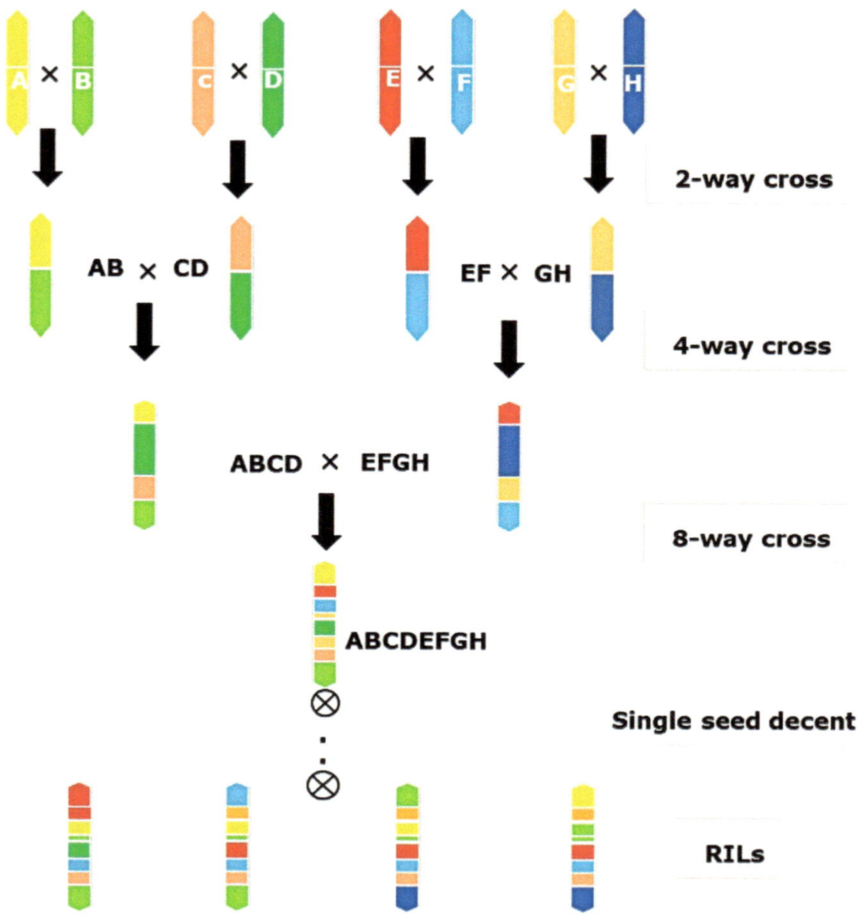

FIGURE 2.4(B) Crossing scheme for MAGIC population using eight parents.

Source: Modified from: Stadlmeier et al. (2018).

resource for generating high-density maps using wild or exotic germplasm. They have their own advantages and disadvantages compared to bi-parental mapping populations but also have many distinct applications. For example, a large set of RILs (>1 k) can be used to identify the epistatic interactions to generate information on the complexity of many traits with respect to the number and the type of genes involved. They will also help in the screening of definite subcategories of the population for important traits (such as yield under stress situations), where definite crop phenology (ex. flowering time) can be exploited simultaneously whilst still incorporating genetic diversity

TABLE 2.2 Use of MAGIC Population in Some Crops

Crop	Experimental Design	Progeny Studied	Phenotyping	Genotyping	Analyzes	References/Research Institutions
Rice	Half-diallel: 35 MAGIC and RIL funnels	S6:8; DH	Yield, grain quality, drought, salinity tolerance, disease resistance,	GBS-SNP	QTL mapping, Genome-wide association mapping (GWAS)	Bandillo et al. (2013)
	Half-diallel: 35 *japonica* MAGIC funnels	S4:6	–	–	–	
	Half-diallel: 35 funnels, then the funnels were inter-crossed to produce MAGIC GLOBAL population	S7	–	–	–	
Wheat	1 funnel	F_6	Height and weight of plant, coleoptile length	DArTs, SNPs, SSRs	Linkage map construction (LMC), QTL mapping	Rebetzke et al. (2014)
	Half-diallel: 315 funnels	F_7	Yield, flowering time, plant height, yellow rust, fusarium, mildew, awn presence	SNP array	QTL mapping, GWAS	Mackay et al. (2014)
	125 funnels	RILs	–	–	–	CISRO
	Half-diallel	RILs	–	–	–	NIAB
	60 parents randomly inter-crossed	F_3	Heading date	SNP array	Linkage disequilibrium analysis (LDA), GWAS	Thepot et al. (2015)

TABLE 2.2 *(Continued)*

Crop	Experimental Design	Progeny Studied	Phenotyping	Genotyping	Analyzes	References/Research Institutions
Maize	Parents (8nos): 35 funnels	F_6 by SSD	Plant height, ear height, yield, flowering time	SNP chip; genotyping-by-sequencing	–	Pea et al. (2013)
Barley	Landraces (7 nos), improved variety (1 nos), 1 funnel	DH lines	Flowering time	SNP chip	LMC, QTL mapping, GWAS	Sannemann et al. (2015)
Chickpea	Desi lines (8nos.)	F_6	–	Re-sequencing	–	Gaur et al. (2012)
Pigeon pea	7 funnels	S1	–	–	–	ICRISAT
Peanut	14 funnels	S1	–	–	–	ICRISAT
Tomato	Normal and cherry tomato (four each)	S3	Fruit weight	SNP chip	LMC, QTL mapping, GWAS	Pascual et al. (2015)

Source: Modified from: Huang et al. (2015).

(Cavanagh et al., 2008). The multi-parental populations confer numerous advantages like:

- A large number of targeted traits from each parent can be analyzed based on the selection of the parents used to make multi-parental cross.
- QTLs can be detected with increased precision and resolution due to more recombination events.
- MAGIC population enables novel rearrangements of alleles by shuffling of parental genes.
- To examine multiple alleles.
- Generate larger genetic variability.
- Likelihood to get the best combination of desirable genes increases.
- Phenotypic selection (PS) in advanced generations reduces the frequency of harmful or undesirable alleles obtained from the donors.
- It is a permanent mapping population for precise mapping of QTLs.

2.5 MARKER GENOTYPING

The next phase in linkage mapping is identifying polymorphic DNA markers that clearly tell the differences between selected parents. For this, it is vital that sufficient genetic variability exists between the parents (Young, 1994; Yang et al., 2012; Wu et al., 2013). Once the polymorphic genic markers have been identified, they are used to screen the developed mapping population as well as the parents for the desired traits. This part of the mapping program is called 'genotyping' of the population. It includes DNA extraction from each plant of the mapping population and using the markers to identify QTLs. But extracting and analyzing each plant from a mapping population is time consuming and tedious. To overcome this laborious and time-consuming step, the bulk segregant analysis (BSA) method can be used for genotyping and QTL analysis for desired traits.

2.5.1 BULK SEGREGANT ANALYSIS (BSA)

This is frequently used as a 'short cut' for molecular tagging of a gene or QTL mapping in the crop improvement program. The method was first suggested for use in crop genetics and breeding by Michelmore in 1991. It is based on the concept where a DNA bulk obtained from a group of individuals sharing the same phenotype, containing all alleles of the selected population. Consequently, two distinct bulked group of segregating individuals differing for a trait will differ only at the locus that harbors the trait. Thus, in BSA method, first, a segregating population is developed from a biparental cross of the phenotypically contrasting parents. The next two bulks of DNA samples are taken from 10–20 individual plants from the segregating population and compared. Within each DNA bulk, the plants are identical for the locus of interest but are arbitrary for the rest other loci. The genetic markers that are polymorphic between the two DNA pools are presumed to be linked to the trait locus. Once the polymorphic markers are identified, genotyping of the mapping populations along with the parents is performed. It helps in QTL analysis and determining the position of QTL easily and rapidly.

2.6 MAKING A LINKAGE MAP

After genotyping and analyzing QTLs for the trait under study, the data is coded for each molecular marker used on each individual of the test population. Then using some suitable software or computer program linkage analysis is done. The distance between the genes and DNA markers is key to creating a linkage map based on the fact that if the distance between genetic markers is large, then the chance of crossing over (CO) occurring during meiosis is high. Hence, in a linkage map, the distance is measured in terms of the frequency of recombination between genetic markers.

Recombination frequency is calculated as = [the number of recombinants/the total number of progenies] × 100.

A mapping function is used to translate the recombination frequencies between two loci into a map distance or centiMorgan (cM) where 1 cM equals to the distance over which one crossover occurs (on average per meiotic event). For small distances (Collard et al., 2005), a mapping function is simply calculated as map distance (d) = recombination fraction (r). When the distance between loci is small (<10 cM to 20 cM), the map distance is equal to the recombination frequency, but for map distances that are greater than 20 cM, this association does not hold good due to the chance of interference. Hence, two mapping functions are commonly used, i.e., the Kosambi mapping function and the Haldane mapping function.

The Kosambi mapping function (Hartl & Jones, 2001) assumes that the occurrence of recombination events in one location interferes with the occurrence of adjacent recombination events. In contrast, the Haldane mapping function assumes no interference between adjacent crossover events, i.e., all crossover events occur independently of one another (Hartl & Jones, 2001).

The Haldane mapping function (Singh & Singh, 2015) amends the recombination frequencies for several CO events considering there is no interference and incidence of a CO does not affect the probability of another CO occurring in the neighboring regions of the chromosome. In the case of multiple CO, the 'Haldane' genetic distance in Morgans (m), is a function of the frequency of recombination (r) as follows: $m = -50 \ln(1-2r)$. 'Kosambi' mapping function (m_k) considers the phenomenon of interference, i.e., CO at a chromosomal site interferes with another CO in its surrounding regions. Thus, the result is that the observed frequency of two simultaneous COs occurring in the neighboring regions of a chromosome is lower than expected (Singh & Singh, 2015). The mapping function is calculated by taking the ratio of the observed frequency of CO to the expected frequency of CO (which is termed as coincidence). Hence, $m_k = 25 \ln[(1 + 2r)/(1 - 2r)]$.

When it is difficult to verify the independent segregation of two genes, the linkage can be tested with Chi-square test and LOD Score. The linkage between genetic markers can be tested using a logarithm of odds (LOD) value or LOD score, i.e., the ratio of the likelihood of linkage (r=0) versus the likelihood of no linkage (null hypothesis, r = 0.5, i.e., among 1,000 plants, there is one chance of crossover). A LOD score greater than 3 rejects the null hypothesis. A LOD score of greater than three between two genic markers indicates that the occurrence of linkage is 1,000 times more likely, and the QTL position is likely to be present between the two linked markers (Figure 2.5). Hence, LOD values of greater than 3 are used to construct linkage maps.

2.7 QTL ANALYSIS

The mapping population is divided into different genotypic groups based on the presence or absence of a particular marker locus (Collard et al., 2005). This helps to determine whether significant differences exist between the groups with respect to a trait under study. Depending on the marker and population type, a significant difference of phenotypic means between the groups (either 2 or 3 groups), indicates that the marker locus is linked to a QTL for the plant character under study (Young, 1994). A significant P value indicates linkage between the concerned marker and QTL. This

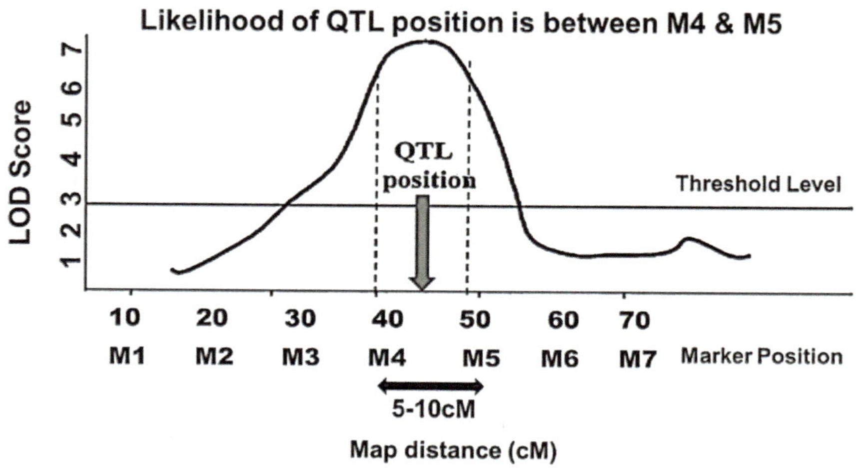

FIGURE 2.5 Relationship between LOD score and marker distance.

Source: Modified from: Collard et al. (2005).

is because the QTL and the marker that is linked to it will be inherited together in the progeny. Hence, the mean of the population having the linked marker will be significantly different from the other group (Collard et al., 2005).

2.7.1 METHODS TO DETECT QTL

The three common methods used for detecting single QTLs are: (i) single-marker analysis (SMA); (ii) simple interval mapping (SIM); and (iii) composite interval mapping (CIM). SMA (also known as 'single-point analysis') is the simplest and most straightforward method that is used for detecting QTLs associated with single genetic markers. The statistical methods used for SMA encompass 't-tests,' 'analysis of variance' (ANOVA), and linear regression (Akond et al., 2019; Brondani et al., 2002; Collard et al., 2005). Of these three, linear regression is the most commonly used method. This is because R^2 (coefficient of determination or regression) from the marker describes the extent of phenotypic variation due to the QTL linked to the marker. The two major disadvantages are that the increase in the distance between QTL and marker significantly decreases the likelihood of detecting a QTL as well as the association of the markers with one or more

QTLs. Some commonly used computer programs for SMA are Q gene and QTX (Collard et al., 2005).

The SIM method helps to make linkage or association maps and simultaneously analyze the distance between adjacent pairs of linked markers. In 1989, Lander & Botstein first suggested the use of SIM in linkage mapping using RFLPs. This method is considered statistically more robust than SMA as it uses linked genetic markers for analyzing the recombination frequency between the markers and the QTLs (Collard et al., 2005). The presence of possible QTL is assessed if the LOD ratios exceeds a critical threshold value of 3 (Figure 2.5).

CIM is now used commonly for mapping QTLs as this method has the advantage of combining interval mapping with linear regression. Also, it can include additional markers in addition to the adjacent pair of linked markers (Lu et al., 1997; Collard et al., 2005; Ramayya et al., 2017). Thus, compared to SMA at mapping QTLs, CIM is more precise and effective. Also, 'multiple interval mapping' (MIM) has also become popular for the mapping of various quantitative genes.

2.7.2 ANALYZING INTERVAL MAPPING RESULTS

An identified QTL can be considered statistically significant when the critical threshold value or maximum map distance transcends a specified significance level. The determination of significance thresholds can be done by applying permutation tests or a LOD score.

The efficiency of RIL, F_2, and BC populations to detect recombinants by using either codominant or dominant markers is known as the amount of information per individual (i_p). This is the inverse of the variance divided by population size. In the case of an RI population information per individual is calculated as $2/r(1 + 2r)^2$. The amount of information per individual is usually represented by the logarithm of i_p and is plotted against the recombination fraction (r). An RI population is very efficient for closely linked markers (Reiter et al., 1992). A number of software can be used for the data curation and QTL analysis (Table 2.3).

To summarize, the major steps involved in QTL mapping are parent selection, development of a suitable mapping population followed by marker phenotyping and genotyping followed by linkage analysis, and finally, map construction (Figure 2.6).

TABLE 2.3 Software Packages Available for QTL Mapping

Software Package	Applicability in the Type of Mapping Population	Functionality
MapMaker/QTL	F_2, BC, RIL	Interval mapping
QTL cartographer	F_2, BC	Composite interval mapping
MapQTL	F_2, BC, RIL, DH, heterozygous F_1	Interval mapping
Mapmanager QTL	F_2, BC	Interval mapping using regression
Q gene	F_2, BC	Linear regression
HAPPY	General MAP	QTL analysis; permutation
R/qtl	MAGIC (4-way cross, 8-way cross, 16-way cross)	Simulation; constructing map; QTL analysis
R/ricalc	MAGIC (by selfing), Sib mating	Simulation; probability calculation
R/mpMap	MAGIC (4-way, 8-way cross) by selfing	Simulation; constructing map; QTL analysis
R/spclust	NAM, MAGIC (4-way, 8-way cross) by selfing	Selective phenotyping
R/mpwgaim	MAGIC (4-way, 8-way cross) by selfing	QTL analysis
Alpha MPSim	General MPP	Simulation

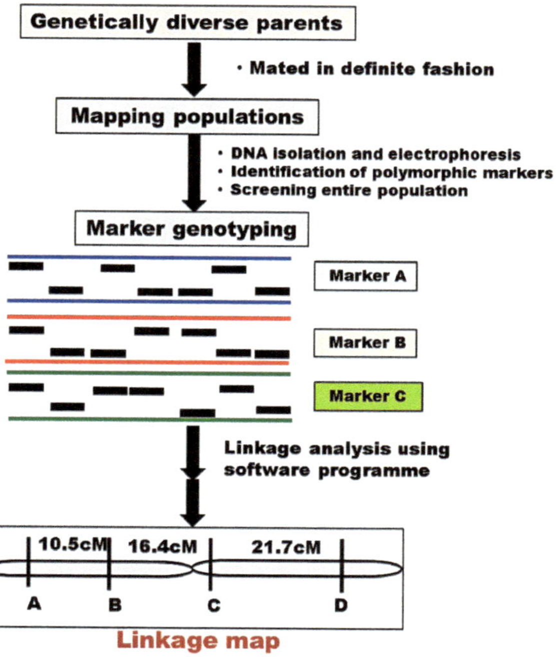

FIGURE 2.6 Major steps in construction of a hypothetical linkage map.

2.8 APPLICATIONS OF QTL MAPPING IN CROP IMPROVEMENT

A plant breeder may not know the precise location of QTL, but it can be introgressed using marker-assisted backcrossing (MABC). In maize, the QTLs with major effects conferring resistance to downy mildew have been identified and transferred into popular varieties. The introgression of QTLs into elite lines/germplasm and MAS for QTLs in crop improvement has been undertaken in many crops like rice, wheat, maize, tomato, pulses, and oilseeds. Much success has been achieved in the development of improved rice varieties by pyramiding or introgressing several genes (QTLs) for various traits employing MAS (Table 2.4). In many crops, including rice, potential QTLs have been mapped using different mapping populations for various growth, physiological, and yield attributing traits (Table 2.5). QTLs so identified for various traits in different crops can be used in crop improvement to enhance the yield by introgression of QTLs for yield attributing traits as well as to develop biotic and abiotic stress-resistant elite lines.

TABLE 2.4 List of Rice Varieties Improved by MAS

Rice Varieties/ Genotypes	Improved Varieties Developed by MAS	Character	Target Genes	References
Pusa basmati 1	Improved Pusa Basmati-1	Bacterial blight	*Xa13*, *Xa21*	Kottapalli et al. (2010)
Samba Mahsuri	Improved Samba Mahsuri	Bacterial blight	*Xa5*, *Xa13*, *Xa21*	
Swarna	Swarna sub-1	Submergence	*Sub1*	Nandi et al. (1997)
IR64	IR64 sub-1	Submergence	*Sub1*	Reddy et al. (2009)
Pusa RH10	PRR78/IRBB60	Rice blast	*Pi54* and *Pix5*	Singh et al. (2011)
Pusa RH 10	Pusa 6A, Pusa 6B	Bacterial blight	*Xa13* and *Xa21*	
KMR3 restorer	KMR3/IRBB60	Bacterial blight	*Xa4*, *Xa5*, *Xa13*, and *Xa21*	Shanti et al. (2010)
Lalat	Improved Lalat	Bacterial blight	*Xa4*, *Xa5*, *Xa13*, and *Xa21*	Dokku et al. (2013)
Tapaswini	Improved Tapaswini	Bacterial blight	*Xa4*, *Xa5*, *Xa13*, and *Xa21*	
Mangeumbye	Improved Mangeumbye	Bacterial blight	*Xa4*, *Xa5*, and *Xa21*	Suh et al. (2013)
PRR78	Improved Pusa RH 10	Rice blast	*Pi54* and *Pix5*	Singh et al. (2013)

TABLE 2.5 List of Important QTLs Identified for Different Characters in Major Crops

Crop	Characters	QTLs (Allele) Identified	References
Rice	Grain yield (under drought stress)	*qDTY 1.1* (RM11943-RM431), *qtl12.1* (RM28048-RM511)	Bernier et al. (2007); Vikram et al. (2011)
	Panicle length	*pl9*	Zhu et al. (2018)
	Plant height	*Ph1* (RZ730-RRZ801), *Ph4* (RZ590-RG214)	Yan et al. (1998)
	1,000 grain weight	*qtgw5.1* (chr05-6059062-chr05-6145897)	Bhatia et al. (2018)
	Grains per panicle	*qGN1c*, *qTSN12.2* (RM28746-RM28753)	Sasaki et al. (2017); Xu et al. (2019)
	Heading dates	*QHd3A*, *QHd8A* (RG20-RG1034)	Li et al. (1995)
	Root length and thickness	*qRT9*	Li et al. (2015)
	Shoot fresh weight (salt stress)	*qSWD2* (RM1287)	Haq et al. (2008)
	Shoot dry weight (salt stress)	*qSFD1* (RM6289), *qSFD2* (RG544)	
	Rice protein content	*qRPC-5* (RG-435-RG 127a), *q RPC-7* (ZG34B-G20)	Hu et al. (2004)
	Rice amylose %	*Wx 1-1* (WX-UIL3), *q AC9.3*	Ando et al. (2010)
	Spikelet fertility	*spf 5* (RM13-RM405), *spf 6* (RG213-RM557), *spf 8* (H4698-1-H3878-1)	Song et al. (2005)
	Pollen fertility	*pf 5* (MRG1010-RM13), *pf 12* (RM19-RM247)	
	Embryo sac fertility	*Ef6* (C11-RM276)	
	Cold tolerance	*q CTS12* (RM5746-RM3103), *q SCT1a* (RM3602-RM5310)	Andaya & Tai (2006); Kim et al. (2014)
	Seedling salt tolerance	*qSH1.2* (RM5389-RM5759); *qSH4* (RM518-RM16535)	Wang et al. (2012)
Wheat	Grain yield	*QGy.ccsu-1D.1* (Xgwm957-Xgwm232), *QGy.ccsu-2D.2* (Xwmc41-Xgwm608)	Kumar et al. (2006)
	Biological yield	*QBy.ccsu-4A.2* (Xgwm959-Xgwm832)	
	Harvest index	*QHi.ccsu-4A.2* (Xgwm959-Xgwm832)	
	Spikelet per spike	*QSps.ccsu-4A.5* (Xgwm959-Xgwm832)	
Maize	Plant height	*W ph-1*, *s-ph 2*, *s-ph 7*	Guo et al. (2008)
	1,000 kernel weight	*w-100 gw1*, *s-100gw2*, *Q qkwei 4*	
	Kernel yield	*W gy-1*	

TABLE 2.5 *(Continued)*

Crop	Characters	QTLs (Allele) Identified	References
Chickpea	Ascochyta blight resistance	*QTLs, Ar 19,* QTL_{AR3}	Rakshit et al. (2003); Iruela et al. (2009)
	Seed per pod	*Spp*	Radhika et al. (2007)
	Double podding	*Sfl*	
	Seed coat thickness	QTL_{Tt}	Cobos et a . (2009)
Common bean	Anthracnose resistance	*Are*	Adam-Blondon et al. (1994)
	Fusarium wilt resistance	*PvPR1*	Schneider et al. (2001)
Pea	Powdery mildew resistance	*Er*	Dirlewanger et al. (1994)
Mung bean	Powdery mildew resistance	*QTLs*	Humphry et al. (2003)
Lathyrus	Ascochyta blight resistance	*QTL*	Skiba et al (2004)
Oilseed rape (*Brassica napus*)	Silique length	*cqSLA-9*	Yang et al. (2012)
	Seed weight	*CqSWA-9*	
	High oleic and low linoleic	*FAD2*	
	1,000 seed weight	*TSWA 7a, TSWA 7b*	Fan et al. (2010)
	Sclerotia stem rot resistance	*QTL SRO6*	Wu et al. (2013)
Ground nut	Pod length	*q PLA 05.7*	Luo et al. (1997)
	100 seed weight	*q HPWA 05.6*	
Soya bean	100 seed weight	*q SW-02-1*	Kulkarni et al. (2017)
Tomato	Plant height	*Ht-6, ht-5b*	De Vicente & Tanksley, (1993)
	Number of flower buds	*Buds 3*	
	Days to first flower	*Dff-3, Dff-5*	
	Fruit shape	*Off s 8.1*	Grandilo et al. (1996)
	Fruit weight	*fw 2.2*	Alpet et al (1995)
	Carpel shape	*Fs 8.1*	Ku et al. (2000)
	Lycopene content	*lyc 12.1*	Kinkade & Foolad, (2003)

2.9 CONCLUSION

QTL mapping helps in dissecting the complex loci and thereby provides greater insight into the molecular architecture of these complexes. As discussed, the initial mapping process involves careful selection of parents exhibiting contrasting values for the said traits. A suitable mapping population is essential for both phenotyping and genotyping. Nowadays, MAGIC population has been given more emphasis over several bi-parental populations due to the use of many parental lines possessing various important traits. A wide array of software for QTL analysis, such as MapMaker, QGene, and QTL cartographer, are freely available. Once the position of markers flanking the target genes, governing several agronomically important traits, are accomplished, MAS can be subsequently employed in improving genotypes for several biotic and abiotic traits. Several varieties have been developed by utilizing this strategy in different crops, particularly rice. Improved Pusa Basmati-1 and Swarna sub-1 are the two most important achievements using MAS in rice. The use of QTL mapping addresses some issues that the breeder came across while using conventional selection procedures and, ultimately, improves the efficiency of selecting suitable genotypes.

KEYWORDS

- **bulk segregant analysis**
- **composite interval mapping**
- **doubled haploids**
- **genetic markers**
- **linkage map**
- **mapping population**
- **near-isogenic lines**
- **QTL analysis**
- **recombinant inbred lines**
- **simple interval mapping**

REFERENCES

Adam-Blondon, A. F., Sévignac, M., Bannerot, H., & Dron, M., (1994). SCAR, RAPD and RFLP markers linked to a dominant gene (Are) conferring resistance to anthracnose in common bean. *Theoretical and Applied Genetics, 88*(6), 865–870.

Akond, Z. M., Alam, J., Hasan, M. N., Uddin, M. S., Alam, M., & Mollah, M. N. H., (2019). A comparison on some interval mapping approaches for QTL detection. *Bioinformation, 15*(2), 90.

Alpert, K. B., Grandillo, S., & Tanksley, S. D., (1995). fw 2.2: A major QTL controlling fruit weight is common to both red-and green-fruited tomato species. *Theoretical and Applied Genetics, 91*(6, 7), 994–1000.

Andaya, V. C., & Tai, T. H., (2006). Fine mapping of the qCTS12 locus, a major QTL for seedling cold tolerance in rice. *Theoretical and Applied Genetics, 113*(3), 467–475.

Ando, I., Sato, H., Aoki, N., Suzuki, Y., Hirabayashi, H., Kuroki, M., & Takeuchi, Y., (2010). Genetic analysis of the low-amylose characteristics of rice cultivars Oborozuki and Hokkai-PL9. *Breeding Science, 60*(3), 187–194.

Bandillo, N., Raghavan, C., Muyco, P. A., Sevilla, M. A. L., Lobina, I. T., Dilla-Ermita, C. J., & Leung, H., (2013). Multi-parent advanced generation inter-cross (MAGIC) populations in rice: Progress and potential for genetics research and breeding. *Rice, 6*(1), 1–15.

Bernier, J., Kumar, A., Ramaiah, V., Spaner, D., & Atlin, G., (2007). A large-effect QTL for grain yield under reproductive-stage drought stress in upland rice. *Crop Science, 47*(2), 507–516.

Bhatia, D., Wing, R. A., Yu, Y., Chougule, K., Kudrna, D., Lee, S., & Singh, K., (2018). Genotyping by sequencing of rice interspecific backcross inbred lines identifies QTLs for grain weight and grain length. *Euphytica, 214*(2), 1–16.

Brondani, C., Rangel, P., Brondani, R., & Ferreira, M., (2002). QTL mapping and introgression of yield-related traits from *Oryza glumaepatula* to cultivated rice (*Oryza sativa*) using microsatellite markers. *Theoretical and Applied Genetics, 104*(6), 1192–1203.

Cavanagh, C., Morell, M., Mackay, I., & Powell, W., (2008). From mutations to MAGIC: Resources for gene discovery, validation and delivery in crop plants. *Current Opinion in Plant Biology, 11*(2), 215–221.

Churchill, G. A., Airey, D. C., Allayee, H., Angel, J. M., Attie, A. D., Beatty, J., & Zou, F., (2004). The collaborative cross, a community resource for the genetic analysis of complex traits. *Nature Genetics, 36*(11), 1133–1137.

Cobos, M. J., Winter, P., Kharrat, M., Cubero, J. I., Gil, J., Millan, T., & Rubio, J., (2009). Genetic analysis of agronomic traits in a wide cross of chickpea. *Field Crops Research, 111*(1, 2), 130–136.

Collard, B. C., Jahufer, M. Z. Z., Brouwer, J. B., & Pang, E. C. K., (2005). An introduction to markers, quantitative trait loci (QTL) mapping and marker-assisted selection for crop improvement: The basic concepts. *Euphytica, 142*(1), 169–196.

DeVicente, M. C., & Tanksley, S. D., (1993). QTL analysis of transgressive segregation in an interspecific tomato cross. *Genetics, 134*(2), 585–596.

Ding, J. Q., Ma, J. L., Zhang, C. R., Dong, H. F., Xi, Z. Y., Xia, Z. L., & Wu, J. Y., (2011). QTL mapping for test weight by using F 2: 3 population in maize. *Journal of Genetics, 90*(1), 75–80.

Dirlewanger, E., Isaac, P. G., Ranade, S., Belajouza, M., Cousin, R., & De Vienne, D., (1994). Restriction fragment length polymorphism analysis of loci associated with disease resistance genes and developmental traits in *Pisum sativum* L. *Theoretical and Applied Genetics, 88*(1), 17–27.

Dokku, P., Das, K. M., & Rao, G. J. N., (2013). Genetic enhancement of host plant-resistance of the Lalat cultivar of rice against bacterial blight employing marker-assisted selection. *Biotechnology Letters, 35*(8), 1339–1348.

Douaihy, B., Sobierajska, K., Jasińska, A. K., Boratyńska, K., Ok, T., Romo, A., & Boratyński, A., (2012). Morphological versus molecular markers to describe variability in *Juniperus excelsa* subsp. *excelsa* (*Cupressaceae*). *AoB Plants, 2012*.

ENDo, T. R., (1986). Complete identification of common wheat chromosomes by means of the C-banding technique. *The Japanese Journal of Genetics, 61*(1), 89–93.

Fan, C., Cai, G., Qin, J., Li, Q., Yang, M., Wu, J., & Zhou, Y., (2010). Mapping of quantitative trait loci and development of allele-specific markers for seed weight in *Brassica napus*. *Theoretical and Applied Genetics, 121*(7), 1289–1301.

Gaur, P. M., Jukanti, A. K., & Varshney, R. K., (2012). Impact of genomic technologies on chickpea breeding strategies. *Agronomy, 2*, 199–221.

Gill, B. S., Friebe, B., & Endo, T. R., (1991). Standard karyotype and nomenclature system for description of chromosome bands and structural aberrations in wheat (*Triticum aestivum*). *Genome, 34*(5), 830–839.

Goel, S., Singh, K., Singh, B., Grewal, S., Dwivedi, N., Alqarawi, A. A., & Singh, N. K., (2019). Analysis of genetic control and QTL mapping of essential wheat grain quality traits in a recombinant inbred population. *PLoS One, 14*(3), e0200669.

Grandillo, S., Ku, H. M., & Tanksley, S. D., (1996). Characterization of fs8. 1, a major QTL influencing fruit shape in tomato. *Molecular Breeding, 2*(3), 251–260.

Guo, J., Su, G., Zhang, J., & Wang, G., (2008). Genetic analysis and QTL mapping of maize yield and associate agronomic traits under semi-arid land condition. *African Journal of Biotechnology, 7*(12).

Gupta, P. K., Varshney, R. K., Sharma, P. C., & Ramesh, B., (1999). Molecular markers and their applications in wheat breeding. *Plant Breeding, 118*(5), 369–390.

Haq, T. U., Akhtar, J., Gorham, J. O. H. N., Steele, K. A., & Khalid, M., (2008). Genetic mapping of QTLs, controlling shoot fresh and dry weight under salt stress in rice (*Oryza sativa* L.) cross between CO39× moroberekan. *Pakistan Journal of Botany, 40*(6), 2369–2381.

Hartl, D. L., & Jones, E. W., (2009). *Genetics: Analysis of Genes and Genomes*. Jones & Bartlett Learning.

Hittalmani, S., Huang, N., Courtois, B., Venuprasad, R., Shashidhar, H. E., Zhuang, J. Y., & Khush, G. S., (2003). Identification of QTL for growth-and grain yield-related traits in rice across nine locations of Asia. *Theoretical and Applied Genetics, 107*(4), 679–690.

Hu, J., Guo, C., Wang, B., Ye, J., Liu, M., Wu, Z., & Liu, K., (2018). Genetic properties of a nested association mapping population constructed with semi-winter and spring oilseed rapes. *Frontiers in Plant Science, 9*, 1740.

Hu, Z. L., Li, P., Zhou, M. Q., Zhang, Z. H., Wang, L. X., Zhu, L. H., & Zhu, Y. G., (2004). Mapping of quantitative trait loci (QTLs) for rice protein and fat content using doubled haploid lines. *Euphytica, 135*(1), 47–54.

Huang, B. E., Verbyla, K. L., Verbyla, A. P., & Cavanagh, C. R., (2015). MAGIC populations in crops: Current status and future prospects. *Theoretical and Applied Genetics, 128*, 999–1017.

Humphry, M. E., Magner, T., McIntyre, C. L., Aitken, E. A., & Liu, C. J., (2003). Identification of a major locus conferring resistance to powdery mildew (*Erysiphe polygoni* DC) in mungbean (*Vigna radiata* L. Wilczek) by QTL analysis. *Genome, 46*(5), 738–744.

Iruela, M., Pistón, F., Cubero, J. I., Millán, T., Barro, F., & Gil, J., (2009). The marker SCK13 603 associated with resistance to *Ascochyta* blight in chickpea is located in a region of a putative retrotransposon. *Plant Cell Reports, 28*(1), 53–60.

Jones, N., Ougham, H., & Thomas, H., (1997). Markers and mapping: We are all geneticists now. *The New Phytologist, 137*(1), 165–177.

Joshi, C. P., & Nguyen, H. T., (1993). RAPD (random amplified polymorphic DNA) analysis based intervarietal genetic relationships among hexaploid wheats. *Plant Science, 93*(1, 2), 95–103.

Joshi, S. P., Ranjekar, P. K., & Gupta, V. S., (1999). Molecular markers in plant genome analysis. *Current Science*, 230–240.

Khan, S., & Spoor, W., (2001). Use of molecular and morphological markers as a quality control in plant tissue culture. *Pakistan Journal of Biological Sciences, 4*(4), 479–482.

Kim, S. M., Suh, J. P., Lee, C. K., Lee, J. H., Kim, Y. G., & Jena, K. K., (2014). QTL mapping and development of candidate gene-derived DNA markers associated with seedling cold tolerance in rice (*Oryza sativa* L.). *Molecular Genetics and Genomics, 289*(3), 333–343.

Kinkade, M. P., & Foolad, M. R., (2013). Validation and fine mapping of lyc12. 1, a QTL for increased tomato fruit lycopene content. *Theoretical and Applied Genetics, 126*(8), 2163–2175.

Kottapalli, K. R., Narasu, M. L., & Jena, K. K., (2010). Effective strategy for pyramiding three bacterial blight resistance genes into fine grain rice cultivar, Samba Mahsuri, using sequence tagged site markers. *Biotechnology Letters, 32*(7), 989–996.

Ku, H. M., Grandillo, S., & Tanksley, S. D., (2000). fs8. 1, a major QTL, sets the pattern of tomato carpel shape well before anthesis. *Theoretical and Applied Genetics, 101*(5, 6), 873–878.

Kulkarni, K. P., Asekova, S., Lee, D. H., Bilyeu, K., Song, J. T., & Lee, J. D., (2017). Mapping QTLs for 100-seed weight in an interspecific soybean cross of Williams 82 (*Glycine max*) and PI 366121 (Glycine soja). *Crop and Pasture Science, 68*(2), 148–155.

Kumar, N., Kulwal, P. L., Balyan, H. S., & Gupta, P. K., (2007). QTL mapping for yield and yield contributing traits in two mapping populations of bread wheat. *Molecular Breeding, 19*(2), 163–177.

Kumawat, G., Kumawat, C. K., Chandra, K., Pandey, S., Chand, S., Mishra, U. N., & Sharma, R., (2020). Insights into marker assisted selection and its applications in plant breeding. In: *Plant Breeding-Current and Future Views*. IntechOpen.

Li, J., Han, Y., Liu, L., Chen, Y., Du, Y., Zhang, J., & Zhao, Q., (2015). qRT9, a quantitative trait locus controlling root thickness and root length in upland rice. *Journal of Experimental Botany, 66*(9), 2723–2732.

Li, L., Zhao, S., Su, J., Fan, S., Pang, C., Wei, H., & Yu, S., (2017). High-density genetic linkage map construction by F_2 populations and QTL analysis of early-maturity traits in upland cotton (*Gossypium hirsutum* L.). *PLoS One, 12*(8), e0182918.

Li, S., Jia, J., Wei, X., Zhang, X., Li, L., Chen, H., & Li, L., (2007). A intervarietal genetic map and QTL analysis for yield traits in wheat. *Molecular Breeding, 20*(2), 167–178.

Li, Z., Pinson, S. R. M., Stansel, J. W., & Park, W. D., (1995). Identification of quantitative trait loci (QTLs) for heading date and plant height in cultivated rice (*Oryza sativa* L.). *Theoretical and Applied Genetics, 91*(2), 374–381.

Liu, B. H., (1991). Development and prospects of dwarf male-sterile wheat. *Chinese Science Bulletin, 36*(4), 306.

Lu, C. F., Shen, L. S., Tan, Z., Xu, Y., He, P., Chen, Y., & Zhu, L., (1996). Comparative mapping of QTLs for agronomic traits of rice across environments using a doubled haploid population. *Theoretical and Applied Genetics, 93*(8), 1211–1217

Luo, H., Ren, X., Li, Z., Xu, Z., Li, X., Huang, L., & Jiang, H., (2017). Co-localization of major quantitative trait loci for pod size and weight to a 3.7 cM interval on chromosome A05 in cultivated peanut (*Arachis hypogaea* L.). *BMC Genomics, 18*(1), 1–12.

Ma, J., Pei, W., Ma, Q., Geng, Y., Liu, G., Liu, J., & Yu, J., (2019). QTL analysis and candidate gene identification for plant height in cotton based on an interspecific backcross inbred line population of *Gossypium hirsutum* × *Gossypium barbadense*. *Theoretical and Applied Genetics, 132*(9), 2663–2676.

Mackay, I. J., Bansept-Basler, P., Barber, T., Bentley, A. R., Cockram, J., Gosman, N., & Howell, P. J., (2014). An eight-parent multiparent advanced generation inter-cross population for winter-sown wheat: Creation, properties, and validation. *G3: Genes, Genomes, Genetics, 4*(9), 1603–1610.

McCough, S. R., & Doerge, R. W., (1995). QTL mapping in rice. *Trends in Genetics, 11*(12), 482–487.

Meena, A. K., Ramesh, M., Nagaraju, C., & Kumhar, B. L., (2017). A review of QTL mapping in cotton: Molecular markers, mapping populations and statistical methods. *International Journal of Current Microbiology and Applied Sciences, 6*, 3057–3080.

Meksem, K., & Kahl, G., (2006). *The Handbook of Plant Genome Mapping: Genetic and Physical Mapping*. John Wiley & Sons.

Mohan, M., Nair, S., Bhagwat, A., Krishna, T. G., Yano, M., Bhatia, C. R., & Sasaki, T., (1997). Genome mapping, molecular markers and marker-assisted selection in crop plants. *Molecular Breeding, 3*(2), 87–103.

Nandi, S., Subudhi, P. K., Senadhira, D., Manigbas, N. L., Sen-Mandi, S., & Huang, N., (1997). Mapping QTLs for submergence tolerance in rice by AFLP analysis and selective genotyping. *Molecular and General Genetics MGG, 255*(1), 1–8.

Pascual, L., Desplat, N., Huang, B. E., Desgroux, A., Bruguier, L., Bouchet, J. P., Le, Q. H., et al., (2015). Potential of a tomato MAGIC population to decipher the genetic control of quantitative traits and detect causal variants in the resequencing era. *Plant Biotechnology Journal, 13*(4), 565–577.

Paterson, A. H., Damon, S., Hewitt, J. D., Zamir, D., Rabinowitch, H. D., Lincoln, S. E., & Tanksley, S. D., (1991). Mendelian factors underlying quantitative traits in tomato: Comparison across species, generations, and environments. *Genetics, 127*(1), 181–197.

Paterson, A. H., Lander, E. S., Hewitt, J. D., Peterson, S., Lincoln, S. E., & Tanksley, S. D., (1988). Resolution of quantitative traits into Mendelian factors by using a complete linkage map of restriction fragment length polymorphisms. *Nature, 335*(6192), 721–726.

Pea, G., Dell'Acqua, M., Hlaing, A. L. L., & Pe, M. E., (2013). *From Mice to Maize: A Multiparental Population for Fine Mapping in Zea Mays.* MAGIC populations workshop. http://openwetware.org/images/e/e6/MatteoDellAcqua_MaizePoster.pdf (accessed on 04 January 2022).

Perpiñá, G., Esteras, C., Gibon, Y., Monforte, A. J., & Picó, B., (2016). A new genomic library of melon introgression lines in a cantaloupe genetic background for dissecting desirable agronomical traits. *BMC Plant Biology, 16*(1), 1–21.

Radhika, P., Gowda, S. J. M., Kadoo, N. Y., Mhase, L. B., Jamadagni, B. M., Sainani, M. N., & Gupta, V. S., (2007). Development of an integrated intraspecific map of chickpea (*Cicer arietinum* L.) using two recombinant inbred line populations. *Theoretical and Applied Genetics, 115*(2), 209–216.

Rakshit, S., Winter, P., Tekeoglu, M., Muñoz, J. J., Pfaff, T., Benko-Iseppon, A. M., & Kahl, G., (2003). DAF marker tightly linked to a major locus for *Ascochyta* blight resistance in chickpea (*Cicer arietinum* L.). *Euphytica, 132*(1), 23–30.

Ramya, P., Chaubal, A., Kulkarni, K., Gupta, L., Kadoo, N., Dhaliwal, H. S., & Gupt, V., (2010). QTL mapping of 1000-kernel weight, kernel length, and kernel width in bread wheat (*Triticum aestivum* L.). *Journal of Applied Genetics, 51*(4), 421–429.

Rebetzke, G. J., Verbyla, A. P., Verbyla, K. L., Morell, M. K., & Cavanagh, C. R., (2014). Use of a large multiparent wheat mapping population in genomic dissection of coleoptile and seedling growth. *Plant Biotechnology Journal, 12*(2), 219–230.

Reddy, C. S., Babu, A. P., Swamy, B. M., Kaladhar, K., & Sarla, N., (2009). ISSR markers based on GA and AG repeats reveal genetic relationship among rice varieties tolerant to drought, flood, or salinity. *Journal of Zhejiang University Science B, 10*(2), 133–141.

Reiter, R. S., Williams, J. G., Feldmann, K. A., Rafalski, J. A., Tingey, S. V., & Scolnik, P. A., (1992). Global and local genome mapping in *Arabidopsis thaliana* by using recombinant inbred lines and random amplified polymorphic DNAs. *Proceedings of the National Academy of Sciences, 89*(4), 1477–1481.

Ribaut, J. M., & Hoisington, D., (1998). Marker-assisted selection: New tools and strategies. *Trends in Plant Science, 3*(6), 236–239.

Sannemann, W., Huang, B. E., Mathew, B., & Leon, J., (2015). Multi-parent advanced generation inter-cross in barley: High-resolution quantitative trait locus mapping for flowering time as a proof of concept. *Molecular Breeding, 35*, 86.

Sasaki, K., Fujita, D., Koide, Y., Lumanglas, P. D., Gannaban, R. B., Tagle, A. G., & Ishimaru, T., (2017). Fine mapping of a quantitative trait locus for spikelet number per panicle in a new plant type rice and evaluation of a near-isogenic line for grain productivity. *Journal of Experimental Botany, 68*(11), 2693–2702.

Schneider, K. A., Grafton, K. F., & Kelly, J. D., (2001). QTL analysis of resistance to *Fusarium* root rot in bean. *Crop Science, 41*(2), 535–542.

Shanti, M. L., Shenoy, V. V., Devi, G. L., Kumar, V. M., Premalatha, P., Kumar, G. N., & Freeman, W. H., (2010). Marker-assisted breeding for resistance to bacterial leaf blight in popular cultivar and parental lines of hybrid rice. *Journal of Plant Pathology*, 495–501.

Shi, Z., Zhang, R., Xing, J., Duan, M., Wang, Y., Su, A., & Zhao, J., (2018). QTL mapping of three ear traits using a doubled haploid population of maize. *Plant Breeding, 137*(5), 706–713.

Singh, A. K., Gopalakrishnan, S., Singh, V. P., Prabhu, K. V., Mohapatra, T., Singh, N. K., & Marathi, B., (2011). Marker assisted selection: A paradigm shift in basmati breeding. *Indian Journal of Genetics and Plant Breeding, 71*(2), 120.

Singh, B. D., & Singh, A. K., (2015). *Marker-Assisted Plant Breeding: Principles and Practices*. Springer (India) Pvt. Ltd.

Singh, V. K., Singh, A., Singh, S. P., Ellur, R. K., Singh, D., Gopala, K. S., & Singh, A. K., (2013). Marker-assisted simultaneous but stepwise backcross breeding for pyramiding blast resistance genes Piz5 and Pi54 into an elite basmati rice restorer line 'PRR 78'. *Plant Breeding, 132*(5), 486–495.

Skiba, B., Ford, R., & Pang, E. C. K., (2004). Construction of a linkage map based on a *Lathyrus sativus* backcross population and preliminary investigation of QTLs associated with resistance to *Ascochyta* blight. *Theoretical and Applied Genetics, 109*(8), 1726–1735.

Song, X., Qiu, S. Q., Xu, C. G., Li, X. H., & Zhang, Q., (2005). Genetic dissection of embryo sac fertility, pollen fertility, and their contributions to spikelet fertility of intersubspecific hybrids in rice. *Theoretical and Applied Genetics, 110* (2), 205–211.

Stadlmeier, M., Hartl, L., & Mohler, V., (2018). Usefulness of a multiparent advanced generation intercross population with a greatly reduced mating design for genetic studies in winter wheat. *Frontiers in Plant Science, 9*, 1825.

Suh, J. P., Jeung, J. U., Noh, T. H., Cho, Y. C., Park, S. H., Park, H. S., & Jena, K. K., (2013). Development of breeding lines with three pyramided resistance genes that confer broad-spectrum bacterial blight resistance and their molecular analysis in rice. *Rice, 6*(1), 1–11.

Thépot, S., Restoux, G., Goldringer, I., Hospital, F., Gouache, D., Mackay, I., & Enjalbert, J., (2015). Efficiently tracking selection in a multiparental population: The case of earliness in wheat. *Genetics, 199*(2), 609–623.

Vikram, P., Swamy, B. M., Dixit, S., Ahmed, H. U., Cruz, M. T. S., Singh, A. K., & Kumar, A., (2011). qDTY 1.1, a major QTL for rice grain yield under reproductive-stage drought stress with a consistent effect in multiple elite genetic backgrounds. *BMC Genetics, 12*(1), 1–15.

Wang, J., Zhang, X., & Lin, Z., (2018). QTL mapping in a maize F2 population using genotyping-by-Sequencing and a modified fine-mapping strategy. *Plant Science, 276*, 171–180.

Wang, L., Cheng, Y., Ma, Q., Mu, Y., Huang, Z., Xia, Q., & Nian, H., (2019). QTL fine-mapping of soybean (*Glycine max* L.) leaf type associated traits in two RILs populations. *BMC Genomics, 20*(1), 1–15.

Wang, Z., Cheng, J., Chen, Z., Huang, J., Bao, Y., Wang, J., & Zhang, H., (2012). Identification of QTLs with main, epistatic and QTL× environment interaction effects for salt tolerance in rice seedlings under different salinity conditions. *Theoretical and Applied Genetics, 125*(4), 807–815.

Winter, P., & Kahl, G., (1995). Molecular marker technologies for plant improvement. *World Journal of Microbiology and Biotechnology, 11*(4), 438–448.

Wu, J., Cai, G., Tu, J., Li, L., Liu, S., Luo, X., & Zhou, Y., (2013). Identification of QTLs for resistance to sclerotinia stem rot and BnaC. IGMT5. a as a candidate gene of the major resistant QTL SRC6 in *Brassica napus*. *PloS One, 8*(7), e67740.

Xiao, J., Li, J., Grandillo, S., Ahn, S. N., Yuan, L., Tanksley, S. D., & McCouch, S. R., (1998). Identification of trait-improving quantitative trait loci alleles from a wild rice relative, *Oryza rufipogon*. *Genetics, 150*(2), 899–909.

Xu, Y., (2010). *Molecular Plant Breeding*. CABI.

Xu, Z., Miao, Y., Chen, Z., Gao, H., Wang, R., Zhao, D., & Liu, Q., (2019). Identification and fine mapping of qGN1c, a QTL for grain number per panicle, in rice (*Oryza sativa*). *Molecular Breeding, 39*(9), 1–12.

Yan, J., Zhu, J., He, C., Benmoussa, M., & Wu, P., (1998). Molecular dissection of developmental behavior of plant height in rice (*Oryza sativa* L.). *Genetics, 150*(3), 1257–1265.

Yang, P., Shu, C., Chen, L., Xu, J., Wu, J., & Liu, K., (2012). Identification of a major QTL for silique length and seed weight in oilseed rape (*Brassica napus* L.). *Theoretical and Applied Genetics, 125*(2), 285–296.

Young, N. D., (1994). Constructing a plant genetic linkage map with DNA markers. In: *DNA-Based Markers in Plants* (pp. 39–57). Springer, Dordrecht.

Yu, J., & Buckler, E. S., (2006). Genetic association mapping and genome organization of maize. *Current Opinion in Biotechnology, 17*(2), 155–160.

Yu, J., Holland, J. B., McMullen, M. D., & Buckler, E. S., (2008). Genetic design and statistical power of nested association mapping in maize. *Genetics, 178*(1), 539–551.

Zhu, Z., Li, X., Wei, Y., Guo, S., & Sha, A., (2018). Identification of a novel QTL for panicle length from wild rice (*Oryza minuta*) by specific locus amplified fragment sequencing and high density genetic mapping. *Frontiers in Plant Science, 9*, 1492.

CHAPTER 3

Use of Cutting-Edge Technologies for the Improvement of Pulse Crops

BENJAMIN KARIKARI,[1,2] JOHN K. AHIAKPA,[2,3] JAVAID A. BHAT,[4] and SYLVESTER N. T. T. ADDY[5]

[1]Department of Agricultural Biotechnology, Faculty of Agriculture, Food and Consumer Sciences, University for Development Studies, Nyankpala Campus, Tamale, Ghana

[2]Organization of African Academic Doctors (OAAD), Off Kamiti Road, Nairobi, Kenya

[3]Research Desk Consulting Limited, Kwabenya-Accra, Ghana

[4]International Genome Center, Jiangsu University, Zhenjiang, China

[5]Legumes and Oil Seed Improvement Department, Council for Scientific and Industrial Research-Crops Research Institute (CSIR-CRI), Fumesua-Kumasi, Ghana

ABSTRACT

Pulses have significant potential in the battle against malnutrition, particularly hidden hunger. Aside from that, they serve as food security crops through their symbiotic associations with nitrogen-fixing bacteria, which lessens the need for farmers to apply nitrogen-based fertilizers to their fields. As a result of the application of next-generation sequencing (NGS) and high-throughput (HT) technologies in molecular biology, genotyping by sequencing (GBS) has substantially lowered sequencing costs and made the development of genomic and molecular markers more accessible and achievable at a low cost and investment. The application of GBS in pulse crops (lentil, common

Smart Breeding: Molecular Interventions and Advancements for Crop Improvement. Kailash Chandra, Subhash Chand, Ravi Prakash Saini, and Rohit Sharma (Eds.)

bean, field pea, and mung bean), such as linkage mapping, genome-wide association studies (GWAS), and genomic selection (GS) is discussed in this chapter. Gene-based single nucleotide polymorphism (SNP) markers derived from GBS have been developed and used for linkage and GWAS mappings in pulses to elucidate genetic bases for a number of economically, nutritionally, and environmentally significant climate-resilient traits. The markers associated with these significant features could be employed in marker-assisted breeding to improve the breeding of these traits. Genes associated with each of these traits were discovered by mining the genomic areas connected together. These genes were then functionally confirmed to determine their functional roles in the specific traits. GBS-based GS studies in pea and lentils have been reported by several researchers. GS was used to increase grain yield in pea recombinant inbred populations under severe terminal drought escape and intrinsic drought tolerance (DT). GS outperformed both MAS and phenotypic selection (PS) in terms of selection efficiency, with at least 18% greater selection efficiency. It is anticipated that the information offered in this chapter will be useful in future meta-QTL analyzes.

3.1 INTRODUCTION

The world's population is expected to reach 10 billion by 2050 (Hickey et al., 2019). The predicted increase in the global population, along with a decline in arable lands for agriculture, climate change, and an increase in biotic and abiotic pressures, pose a severe danger to food and nutritional security. Approximately 2 billion people worldwide do not have enough food to consume, and nearly a billion people suffer from malnutrition (Von Grebmer et al., 2019). Due to the fact that more than half of the world's population suffers from malnutrition, particularly hidden hunger (Gödecke et al., 2018; Kaur et al., 2020), pulses have significant potential in the fight against malnutrition (Hall et al., 2017). Pulses are a good source of plant-based protein since they are low in fat and contain no cholesterol. They are also high in fiber, a good source of folate, and a good supply of minerals such as potassium (K), iron (Fe), magnesium (Mg), calcium (Ca), and zinc (Zn). Moreover, they are low in glycemic index, which can aid in the stabilization of blood sugar and insulin levels, making them ideal for persons with diabetes and perfect for weight management (McCrory et al., 2010; Ramdath et al., 2016). When compared to other crops, the seeds of these pulses may be stored for an extended period of time. Pulses contribute to food security *via* their symbiotic associations with nitrogen-fixing bacteria, which reduces the application of nitrogen-based fertilizers on farmers' fields (Foyer et al., 2016). It reduces the cost of production and ensures sustainable agriculture through farming practices, such as crop rotation, which replenishes soil fertility.

It is projected that the world's population expansion requires an increase in food production by 100–110% to meet the needs of this frightening demographic and industrial demands (Ray et al., 2013; Tilman et al., 2011). Plant breeding is a strategic technique for producing greater food output to satisfy the world's predicted population expansion. Plant breeding is a part of science that deals with modifying the genetic makeup of plants to generate plants with desired characteristics for end-users. Plant breeding is performed utilizing conventional and molecular approaches. A key shortcoming of conventional plant breeding methods is the longer duration required for the growth of numerous generations to select and assess desirable genotypes (Tester & Langridge, 2010). It can impede the progress needed toward the global food and nutritional security goal as proposed by the United Nations by 2030. The latter strategy uses less than half of the period needed for the former strategy for breeding and as a result, it offers a more promising approach to breed and produce new varieties with optimum yield, nutritional qualities and tolerance to stress to feed the rising population and meet industrial needs (Goel et al., 2020; Lenaerts et al., 2019; Varshney et al., 2020).

Molecular markers are one of the tools for molecular breeding. Molecular markers are deployed in numerous molecular breeding procedures, including linkage mapping, genome-wide association studies (GWAS), marker-assisted breeding (MAB), and genomic selection (GS). Until recently, genetic and molecular marker data for pulses were limited (Bohra et al., 2020; Varshney et al., 2019a). The introduction of next-generation sequencing (NGS) and high-throughput (HT) technologies in molecular biology has substantially lowered sequencing costs and made the development of genomic and molecular markers more accessible and achievable at low cost and investment. One such technique is genotyping by sequencing (GBS), developed by (Elshire et al., 2011) as a robust, cost-effective, highly multiplexed sequencing strategy for extremely varied species. GBS has been applied to sequence over 20,000 soybean germplasm accessions in the United States Department of Agriculture (USDA) soybean gene bank (Song et al., 2015a), which has been employed by researchers in the soybean research community (Sonah et al., 2015; Wu et al., 2020a; Ziegler et al., 2018). The same method has been applied to several crops, including millet

(Kumar et al., 2016a), rice (Ali et al., 2018; Yadav et al., 2019), and wheat (Chu et al., 2020).

This chapter reviews and summarizes the success made in utilizing GBS in pulses (lentil, common bean, field pea, and mung bean) improvement projects in the areas of linkage, GWAS mapping, and GS. This chapter also sheds light on the potential for plant breeding in the future.

3.2 GBS EFFORTS IN PULSES

In the field of sequencing, several NGS procedures have been created and put into practice. GBS, complexity reduction of polymorphic sequences (CRoPS), double digest RADSeq (ddRADSeq) and its variants (ezRADSeq, DArTSeq), genome reduction on restriction site conservation (GR-RSC), and sequence-based polymorphic marker technology are some of the techniques used. Reduced representation libraries (RRLs) are one type. It is possible to employ these technologies even if you do not have prior information of the whole genome sequence (WGS) (Ray & Satya, 2014; Toonen et al., 2013). The dramatic decrease in sequencing costs and the availability of reference genomes for the majority of species in the last decade, it has become much more possible to generate SNPs/markers for linkage mapping, and GWAS.

Following the development of its protocol by (Elshire et al., 2011), GBS has emerged as the most robust, HT, cost-effective, and straightforward technology for obtaining thousands of markers from large numbers of individuals among the above-mentioned NGS technologies. Illumina platforms based on sequencing-by-synthesis technology continue to be the most used NGS system for GBS (Bohra et al., 2020), with an average read length ranging from 125 to 300 bp (Hu et al., 2018). However, there is typically a substantial percentage of data that is always missing. It has been possible to create hundreds of single nucleotide polymorphisms (SNPs) using GBS from Illumina platforms for a number of pulses (Table 3.1). These markers have been employed in huge breeding programs for a variety of purposes, including linkage mapping, GWAS, genomic selection (GS), genetic diversity, haplotype, and allele mining, germplasm characterization, and quality control. According to Campa et al. (2018), the genetic diversity, population structure, and linkage disequilibrium (LD) in 308 Spanish common beans with 3,099 SNPs by GBS were all investigated by GBS.

TABLE 3.1 Platforms Used for Genotyping by Sequencing in Selected Pulse Crops

Pulse	Genotyping Platform	Number of SNPs/InDels	References
Lentil	Illumina Hiseq 2500	2,66,356	Wong et al. (2015)
Pea	Illumina HiSeq 2000	95,740	Annicchiarico et al. (2017)
Faba bean	Illumina iScan reader (Illumina Inc.)	22,217	Sudheesh et al. (2019)
Adzuki bean	Illumina HiSeq 2000	2,347	Kang et al. (2015)
Mung bean	Illumina Hiseq 2000	1,993	Kang et al. (2014)
Common bean	Illumina HiSeq 2500	7,530	Schröder et al. (2016)
Common bean	Illumina HiSeq 2000	77,595	Ariani et al. (2016)
Pea	HiSeq 2000	64,263	Boutet et al. (2016)
Common bean	Illumina HiSeq	1,112	Bhakta et al. (2015)
Common bean	HiSeq 2500	84,416	Wen et al. (2019)
Field Pea	Illumina HiSeq™ 2500 platform	16,877	Gali et al. (2019)
Lentil	Illumina HiSeq 2500	6,693	Pavan et al. (2019)
Mung bean	Illumina HiSeq 4000	76,160	Reddy et al. (2020)

3.2.1 *HIGH-DENSITY LINKAGE MAPS IN PULSES*

It has been possible to gain a complete and comprehensive understanding of the genetic foundation underpinning complex quantitative traits in numerous crops through the construction of high-density genetic maps and their use in the detection of quantitative trait loci (QTLs) and genes (Geleta et al., 2020; Gong et al., 2019; Karikari et al., 2019; Khan et al., 2020; Li et al., 2019; Su et al., 2020; Wang et al., 2020; Zhang et al., 2019a). A high-density SNP map, such as that employed by Khan et al. (2020), made it possible to refine the mapping of QTLs and candidate genes for *Aspergillus flavus* resistance in peanuts (*Arachis hypogaea*). Their research identified two QTLs, *qRAF-3-1* and *qRAF-14-1*, which were fine mapped within a physical distance of 1.44 Megabase pair (Mbp) and 2.22 Mbp, respectively, and associated with 67 and 137 model genes, respectively. They discovered six genes with the same function in both QTL regions as a result of this analysis. High-density genetic linkage maps are incredibly significant tools for genomic and genetic studies, and they serve as the foundation for the genetic improvement of pulses.

Significant progress has been made in pulses by using high-value density maps in the breeding programs (Table 3.1). High-density genetic maps have been produced for the legume mung bean Hwang (2015), with 798 SNPs spanning a total length of 757.39 centimeters (cM) and an average distance between markers of 0.95 cM. In their study, Ma et al. (2020) developed a high-density genetic map that covered 978.1 cM of the complete genome and had 2865 SNPs at the density of 2.9 markers per cM, covering a total of 978.1 cM in the genome. As an example, Aldemir et al. (2017) found a total of 4177 SNPs on seven linkage groups (LGs) spanning 497.1 cM, with an average spacing between markers of 0.12 cM between each marker. Gali et al. (2018) constructed a high-density genetic map of dry field bean for three different recombinant inbred lines (RILs): PR-02 (Orb CDC Striker), PR-07 (Carerra CDC Striker), and PR-15 (1-2347-144 CDC Meadow), with 1560, 2990, and 3077 SNPs derived from GBS, respectively. Smitchger (2017) used field bean in the construction of a high-density genetic map, which included 4753 SNP markers and described a genetic map of an 8-way multiparent advanced generation intercross (MAGIC) population derived from eight common bean breeding lines that was just released (*SXB412*, *INB827*, *ALB213*, *SEN56*, *SCR2*, *MIB778*, *SCR9*, and *INB841*). The genetic map for the MAGIC population was generated with a total length of 857 cM, with a recombination rate of 2,407 kbp/cM in the pericentromeric region and 291 kbp/cM in the euchromatic regions of the chromosomes. The pericentromeric region had a recombination rate of 2,407 kbp/cM, while the euchromatic regions had a recombination rate of 603 kbp/cM. The mean number of recombination events per chromosome and the mean number of recombination events per RIL in the complete genome were 3.22% and 44%, respectively (Smitchger, 2017). Other authors have worked with the same approach, creating high-density genetic maps for a variety of pulses, including the common bean (Diaz et al., 2020; Murube et al., 2020; Tock et al., 2017; Vasconcellos et al., 2017), lentil (Bhadauria et al., 2017), mung bean (Schafleitner et al., 2016; Sokolkova et al., 2020), and lima bean (Mhora et al., 2016).

3.2.2 GBS PROTOCOLS AND ANALYTICAL PIPELINES FOR PULSES

Following the discovery of GBS by Elshire et al. (2011), a number of methods and analytical pipelines have been developed (Glaubitz et al., 2014; Poland et al., 2012; Poland & Rife, 2012; Sonah et al., 2013). The eight-step method presented in this section provides a summary of the protocols and analytical pipelines used in pulses (Figure 3.1). Isolation of DNA and a quality control test are the first two processes. A restriction enzyme is then used to digest the extracted DNA. This step is primarily accomplished using a single common restriction enzyme, ApeKI, in accordance with the procedure outlined by Elshire et al. (2011). However, a two-enzyme system involving PstI and MspI has been used in lentil (Bhadauria et al., 2017; Ma et al., 2020; Wong et al., 2015), field pea (Gali et al., 2019; Jha et al., 2020), and common bean with MseI and TaqI enzymes (Schröder et al., 2016). This procedure, devised by Poland et al. (2012), has been shown to result in a better degree of complexity reduction as well as a more uniform library for sequencing than the first enzyme restriction protocol with ApeKI (He et al., 2014).

Using restriction enzyme(s) to digest DNA results in fragments that are ligated to two adaptors: one adaptor with a forward primer for amplification, sequencing platform primer sites, and a unique DNA barcode that allows sample multiplexing (Figure 3.1). Pooling, randomly shearing, and size selection are used to create a uniform collection of DNA fragments that are all the same size. A second adaptor is added to the fragments, which allows only fragments that have been ligated to the first adaptor to amplify. The sixth stage entails multiplex enrichment of polymerase chain reaction (PCR) products (a set of SNPs has been defined for a section of the genome with PCR primers designed to amplify the areas of interest). In the following step, the PCR-amplified library is cleaned and assessed using an automated capillary sizing device. After that, the sequencing process, which is generally done on the Illumina HiSeq platform, takes place as the penultimate step in producing the FASTQ file (Figure 3.1). A handful of bioinformatic analyzes are required in the last step, which is outside the scope of this chapter.

3.3 GBS-BASED QTL MAPPING EFFORTS IN PULSES

Pulses are threatened by both abiotic and biotic stressors, which make their production on a global scale unsustainable. As a result, significant progress has been made in recent years in unraveling the genetic architecture of important abiotic and biotic stressors, yield, and its components, as well as some seed quality parameters, using a GBS-based linkage mapping technique (Table 3.2).

3.3.1 *BIOTIC STRESSES*

There are different types of biotic stressors that affect pulses, the majority of which are pests and diseases. There are 20 diseases caused by viruses, fungi, and bacteria that damage the mung bean plant, as well as 84 pests that

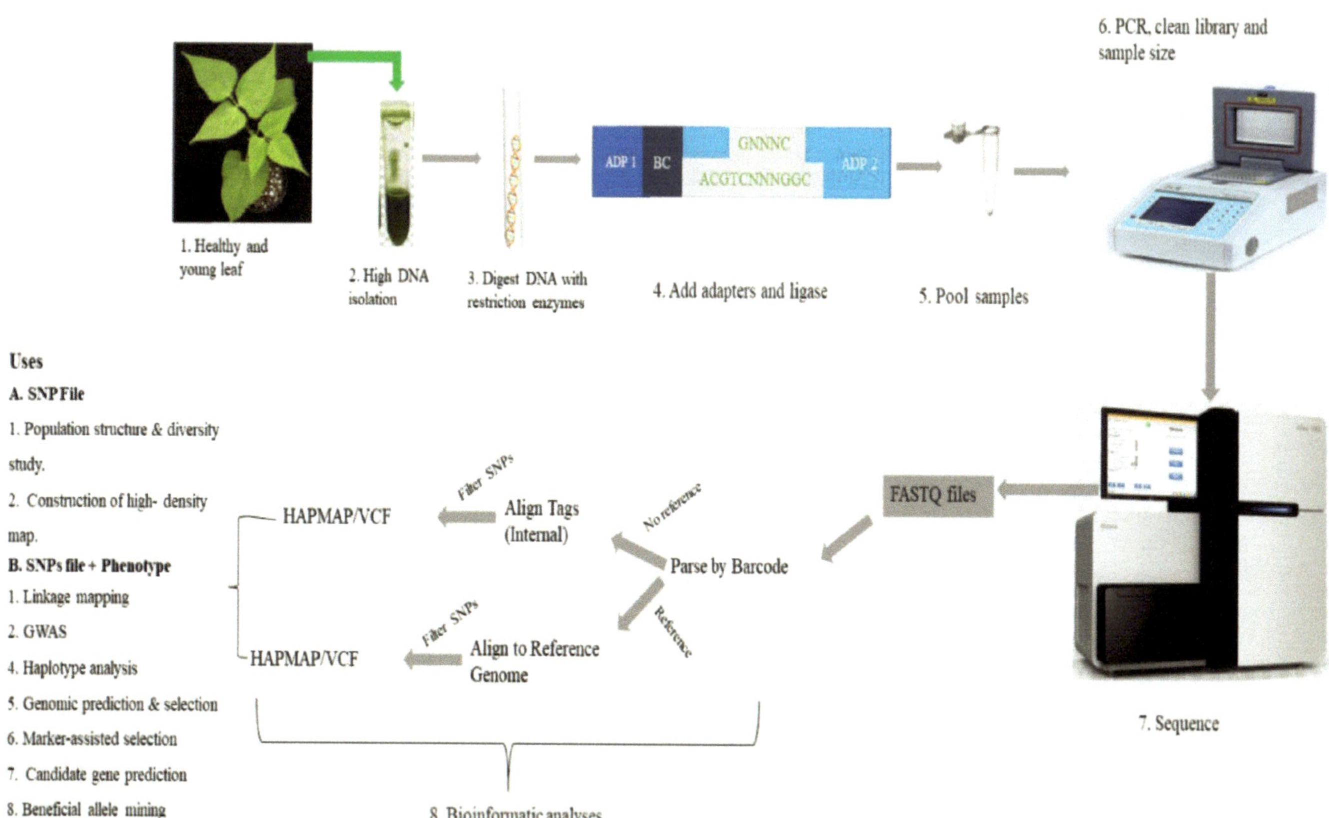

FIGURE 3.1 Schematic flow of GBS protocol and analytical procedure in pulses.

attack it (Mogali and Hegde, 2020). Hwang (2015) delineated one fungal disease, *Cercospora* leaf spot (CLS), for resistance breeding and discovered two QTLs, *CSL 5-1* (97_756706–17_3170881) and *CSL 11-1* (15_2959889–15_2508102), on chromosomes 5 and 11, with 3.34 and 17.72 logarithms of odds (LOD) and phenotypic variation explained (R^2) by a QTL (5.14%).

Schafleitner et al. (2016) successfully mapped resistance to bruchids in wild mung bean using two closely related RIL populations (TC1966-NM92 and V2802-NM94). A bruchid resistance gene map was created for chromosomes 1 (26370595–28364419), 2 (23741639–24208814), 3 (10257752–11651817), 4 (15495366–15684641), 5 (5179402–1541296), and 7 (13713780. Subsequently, markers linked to the genomic area on chromosome 10 could be verified for use in marker-assisted selection (MAS) to improve Bruchid resistance in cattle. A total of 10 (2772296–3814194) with a LOD in the range of 10–52.3 were reported. The genomic region on chromosome 10 was one of those that was consistently found in both populations. If not controlled, this insect can cause 30–40% losses in a short period of three months, and if not controlled, it can cause 100% losses (Sekar & Nalini, 2017a, b).

Lentil root rot (ARR) disease, caused by the fungus *Aphanomyces euteiches* Drechs, is one of the most severe diseases in the production of lentils. It can cause output losses of up to 80% in some cases. A wide variety of plants in the Fabaceae family are affected by this disease, with pea, lentil, faba bean, and dry bean being the most common hosts (Gaulin et al., 2007; Malvick et al., 1998; Wu et al., 2018). In spite of the fact that there are no lentil cultivars that are resistant to ARR, Ma et al. (2020) used a breeding line (K192-1) that had high partial resistance and a susceptible breeding line (K191-2) to develop 189 RIL populations in an attempt to understand the genetic architecture of ARR. They used both manual and image-based phenotyping under two different situations: controlled conditions and field conditions (*see* Figure 3.1). Root rot index (RRI), shoot dry weight loss per plant (SDL), and root dry weight loss per plant (RDWLP) were also studied under both conditions (RDL). The following parameters were measured: the average intensity of the blue channel acquired from an RGB camera (RGB.blue), the standard deviation of the saturation channel acquired from an RGB camera (RGB.saturation), the number of pixels loss per plant in shoot acquired from an RGB camera (RGB.SPL), the number of pixels loss per plant in root acquired from an RGB camera (RGB.RPL), the above-ground index (AGI), and the canopy area acquired from a multispectral camera (Multispectral.can (Multispectral.NDVI). Specifically, they discovered 20

QTLs on chromosomes 2, 5, and 7, which were divided into three groups of 6, 4, and 3. QTLs were also found on chromosomes 3, 4, and 6, with two each on chromosomes 3, 4, and 6. It was found that the LOD and R^2 ranged from 2.6–5.6% and 5.2–12.1%, respectively, for the samples. One repetitive marker on chromosome 2 at position 10483056 bp underlined the RRI (*Q.RRI-Lc2.1*), RGB.blue (*Q.BLU.Lc2.1*), AGI (*Q.AGI.Lc2.1*), and multispectral canopy (*Q.CAN-Lc2.1*) with beneficial alleles from either of the parents. Another repetitive marker on chromosome 2 at position 10483056 bp underlined the AGI (*Q.AGI*). This site is considered a hotspot for ARR and could be a potential target for molecular breeding to improve root rot resistance in lentils.

Most major lentil production areas around the world, particularly in Africa, Eastern Europe, North America, and South Asia, are affected by fungal diseases such as anthracnose and *Stemphylium* blight (Gupta et al., 2020). *Colletotrichum lentis* and *Stemphylium botryosum* are the pathogens that cause these disorders to manifest themselves. Bhadauria et al. (2017) used two wild relatives of lentil (*Lens ervoides*), L01-827A and IG 72815, that had superior resistance to anthracnose, Ascochyta blight, and *Stemphylium* blight, to develop 94 RIL populations to discover the genetic factors that underlie genetic resistance in the cultivated lentil. A semi-quantitative rating scale was used to assess the responses of the RILs and their parents, as well as two controls (Lens culinaris Eston and CDC Robin). The RILs and their parents were inoculated with two isolates of *C. lentis* (CT-21 as race 1 and CT-30 as race 0), and *S. botryosum* SB19, and their responses were assessed using a semi-quantitative rating scale seven days after the inoculation (dpi). Around 14 QTLs, with LOD (4.08–8.67) and R^2 (9.91–24.75), were discovered on chromosomes 2, 3, 5, and 7. These QTLs were associated with resistance to *C. lentis* race 0, *C. lentis* race 1, and *S. botryosum* and were located on chromosomes 2, 3, 5, and 7. Of them, *qANTH0-3* (54.2–63.1 cM) and *qANTH1-3-2* (55.5–60 cM) shared chromosome 3 with the two isolates of *C. lentis* and overlapped with each other. Furthermore, there were two pairs of QTLs on chromosome 5 that overlapped for resistance to the two isolates of *C. lentis*: *qANTH0-5-1* (37.6–42 cM) and *qANTH1-5-1* (38.7–43.1 cM), and *qANTH0-5-2* (29.8–33.5 cM) and qANTH1-5-2 (29.8–33.5 cM) (32.4–35.9 cM).

Common bean output and quality are severely hampered by white mold, which is caused by the fungus *Sclerotinia sclerotiorum* (Lib.) (Vasconcellos et al., 2017). White mold is a serious disease that affects common bean productivity and quality around the world. It has the potential to result in a total yield loss of 100% (Schwartz & Singh, 2013). A meta-QTL study for resistance to white mold in common with 14 RIL populations was performed by Vasconcellos et al. (2017). Two of the 14 populations (AO derived from the cross A195 OSU6137, and the other, GW, was generated from the cross G122 × WMG904-20-3) and were genetically genotyped by GBS. QTLs on chromosome 1 were found in AO RILs, two of which were on chromosome 1 (*WM1.1* and *WM1.2*), and one QTL on each of the following chromosomes: 3, 5, 7, 8, 9, 10, and 11. *WM7.4* was discovered in the AO population within a physical interval of 41.80–42.41 Mb, and it was also discovered in two other populations, O12 (Orion USPT-WM-12) and XC (Xana Cornell 49242), within a physical interval of 36.47–37.15 Mb and 39.3–44.95 Mb, respectively. A QTL (*WM8.1*) was discovered in GW on chromosome 8 with a low level of evidence (LOD) and 3.7 and 19.3% (R^2), respectively. A total of nine meta-QTLs were discovered after the nine QTLs identified in the two populations were combined with QTLs from the other 12 RIL populations.

Furthermore, the bacterial disease *Pseudomonas syringae* pv. *phaseolicola* (Psph), which is caused by a bacterium, is a disease of significant economic importance that has the potential to inflict yield losses of up to 45% worldwide (Félix-Gastélum et al., 2016; Schwartz & Singh, 2013). Psph resistance mapping was performed by Tock et al. (2017) using three RIL populations of common bean, namely SE (SOA-BN EDT Mund), CP (Canadian Wonder PI 150414) and RC (Rojo CAL 143), to map for resistance to Halo blight disease with nine races of Psph. They were isolated from larger areas of common bean production and are as follows: UK-725A (race 1), USA-882 (race 2), Tanzania-1301A (race 3), Kenya-1375A (race 5), UK-716B (race 6), Rwanda-1449B/Tanzania-1299 (race 7), Lesotho-2656A (race 8), and Malawi-2709A (race 9). The short arm of chromosome 4 was discovered to include one major impact locus for resistance to races 1 and 6 in the SE population, as well as to other races in the CP and RC populations (Tock et al., 2017). This hotspot was classified as HB4.2 to represent the location of the major effect locus.

Resistance to downy mildew in lima bean was mapped using bulked segregant analysis (BSA) combined with GBS on 216 lima bean progeny segregation for a dominant race F (Mhora et al., 2016). In the region of chromosome 4 between two markers that were approximately 4.88 cM apart, the researchers discovered seven markers that were substantially related with race F. Using these seven linked markers, lima bean panels consisting of 256 landraces, cultivars, and wild germplasm were screened and identified as promising. Using two of the seven linked markers, the panel was clustered into a haplotype, verifying the correctness of race F resistance in the sample.

By using three closely related RIL populations, Gali et al. (2018) were able to deconstruct the genetic regulation of *Mycosphaerella* resistance in field peas. They discovered a large and stable QTL (designated *qMR-LG3b-1*) in the PR-07 RIL population with LOD and R^2 values of 3.54–8.85 and 10.7–26.29%, respectively, on linkage group (LG) 3b. An additional environment-specific and large QTL on LG3b, designated *qMR-LG3b-2*, with LOD (3.89) and R^2 (10.46%), was discovered and mapped.

3.3.2 ABIOTIC STRESSES AND CLIMATE CHANGE-RELATED TRAITS

Climate change and human activities have exacerbated abiotic pressures such as high or low temperature, salinity, drought, flooding, and heavy metal contamination, which are among the key factors limiting the productivity and sustainability of pulses. Lodging is one of the most significant limitations to pulses, and it occurs as a result of the plant's architecture and windy conditions. It has a substantial impact on the yield and quality of pulse seeds. It also makes plants more susceptible to the development of fungal infections, as well as reducing their photosynthetic capabilities and harvest efficiency. Lodging could result in yield losses of up to 74% (Amelin & Parakhin, 2003).

Breeding for lodging resistance is a primary goal of pulse improvement programs. GBS-based linkage mapping has been used in mung bean, field pea, and dry field bean to clarify their genetic architectures for MAS against lodging because of its significance (Hwang, 2015). It was revealed that plant height and the number of branches affect lodging resistance in mung bean. QTL (LD 10-1) was found to be associated with markers (32_2548609–32_2193266) on chromosome 10 with a LOD and R^2 of 3.4 and 8.3%, respectively. Alternatively, two QTLs (*PH 3-1* and *PH 4-1*) were mapped on chromosomes 3 and 4 using the LOD and R^2 methods (3.11 vs. 6.01% and 12.89 vs. 30.02%, respectively). For example, the LOD and R^2 values for two QTL's (*NB 2* and *3*), respectively, were in the range of 2.56–2.67% for NB 3-1 and 5.08–5.6% for *NB 2*. Three closely related RIL populations, namely PR-02 (Orb × CDC Striker), PR-07 (Carerra × CDC Striker), and PR-15 (1-2347-144 × CDC Meadow) with 94 lines and 1560, 2990, and 3077 SNPs each were used by Gali et al. (2019) to unravel the genetic architecture of lodging and plant height in field pea (Table 3.2). Four

QTLs for lodging resistance (*qLR-LG3a-1*, *qLR-LG3a-2*, *qLR-LG3b-1*, and *qLR-LG3b-2*) were discovered in the PR-15 and PR-07 RIL populations with LOD 3.43 and R^2 11.19. With LOD (3.43–13.19) and R^2 (11.19–35.29) observed in PR-07 RIL populations in three different environments, qLR-LG3b-2 was the most prevalent QTL. It was found in PR-07 (2010 Rosthern, 2010 Sutherland, and 2011 Sutherland). A total of eight QTLs (*qPH-LG3a-1*, *qPH-LG3a-2*, *qPH-LG3a-3*, *qPH-LG3b-1*, *qPH-3b-2*, *qPH-LG3b-3*, *qPH-LG3b-4*, and *qPH-LG5a-1*) with LOD (R^2 ranges of 3.47–20.52 and 8.02–56.74%, respectively) were found in any of the three populations (Table 3.2). In PR-02 RILs, *qPH-LG3b-4* was detected in 2011 Rosthern, 2011 Sutherland, 2012 Rosthern, and 2012 Sutherland, while *qPH-LG3a-1* was found in three different environments. Three more (*qPH-LG3b-2*, *qPH-LG3b-3*, and *qPH-LG5a-1*) were found in PR-02, PR-07, and PR-02 RILs. Resistant plants have a higher likelihood of lodging (Jha et al., 2017), which has been linked to *Mycosphaerella* resistance. In order to increase lodging and *Mycosphaerella* resistance, *qLR-LG3b-2* and *qMR-LG3b-1* could be targeted for QTL pyramiding. Alternatively, *qPH-LG3b-4* or *qPH-LG3a-1* could be used to achieve optimal plant height.

Smitchger (2017) employed the GBS technique to obtain a total of 4,753 markers for 94 RILs of dry field beans from 254 lines used in their analysis. The resistance to lodging was used to map both major and minor QTLs together. In each of the three LGs, there were three QTLs, one of which was a significant QTL mapped to the LG6 region (Table 3.2). For the production of legumes and pulses, phosphorus is a critical limiting soil nutrient (P). Nodulation, photosynthetic energy, redox reactions, and phosphorylation and dephosphorylation are all processes that require more P in legumes and pulses than in other organisms. For this reason, Reddy et al. (2020) mapped for P use efficiency (PUE) with various attributes at the seedling stage in mung bean under three different P regimes. More than 130 SNPs were found to be associated with the 12 variables studied by two popular GWAS methods (GLM and MLM). Additional potential genes were also discovered.

Flowering time and maturity of pulses can be modified to mitigate different pressures by changing key features (Chaudhary et al., 2019). Progress in pulse phenology, such as days to first flowering (DTFF), days to 50% flowering (DTF), and days to maturity (DTM), is documented in this section. The mapping of DTFF and DTF in mung bean by Hwang (2015) included five chromosomes, 3, 4, 5, 6, and 8, with a LOD and R^2 ranging from 3.96–20.53 and 3.49–33.14%, respectively (Table 3.2). Additional findings include the discovery of 14 DTF QTLs on five chromosomes in three closely related field pea populations evaluated in various conditions by Gali et al. (2018). The most prominent and persistent QTL (*qDTF-LG6-1*) was mapped in PR-02 RILs across four environments in the genetic interval of 9.16–12.84 cM with LOD and R^2 in the range of 9.16–12.84 and 17.88–47.9%, respectively. Except for *qDTF-LG5-1*, all of the other 12 QTLs were context-dependent (Table 3.2). A few QTLs (*qDTM-LG3a-3*, *qDTM-LG5-1*, and *qDTM-LG6-1*, respectively) were found in PR-15 RILs and PR-02 RILs by Gali et al. (2018).

In the PR-15 and PR-02 RIL populations, 10 DTM QTLs were shown to be environment-dependent. Four SNPs in 4010542–4083513 bp and one at 2146621 bp on chromosomes 2 and 8 were shown to be associated with maturation by Sokolkova et al. (2020). Another SNP (8992242) on chromosome 2 was shown to be associated with hypocotyl color.

The common bean MAGIC population mapped numerous QTLs for drought tolerance (DTF) and days to physiological maturity (DPM) under drought conditions by Diaz et al. (2020). This study found a total of four important QTLs, two for DTF and two for DPM (Table 3.2).

3.3.3 YIELD AND RELATED TRAITS

In most legume crops, the seed yield (SY) is influenced by the number of pods per plant, number of seeds per pod, and the seed weight (single, 100, or 1,000). The number of pods per plant is also influenced by the number of branches. These traits are quantitatively inherited with numerous loci and are largely influenced by the environment. In this section, the single/100/1,000-seed weight designated as SW is discussed (Table 3.2).

Three major loci with cumulative R^2 of 58.9% on chromosomes 3, 7, and 8 were mapped in black common bean SY by Sandhu et al. (2018). In addition, nine loci were mapped for SW comprising 1 locus each on chromosomes 1, 2, 3, 7, 8, and 10 (Sandhu et al., 2018), of which 1 on chromosome 3 at genomic interval 53.8–71 cM was consistently detected across multiple environments, and 2 loci on chromosome 8 within intervals of 55.6–60.5 cM and 96.5–98.8 cM were mapped in at least two environments (Table 3.2).

A total of 20 SY and 13 SW QTLs were discovered in field peas by Gali et al. (2018) among seven different LGs (1b, 3a, 3b, 3c, 4, 6, and 7). Among those detected for SY, *qSY-LG1b-1*, *qSY-LG3b-1*, *qSY-LG3b-2*, *qSY-LG4a-1* and *qSY-LG6-4* were mapped consistently across at least two environments in one RIL population, while only *qSW-LG3c-1* was mapped

TABLE 3.2 GBS-based Linkage Mapping Studies Conducted in Pulses on Agronomic and Quality Traits

Crop	Cross	Trait(s)	LOD	R^2	References
Mung bean	VC1973A (*Seonhwanogdu*) × VC2984 (*Kyung-Ki Jaerae* #5)	*Cercospora* leaf spot resistance, Lodging resistance, plant height, number of branches and days to first flowering.	2.56–20.53	3.29–33.14	Hwang (2015)
Lentil	PI 320937 × Eston	Selenium uptake.	3.16–3.96	6.3–16.90	Ates et al. (2016)
	K192-1 × K191-2	Aphanomyces root rot resistance and some image base phenotypes.	2.6–5.6	5.20–12.10	Ma et al. (2020)
	ILL 8006 × CDC milestone	Fe concentration in seed.	3.00–4.45	5.90–12.20	Aldemir et al. (2017)
	L01-827A and IG 72815 (wild)	*Colletotrichum lentis* CT-30 (race 0 and 1, ANTH –0 and –1) and *Stemphylium botryosum.*	3.40–8.05	9.40–24.75	Bhadauria et al. (2017)
Lima bean	B2C × Bridgeton	Downy mildew resistance.	4	NA	Mhora et al. (2016)
Field pea	PR-02 (Orb × CDC Striker); 94; PR-07 (Carerra × CDC Striker) and PR-15 (1–2347-144 × CDC Meadow)	Plant height, days to 50% flowering and maturity, seed shape, 1,000 seed weight, yield, seed dimpling and seed quality (iron, phytate, selenium, starch, zinc and protein concentrations, acid, and neutral detergent fiber).	3.00–22.31	4.45–31.30	Gali et al. (2018)
Dry field bean	Delta × RER	Lodging	NA	NA	Smitchger (2017)
Common bean	BK004-001 and H68-4	100 SW, seed yield, seed hardness, blue to yellow, green to red, hydration coefficient, lightness component, seed stone percentage and visual score.	3.70–23.40	6.7–58.7	Sandhu et al. (2018)
	SOA-BN × Edmund (SE)	Halo blight resistance-race 1 and 6.	4.88–53.43	3.00–95.62	Tock et al. (2017)
	Rojo × CAL 143 (RC)	-Race 2, 3, 6, and 7.	3.87–69.70	3.35–79.2	
	Xana × BAT93	Seed length, width, perimeter, area, and weight; pod length and perimeter, and number of seeds per pod.	NA	NA	Murube et al. (2020)

TABLE 3.2 *(Continued)*

Crop	Cross	Trait(s)	LOD	R^2	References
	AO (A195 × OSU6137)	White mold disease.	3.60–5.90	13.70–20.10	Vasconcellos et al. (2017)
	GW(G122/ WMG904-20-3)	–	3.7	19.3	–
	MAGIC Pop.	Days to flowering.	7.5–58.85	3.43–35.76	Diaz et al. (2020)
	–	Days to physiological maturity.	7.92–21.70	5.11–15.90	
	–	Pod harvest index.	10.73	7.62	
	–	100 seed weight.	8.36	4.74	
	–	Seed yield.	8.76	2.03	
	–	Seed iron.	6.46–7.33	4.06–8.68	
	–	Seed zinc.	6.81	2.65	
Wild mung bean	TC1966 × NM92	Seed yield.	31.9	NA	Schafleitner et al. (2016)
	V2802 × NM94		16.4	NA	
	TC1966 × NM92		3	NA	
	V2802 × NM94	*Bruchid* resistance	39.90–52.30	NA	
	V2802 × NM92		10	NA	
	TC1966 × NM92 and V2802 × NM94		12.2–12.4	NA	

for SW consistently within the genomic interval of 15.5–18.9 cm across four environments (2010 Sutherland, 2011 Rosthern, 2012 Rosthern and 2012 Sutherland) in PR-07 RILs (Table 3.2).

With the potential contribution of the number of branches to SY and lodging resistance, Hwang (2015) mapped two QTLs, *NB 2-1* and NB 3-1, with LOD and R^2 in the range 2.56–2.67 and 5.6–5.8%, respectively, for mung bean (Table 3.2). On the other hand, Diaz et al. (2020) detected three major QTLs, one for pod harvest index (PHI; thus, the ratio of seed weight to pod weight) (*PHI2.1*), one for SY (*Yd1.1*), and another for SW (*100SdW4.1*) on chromosomes 2, 1, and 4, respectively, in a common bean MAGIC population.

3.3.4 SEED QUALITY TRAITS

Pulses contain micronutrients (zinc, iron, and selenium) as essential components of seed quality traits, such as seed protein concentration/content (SPC, %). Around 21 QTLs for seed iron concentration in lentil were mapped by Aldemir et al. (2017) on six chromosomes; thus, three each on chromosomes 1 and 2, chromosomes 4, 5, 6, and 7 harbored six, four, two, and seven QTLs, respectively (Table 3.2). Ates et al. (2016) mapped four QTLs controlling Se accumulation in lentil seeds using a 96 RIL population obtained from the cross "PI 320937" × "Eston" The RILs and their parents were evaluated across six environments (2 years × 3 locations). Four QTLs (*SeQTL2.1*, *SeQTL5.1*, *SeQTL5.2*, and *SeQTL5.3*) were detected simultaneously across the six environments. *SeQTL2.1* was mapped on chromosome 2 within 796.3–803.6 cM with LOD of 3.24–3.96 and R^2 of 6.60–11.50. The remaining 3 (*SeQTL5.1*, *SeQTL5.2*, and *SeQTL5.3*) were detected on chromosome 5 with LOD and R^2 ranging from 3.24–3.90 and 6.30–16.90%, respectively (Table 3.2).

In pulses, a number of promoters and inhibitors of micronutrient availability have been reviewed by Kumar & Pandey (2020). These include Se as a promoter; and phytic acid and fiber as inhibitors. These traits have been mapped to unravel the genetic architecture and develop/select markers for marker-assisted breeding. The amount of phytic acid in some pulses is below 5 g/100 g (dry weight) (Gupta et al., 2015). In field peas, Gali et al. (2018) mapped a number of seed quality traits in the three related RIL populations: two seed protein concentration (SPC) QTLs were mapped on LG1a (*qSPC-LG1b-1*) and LG4a (*qSPC-LG4a-1*) in the PR-02 population with LOD (3.37–5) and R^2 (10.13–15.85%) across two environments. Another five

QTLs (*qSPC-LG3b-1*, *qSPC-LG3b-2*, *qSPC-LG3b-3*, *qSPC-LG7b-2*, and *qSPC-LG7b-3*) and one QTL (*qSPC-LG7b-1*) detected across the two environments were mapped for SPC in the PR-07 RIL population. *qSPhC-LG5-1* and *qSPhC-LG5-2* were mapped for seed phytate concentration (ppm) in lentil on LG5 of the PR-15 RIL population (Gali et al., 2018). These two mapped consistently in 2 environments with LOD and R^2 in the range of 6.69–12.48 and 16.09–33.19%, respectively (Table 3.2). One stable for seed dumpling detected across six different environments (*qSD-LG5a-1*) was mapped on LG5a at the genetic position 2.2–6.3 cM with LOD (4.12–10.37) and R^2 (12.67–29.19%) in the PR-07 RIL population (Table 3.2). In addition, three environment-specific QTLs (*qSFeC-LG2b-1*, *qSFeC-LG4b-4*, and *qSFeC-LG6-1*) and one QTL (*qSFeC-LG3b-1* across two environments) were mapped for seed Fe concentration with LOD (3.34–10.14) and R^2 (10.33–12.55%) in the PR-02 RIL population. Additionally, in the PR-07 RIL population, two major QTLs, one across two environments (*qSFeC-LG3b-3*) and another across five environments (*qSFeC-LG3b-2*), were mapped with LOD and R^2 in the range of 4.04–10.14 and 10.85–35.52%, respectively (Table 3.2). Five environment-specific QTLs (*qSSeC-LG4a-1*, *qSSeC-LG5a-1*, *qSSeC-LG7-1*, and *qSSeC-LG7-2*) with LOD (3.10–11.60) and R^2 (7.14–18.76%) were mapped in PR-02 for seed selenium concentration (ppm) (Table 3.2). Four (*qSZnC-LG1a-1*, *qSZnC-LG3b-1*, *qSZnC-LG6-1*, and *qSZnC-LG6-2*) and three environment-specific QTLs (*qSZnC-LG3b-3*, *qSZnC-LG3b-4*, and *qSZnC-LG7a-2*) were mapped in PR-02 and PR-07 RIL populations for seed Zn concentration with LOD (3.67–13.71) and R^2 (6.34–25.83%) (Gali et al., 2018). However, three stable QTLs (*qSZnC-LG3b-2*, *qSZnC-LG3b-5*, and *qSZnC-LG7a-1*) were mapped in the PR-07 RIL population with LODs of 4.48, 8.56–17.3, and 4.61–5.85, respectively, and R^2 of 9.06–12.98, 23.92–50.13, and 11.41–11.65, respectively (Table 3.2). Around 12 QTLs comprising 9 environment-specific and 3 stable QTLs were mapped for seed starch concentration in the PR-07 RIL population, for details (*see* Table 3.2). Then again, seven and 12 QTLs were found in the PR-07 RIL population for acid detergent fiber (%) and neutral detergent fiber (%) (Table 3.2). Three QTLs (*qSS-LG3c-1*, *qSS-LG3c-2*, and *qSS-LG7a-1*) and one environment-specific QTL (*qSS-LG3c-3*) were mapped for seed shape (Table 3.2).

Besides SY and SW, Sandhu et al. (2018) mapped seed quality traits such as seed stone percentage (SSP), hydration coefficient (HC), lightness component (L* (L)), green to red seed coat color (a* (a)), blue to yellow seed coat color (b* (b)), and visual score (VSC) in black common bean (Table 3.2). Three QTLs (*qSSP-1-1*, *qSSP-2-1*, and *qSSP-7-2*) were mapped in either five or six environments with LOD (3.3–14.7) and R^2 (9.6–33.5%). Two environment-specific QTLs (*qSSP-5-1* and *qSSP-7-1*) were detected with LOD $\geq$ 5.9 and $R^2 \geq$ 10.5% for SSP. One and major QTL (LOD vs R^2) each was mapped on chromosome 1 (*qHC-1-1*, 4.6–9.3 vs 6.6–14.2), 2 (*qHC-2-1*, 8.3–9.8 vs 14.1–18.6) and 7 (*qHC-7-1*, 15.7–23.4 vs 41.5–53.6) for HC at the genomic interval of 115.6–124 cM, 22.3–28.4 cM and 9.3–11.3 cM, respectively. Two additional environment-specific QTLs were detected on chromosomes 8 (*qHC-8-1*) and 10 (*qHC-10-1*). One QTL for a *(a) (qa-7-1), b* (b) (qb-7-2) and VSC (*qVSC-7-1*) co-localized in an overlapping genomic region (9.5–11.5 cM) on chromosome 7, implying that this region controls more than one or more seed coat color and its related traits. Similarly, one major environment-specific QTL (qb-4-1) at a 38.6–46.6 cM genomic interval overlapped with *qVSC-4-1* at 45.1–47.2 cM on chromosome 4. Furthermore, three environment-specific QTLs (*qL-3-1*, *qb-7-1* and *qVSC-10-1*) were mapped for L* (L), b* (b) and VSC, respectively.

Recently, Murube et al. (2020) mapped QTLs associated with pod and seed size in common bean with two nested RIL populations (XC and XB derived from Xana × Cornell 4924 and Xana × BAT93, respectively). A total of 13 QTLs for pod traits were detected in the XC population and 18 QTLs in the XB population. Conversely, 21 and 25 QTLs for seed traits were mapped in the XC and XB RIL populations, respectively. Among these QTLs for both traits and population, 20 and 27 significant epistatic interactions between QTLs were detected in the XC and XB populations. Also, several genomic regions overlapped in both populations for pod and seed size on chromosomes 1, 5, 6, 7, 8, and 11. These results show the importance of these chromosomes in controlling the genetic makeup of pods and seed size in both populations. Some of the QTLs overlapped with narrow intervals than previously reported genomic regions (Blair et al., 2006; da Silva et al., 2018; González et al., 2016; Pérez-Vega et al., 2010; Yuste-Lisbona et al., 2014). For instance, $PWI5^{XB*}$ was found at 56.8–62.8 cM on chromosome 5, which overlapped with SW-5MA in the study by González et al. (2016) at 55.8–68.2 cM; thus, the GBS-based linkage map reduced the interval to less than half of other maps derived from different genotyping techniques. Additionally, three major QTLs, two QTLs (*SdFe6.1* and *6.2*) and one QTL (*SdZn8.1*), were mapped for Fe and Zn in the seed of the common bean MAGIC population (Diaz et al., 2020).

3.4 GENOME SEQUENCING AND RE-SEQUENCING OF PULSES

With the availability of reference genomes for most pulses and a reduction in the cost of sequencing, more genotypes have been sequenced. Re-sequencing via whole-genome re-sequencing (WGRS) is the precursor for deciphering genetic diversity, population structure, domestication pattern, LD, and the unexploited genetic potential in crop plants (Varshney et al., 2019b). For example, Lioi et al. (2019) revealed by GBS that Mesoamerican landraces of common bean had a higher level of genetic diversity compared to the Andean landraces. In the same direction, Kang et al. (2014) sequenced 22 of 18 *Vigna* species with a total length of 421 Mb (80% of the reference genome) and provided empirical evidence on the divergence of what was eventually domesticated while the sampled wild mung bean species appears to have predated domestication. Similarly, Noble et al. (2018) sequenced 466 cultivated and 16 wild mung bean, generating 22,000 polymorphic genome-wide SNPs, which were subsequently used to reveal that polymorphism information content (PIC) and LD in cultivated genotypes were 0.174 and ~100 kb, whereas in the 0.305 and ~ 60 kb were estimated for wild accessions, respectively. Reddy et al. (2020) resequenced 144 mung bean genotypes (40 released varieties, 42 advanced breeding lines and 62 exotic germplasm lines) on the 144-plex GBS library and sequenced (2 × 150-bp single end) using the Illumina HiSeq 4000 platform. A total of 76,160 SNPs of which 19%, 23%, 27%, and 31% were in the scaffold, intragenic, intergenic, and regulatory regions.

Ferreira et al. (2017) identified genomic regions in introgression lines that harbor resistance alleles for anthracnose (Co-2 and Co-3), bean common mosaic virus (BCMV), and bean common mosaic necrosis virus (BCMNV, 1 and bc-3) in common bean. Wen et al. (2019) sequenced 363 common bean accessions comprising the Mesoamerican and the Andean gene pools with GBS on the HiSeq 2500 platform and generated a total of 84,416 high-quality SNPs in comparison to *P. vulgaris* G19833 as the reference genome, which was distributed on 11 chromosomes with an average of 7674 SNPs per chromosome. These SNPs were used for further analyzes. Similarly, Tock et al. (2017) genotyped 384 accessions of cultivated common bean from the Andean gene pool by GBS on an Illumina HiSeq 2500 platform and generated 17,750 SNPs (MAF $\geq$ 0.05, for which $\geq$ 0.8 of samples were genotyped). Oladzad et al. (2019a) also sequenced 482 and 325 accessions from Mesoamerican and the Andean gene pools, respectively, with a read length of 230 nucleotides on a HiSeq 2500 platform, and after filtering and imputation, 205,293 and 260,670 SNPs were obtained in the same order. In addition, Campa et al. (2018) genotyped 308 common bean accessions by GBS on the Illumina HiSeq 4000 platform and obtained 3,099 clean and high-quality SNPs. Cortés & Blair, (2018) sequenced 86 wild common beans and obtained 2,845 high-quality SNP. Diaz et al. (2020) conducted whole-genome sequencing of eight breeding lines of common bean (SXB412, INB827, ALB213, SEN56, SCR2, MIB778, SCR9, and INB841) to develop an 8-way MAGIC population. A total of 20,615 polymorphic markers were generated in the MAGIC population (Diaz et al., 2020).

Gali et al. (2019) sequenced 135 cultivated pea accessions obtained from 23 breeding programs in Africa, Asia, Australia, Europe, and North America on a HiSeq 2500 platform. It generated 60.5% of the reads per accession aligned to the reference sequence, and 91.9% of the aligned sequences were uniquely aligned. After trimming of SNPs with strict criteria of minimum allele frequency (MAF) of 0.05 and 5 minimum read depth, a total of 16,877 high-quality SNPs were used for population and marker-trait association (MTA) analyzes. Pavan et al. (2019) sequenced 349 landraces of lentil collected from 1971–1993, four lines of the self-pollinated cultivar 'Laird,' and generated 410,637 SNP sites 6,693 high-quality SNPs were obtained after filtering the genotypes and SNPs.

3.5 GBS-BASED GWAS EFFORTS IN PULSES

GWAS has gained popularity for unraveling the genetic architecture of several agronomic traits of economic importance compared to linkage mapping (Ibrahim et al., 2020). This popularity is due to higher mapping resolution attributable to high recombination even in the natural population, less time needed to assemble panel Singh et al. (2020). The effectiveness and efficiency power in association mapping (AM) are affected by marker density, population type, sample size, population structure and statistical model(s) (Gupta et al., 2005; Ibrahim et al., 2020; Josephs et al., 2017). The marker density and population-related issues have been easily resolved in recent years with the wide application of NGS techniques at a reduced cost of sequencing (Varshney et al., 2020), and HT GBS-based SNPs have been discovered for GWAS in many pulses. In addition, a number of statistical models have been developed to overcome the limitations of traditional single-locus models (mixed linear model (MLM), general linear model (GLM), compression MLM (CMLM)), and multiple test correction for threshold value of significance and mapping power leads to the exclusion of important loci (Wen et al., 2017). Hence, to overcome these limitations in single-locus models, a number of multiple-site models have been developed, including those reported by Zhang et al. (2019b), some of which are currently used as state-of-the-art procedures to identify quantitative trait nucleotides (QTNs) in some pulses (Oladzad et al., 2019a; Pavan et al., 2019). This section summarizes the progress made in the SNP-derived from GBS-based GWAS applications in pulses.

3.5.1 BIOTIC STRESSES

Soybean cyst nematode (SCN), which is associated with *Heterodera glycines* (HG) in common bean, was mapped via GWAS by Wen et al. (2019) with two types of HG viz, SCN HG types 2.5.7 and 1.2.3.5.6.7, in 363 common bean accessions genotyped by GBS, resulting in a total of 84,416 SNPs. The study identified 5 SNPs on chromosome 1 (1_10061925, 1_18388378, 1_18388392, 1_18388403 and 1_18388408) and 1 SNP on chromosome 9 (9_35068146) linked to HG type 2.5.7 resistance (Table 3.3). Of these, those on chromosome 1 were in the syntenic region of the soybean *Rhg1* locus for SCN resistance. In that same study, 1 SNP (7_44761605) significantly associated with HG 1.2.3.5.6.7 SCN resistance at *P*-value = $9.57 \times 10^{-7} \approx 6.02$ (–log*P*-value) and accounted for 21.7% of R^2. Wen et al. (2019) found three genes (*PHAVU_001G24800g*, *PHAVU_001G247900g*, and *PHAVU_001G247700g*), which are orthologs to three genes on *the Rhg1* locus (*Glyma18g02580*, *Glyma18g02590*, and *Glyma18g02610*, respectively), with each pair having at least 88% similarity. Another study mapped *Psph* race 6 resistance with 504 common bean accessions, mainly from the genetic diversity of cultivated common bean from the Andean gene pool and found three SNPs (S5_38725023, S5_38885617 and S5_38866881, named the locus *HB5-1*) on chromosome 5, which was tightly linked to *Psph* race 6 resistance (Tock et al., 2017) (Table 3.3). Two of these SNPs (*S5_38725023* and *S5_38866881*) were also significantly associated with yield. The locus *HB5-1* could explain the genetic basis for yield reduction under severe halo blight pressure. A total of 13 model genes were within a 162-kb mapping interval of locus *HB5-1*, nine of these encoded for transmembrane receptor-like genes with a leucine-rich repeat domain and a cytoplasmic kinase, usually referred to as RLK.

Oladzad et al. (2019b) used ~ 128 k for the Mesoamerican diversity panel (MDP) and ~ 200 k for the Andean diversity panel (ADP) reported in their earlier study (Oladzad et al., 2019a) to map for *Rhizoctonia solani* resistance in common bean. A total of 8 SNPs peaked with SNP (2_49202443) on chromosome 2 with $-\log_{10}P$ and R^2 of 5.49 and 12.49%, respectively, in ADP, whereas in the MDP, a total of 19 SNPs (2 repetitive SNPs) with $-\log_{10}P$ were detected with R^2 ranging from 4.13–5.40% and 5.00–9.60%, respectively (Table 3.3). A number of candidate genes at ± 100 kb of the position of the SNPs were mined, some encoded for 2-isopropylmalate synthase 1, defensin-like (DEFL) family proteins, TNL disease resistance proteins, leucine-rich repeat protein kinase and NB-ARC domain-containing disease resistance proteins in the two populations (Oladzad et al., 2019b). Moreover, Zitnick-Anderson et al. (2020) used 325 ADP and 469 MDP genotypes with 200 k imputed SNPs for each panel from Oladzad et al. (2019a) to perform MTA as a source of resistance to *Fusarium solani* and with three classes of phenotypic data. In all, 17 SNPs were unevenly distributed on all chromosomes except chromosomes 5, 6, and 10 in the ADP with two repetitive SNPs (*7_38504717* and *8_184683*), while in the MDP, a total of eight SNPs were scattered on chromosomes 1, 3, 4, 7, and 8 (Table 3.3). Several candidate genes were predicted with functions related to plant immune/defense mechanisms, including glucan synthase-like enzyme, senescence-associated genes, NAC domain protein, ethylene response factor 1 and MAC/Perforin domain-containing genes.

In addition, Hart & Griffiths (2015) conducted a case-control GWAS with CMLM in the genome association and prediction integrated tool (GAPIT) R package and identified 44 SNPs tightly associated with *By-2* potyvirus resistance in the physical region of 4799175–8965798 bp on chromosome 2 in common bean (Table 3.3). Oladzad et al. (2019a) mapped for *Macrophomina phaseolina* infection in a subpopulation of common bean named BASE_120 under heat and/or drought conditions; one SNP on chromosome 4, S04_4665828, was tightly linked to *M. phaseolina* infection under either one or both conditions with the $-\log_{10}(P) = 4.05–4.47$ and $R^2 = 7.8–8.9\%$ (Table 3.3). Another SNP on chromosome 10 was associated with *M. phaseolina* infection under both heat and drought combined environment with $-\log_{10}(P)$ and R^2 of 4.06 and 7.9, respectively (Table 3.3).

A study by Dramadri et al. (2020) mapped for resistance to *Pythium ultimum* in common bean with isolate MS61 in Uganda with 260 k GBS-based and 6 k BeadChip SNPs from Oladzad et al. (2019a); and Song et al. (2015b), respectively. The pathogen *P. ultimum* (var. ultimum) causes

TABLE 3.3 GBS-based GWAS in Selected Pulses

Crop	Trait	Significant SNPs	P-Value/–log(P)	R^2 (%)	References
Common bean	*Heterodera glycines* (HG) 2.5.7 and 1.2.3.5.6.7 SCN resistance, seed coat color and weight.	7	5.31–7.10	16.80–64.50	Wen et al. (2019)
	Days to flowering and maturity; growth habit with determinate and indeterminate genotypes, lodging, canopy height, and seed weight.	43	3.10–14.80	4.00–26.00	Moghaddam et al. (2016)
	Halo blight resistance (race 6).	5	6.20–6.82	–	Tock et al. (2017)
	Resistance to *Fusarium solani* (quantitative score).	36	4.03–6.18	8.00–16.00	Zitnick-Anderson et al. (2020)
	Macrophomina phaseolina infection, heat, and drought tolerance traits.	64	2.32–7.89	5.40–13.70	Oladzad et al. (2019a)
	Rhizoctonia solani resistance.	29	4.13–6.60	5.00–14.43	Oladzad et al. (2019b)
	Days to flowering and physiological maturity, pod harvest index, 100 SW, seed yield, seed iron, and zinc.	17	5.80–34.98	NA	Diaz et al. (2020)
	Germination rate, hypocotyl length, total weight, and SPAD index.	45	3.62–6.00	6.02–18.02	Soltani et al. (2017)
	By-2 potyvirus resistance.	44	9.84–10.15	1	Hart & Griffiths (2015)
	Intermittent drought adaptation-leaf biomass, plant height, root biomass, root depth, leaf specific area, soil bulk density, SPAD.	45	5.04–13.52	NA	Berny et al. (2019)
	Bioclimatic-based drought index.	115	7.42–10.91	46.12–63.74	Cortés & Blair, (2018)
Pea	Agronomic and seed quality.	62	2.27–6.48	4.00–16.00	Gali et al. (2019)
	Heat stress responsive traits.	32	2.14–6.30	3.00–15.00	Tafesse et al. (2020)
	Folate and its associated traits-sum of folates.	31	2.01–6.73	NA	Jha et al. (2020)

TABLE 3.3 *(Continued)*

Crop	Trait	Significant SNPs	P-Value/–log(P)	R^2 (%)	References
Lentil	Seed diameter, perimeter, and area.	6	NA	NA	Pavan et al. (2019)
	Aphanomyces root rot resistance and image base traits.	38 SNPs	3.05–9.47	1.40–21.40	Ma et al. (2020)
Mung bean	Macronutrients and micronutrients.	43	3.63–8.88	1.00–39.00	Wu et al. (2020b)
	Resistance to *Pythium ultimum*.	8	5.95–8.02	33.30–37.50	Dramadri et al. (2020)
	Salinity tolerance.	–	3.178–3.675	NA	Breria et al. (2019)
	Phosphorus use efficiency traits.	136	4.01–5.48	10.76–25.40	Reddy et al. (2020)
	Days to maturity and hypocotyl color.	4	NA	NA	Sokolkova et al. (2020)
Mung bean	Seed coat.	9	5.33–7.31	5.70–8.30	Noble et al. (2018)

pythium root rot (PRR) in common bean, which is a common and most important root rot challenge in most parts of East and Central Africa (Dramadri et al., 2020). Three SNPs (S02_25453807, S02_25482756 and S02_25507837) with $-\log_{10}(P)$ = 5.94–8.02 and R^2 = 33.3–39.2% and another unstable 3 SNP (S02_25507946) with $-\log_{10}(P)$ and R^2 of 6.22% and 35.6%, respectively, were linked to *P. ultimum* resistance on chromosome 2. Another SNP (*S04_41963137*) on chromosome 4 with $-\log_{10}P$ (5.94–6.07) and R^2 (35.3–36.4%) and two SNPs (S09_5973960 and S09_6256123) on chromosome 9 with $-\log_{10}P$ and R^2 ranging from 6.02–6.68 and 35.2–37.5%, respectively, were associated with resistance to *P. ultimum*. One SNP (S05_29488066) with $-\log_{10}P$ of 6.7, caused 36.8% phenotypic variation on chromosome 5 in terms of *P. ultimum* resistance.

Ma et al. (2020) detected a total of 34 SNPs on seven chromosomes significantly associated with five ARR-related traits, namely two traditional traits (RRI and RDL) and three image-based traits (RGB.stauration, RGB. blue, and RGB.RPL) with 326 Lentil accessions from 60 countries on four continents. Marker, 1569_6 on chromosome 1 was consistently associated with RRI and RGB blue traits, which accounted for 24.4% R^2 for the two traits at $-\log_{10}P$ of 5.96–9.47 (Table 3.3). Another marker, 5190_31 on chromosome 4, was associated with both RDL and RGB.RPL with $-\log P$ (3.68 and 3.05%) and R2 (3.30% and 1.90%). In addition, two SNPs with the peak one (7154_47) were found from 225823924–226139924 bp on chromosome 4 to be associated with both RRI and RGB saturation traits. Locus 39892874–41416874 bp contained three SNPs on chromosome 6 that significantly influenced three traits (RDL, RGB.RPL, and RGB.saturation) by 13.9% variation in phenotype, with the peak SNP being 9009_9 (Table 3.3). The four repetitive SNPs (1569_6, 5190_31, 7154_47, and 9009_9) associated with at least two ARR-related traits could be validated and used for MAS for resistance to ARR disease. Also, a number of potential candidate genes encoding Leucine-Rich Repeat Receptor-like Kinase (LRR-RLK), cytochrome P450 family 71 protein (CYP71), ATP binding cassette transporter A family protein (ABCA), pectinesterase (PE), and chalcone-flavanone isomerase family protein (CHI) need further functional validation to unravel their involvement in ARR resistance.

3.5.2 ABIOTIC STRESSES AND CLIMATE CHANGE-RELATED TRAITS

Water deficiency affects more than a third of the world's population. As a source of genetic diversity, crop wild relatives (CWR) could be used for plant breeding to generate varieties that are resistant to a wide range of challenges, particularly those connected to climate change (van Treuren et al., 2017). Cortés & Blair (2018) mapped the bioclimatic-based drought index using 86 wild common beans with 22,845 SNPs. At the Bonferroni-corrected significance threshold of 7.36, 115 SNPs were connected with bioclimatic-based drought index and each accounted for R^2 (46.38–63.735%) on 11 chromosomes (Table 3.3). This study found that chromosome 8 has the most SNPs (32), followed by chromosome 3 (21), and chromosome 11 (Table 3.3). A similar approach was used by Josephs et al. (2017); and Song et al. (2015c) in their studies of wild common bean root and shoot variation against intermittent drought adaptation in Mesoamerican wild common bean using the SNP chip platform GWAS. It was determined that a number of attributes might be used to assess the capacity for intermittent drought adaptation in a variety of ways, including leaf biomass, root biomass, total biomass, root depth, specific leaf area, soil-plant analysis development (SPAD), plant height (environmental variables). Overall, 23 and 7 SNPs were associated with the phenotypic and environmental variables (Table 3.3). The highest of the six SNPs were associated with leaf biomass or soil bulk density, and the least number of a SNPs were associated with plant height. A genome scan was performed in the same study to identify genomic regions under natural selection for local adaptation (Berny et al., 2019). A total of 15 selective sweep regions were identified with two SNPs on chromosomes 4, 6, 8, and 10: three SNPs on chromosomes 2 and 11 and the least number of one SNP on chromosome 7 (Table 3.3). Breria et al. (2020) detected two and four SNPs on chromosome 7 (2,696,072 to 2,809,200 bp) and 9 (19,390,227 bp to 20,321,817 bp), respectively, above the threshold of 3 as a signal for salinity tolerance in mung bean with SNPs previously reported Breria et al. (2019). Two other SNPs were found in the contigs for salinity tolerance. One potential candidate gene (*Vradi07g01630*) encoding ammonium transport protein and two (*Vradi09g09510* and *Vradi09g09600*) annotated as OsGrx_S16-glutaredoxin subgroup II and DnaJ domain proteins, respectively) were mined with the hotspots on chromosomes 7 and 9, respectively.

As elucidated previously, Chaudhary et al. (2019); Oladzad et al. (2019a) mapped DTF in common bean with two sub-populations named BASE_Meso and BASE_120 under either heat or drought conditions. A total of 10 SNPs associated with DTF in either environment (PR/HN in 2 years) or combined environments (PR + HN) in BASE_Meso; of these, two SNPs (S03_47356534 and S03_40504942) on chromosome 3 were detected concurrently in either one or combined environments, while the remaining eight

SNPs were environment-specific (Table 3.3). In the same study, (Oladzad et al., 2019a) leveraged the phenotypic data of Gao et al. (2008) to map for DTF and two pleiotropic traits (DTF and DTM) using multi-trait model analysis. Around 19 SNPs comprising 2, 1, 4, and 12 on chromosomes 1, 2, 3, and 11, respectively, were associated with DTF (Table 3.3). Again, 11 SNPs from 51487987–52607649 bp on chromosome 3, two SNPs (4_46335883 and 4_47068592) on chromosome 4 and one SNP (8_24957350) on chromosome 8 were found to regulate DTF and DTM pleiotropically (Table 3.3). SPAD readings measure the greenness of the plant; hence, it can be used as an indicator of plant tolerance to stress. Oladzad et al. (2019a) also identified S08_18593303, S09_17981137 and S08_18397089 under drought, heat, and drought + heat conditions to be associated with SPAD, respectively, with $-\log_{10}P$ (4.37–5.97) and R^2 (7.0–12.7%). The seven major 7 QTLs detected via linkage mapping (3 for DTF: *DF1.1*, *1.2* and *3.1,* and 4 for DPM: *DPM1.1*, *1.2*, *3.1* and *8.1*) were narrowed down in the AM (Diaz et al., 2020). It implies that markers linked to QTLs/QTNs could be integrated into breeding programs via MAS.

A total of 135 cultivated pea accessions with 16,877 SNPs were used by Gali et al. (2019) to dissect the genetic architecture of DTF, DTM, lodging, and plant height. Nine, five, four, and five SNPs were linked to DTF, DTM, lodging, and plant height, respectively, with an uneven distribution on the six chromosomes and three scaffolds (Sc00936_29805, Sc01142_238, and Sc03817_83023) (Table 3.3). In order to map for heat stress-responsive traits (SPAD, photochemical reflectance index (PRI), canopy temperature, reproductive stem length, internode length, and pod number across five diverse environments under control and heat stress conditions, Tafesse et al. (2020) used 16,877 SNPs with known high-quality and 125 accessions of field peas from Gali et al. (2019). In at least three of the five environments studied, six SNPs on chromosome 5 (Chr5Lg3 150942510; Chr5Lg3 446272814; Chr5Lg3 449362407; Chr5Lg3 569788697; and Chr5Lg3 572899434) were stably linked with SPAD (Table 3.3). Chr6LG2 469101917 and Chr7LG7 263964018 were associated with PRI and two SNPs for canopy temperature (Chr4LG4 415994869 and Chr5LG3 309595819) in three environments with –log10P and R2 ranging from 2.29–4.05 to 4.0–11.0%, respectively (Table 3.3). A total of 21 SNPs were found on chromosomes 2, 3, 4, 5, 6, and 7, including one repeating marker (Chr5LG3 566189271), six SNPs for reproductive stem length, six SNPs for internode number, and nine for pod number (Sc04062 32372) (Table 3.3).

3.5.3 YIELD AND RELATED TRAITS

Oladzad et al. (2019a) demonstrated that various yield components, such as seed weight and number of seeds per pod in common bean under heat conditions in the BASE Meso subpopulation are complex quantitative traits that are significantly affected by environment, genotype, and their interactions. Four SNPs consisting of 1 (S03_41096424) on chromosome 3, 1 (S08_9135122) on chromosomes 8 and 2 (S11_41873950 and S11_47305350) on chromosome 11 were detected consistently in three environments (HN15 + 16 and PR-16) with –log10*P* and R2 ranging from 4.76–7.04 and 7.3–13.7%, respectively. Similarly, Wen et al. (2019) detected five and six SNPs on chromosomes 2 and 3 and one SNP each on chromosomes 7 and 11 for seed weight in common bean (Table 3.3). Moreover, the AM results with the common bean MAGIC population detected the *PHI2.1*, *Yd1.1*, and *100SdW4.1* mapped in the linkage mapping (Diaz et al., 2020).

SY and seed weight in field peas were mapped by Gali et al. (2019). Two SNPs (1_366513463 and 1_57305683) on chromosome 1, one SNP (4_373933955) on chromosome, one SNP (*7_488770913*) on chromosome 7 and one SNP (SSc4454_324798) on the scaffold was associated with SY (Table 3.3). One SNP (1_176606388) on chromosome 1 and 3 SNPs (Sc00398_17041, Sc01126_54371, and Sc01886_124838) on the scaffold were associated with seed weight.

3.5.4 SEED QUALITY TRAITS

A range of seed quality traits, including seed coat color, seed composition, seed size, and seed shape, have been identified in various pulses by GWAS. In common bean, 10 SNPs have been found to be associated with seed coat color (Wen et al., 2019). Of these, one and five SNPs were on chromosomes 1 and 9, respectively, and two SNPs were mapped on chromosomes 3 and 6 (Table 3.3). Also, Noble et al. (2018) detected nine SNPs on three chromosomes in the genome of common bean for seed coat color. There were five SNPs on chromosome 4, followed by two on chromosome 5, and the smallest number of SNPs were observed on chromosomes 3 and 7, which had one each. Diaz et al. (2020) also detected two SNPs linked *to SdFe6.1* and *6.2* on chromosome 2 and another on chromosome 8 linked to *SdZn8.1*, which were mapped via linkage mapping (Table 3.3).

Pavan et al. (2019) used 6,693 high-quality SNPs from a core set of 184 accessions of lentil, including those from the Nile Valley of Egypt to Ethiopia, to dissect genetic factors regulating seed area, perimeter, and diameter. A SNP (TP6160) was shown to be strongly associated with all three seed traits, another SNP (TP186539) was found to be connected with both seed area and perimeter, and the last SNP (TP418564) was found to be associated with only seed area. In the same vein, Gali et al. (2019) detected 7 SNPs on three chromosomes (3_197482300), 6 (6_372463590 and 6_68264764) and chromosome 7 (7_7724682) for seed shape in pea and found two SNPs (1_100615820 and 1_46289124) on chromosome 1 and 1 SNP (3_13738628) on chromosome 3 (Table 3.3).

In addition, specific elements targeted at specific promoters to enhance micronutrient bioavailability via biofortification have been studied in pulses (Jha & Warkentin, 2020; Kumar et al., 2016b; Kumar & Pandey, 2020). Micronutrient-related disorders and deficiencies impact one in three people, and this strategy provides a remedy to hidden hunger. Wu et al. (2020b) recently identified several SNPs for two micronutrients (Fe and Zn), two macronutrients (K and phosphorus (P)) and two essential elements, manganese (Mn) and sulfur (S) in mung bean. Three SNPs (SVR01_1387204, SVR01_1392172, and SVR01_1401760) on chromosome 1 and one SNP (SVR05_25211011) on chromosome 5 had pleiotropic effects on both K and P (Table 3.3). Also, one SNP (SVR07_49754942) on chromosome 7 affected both P and S simultaneously ($-\log_{10}P$ (4.22–5.46) and R^2 (18–23%)) (Table 3.3). Furthermore, a total of 33 SNPs were linked to specific nutrients with 3–8 SNPs per nutrient, that is, least and maximum three and eight SNPs for P and S, respectively (Table 3.3).

Gali et al. (2019) elucidated the genetic architecture of two fiber components (acid and neutral detergent fibers), starch, and protein concentrations in peas. One SNP (6_372463590) on chromosome 6 was linked to the two fiber contents. Additionally, SNP (*2_457185*) on chromosome 2 exhibited pleiotropic effects on both neutral detergent fiber (NDF) and starch concentrations, and another SNP (Sc03839_38033) on the scaffold region affected both acid detergent fiber and starch concentration. For the acid detergent fiber, NDF, starch, and protein concentrations with three SNPs each, 17 trait-specific SNPs were discovered. Three SNPs for the NDF and five SNPs were discovered for the starch and protein concentrations. Jha et al. (2020) utilized 109 (85 + 24) accessions of pea from Gali et al. (2019) to identify SNPs linked to folates and its three components, tetrahydrofolate (THF), 5-methltetrahydrofolate (5-MTHF) and 5-formltetrahydrofolate (5-FTHF).

Around 29 SNPs on the scaffold regions were found to regulate folates, THF, 5-MTHF and 5-FTHF in the 85 accessions. Of these, one marker (Sc_18_1682635) on scaffold 18 affected folates and 5-MTHF pleiotropically. SNP (Sc_3060_11265) on scaffold 3060 regulated both folates and 5-FTHF (Table 3.3). To validate the results from the 85 accessions, another 24 core accessions were used to conduct GWAS, out of which two SNPs (Sc_6992_86348 and Sc_18_1682635) were aligned to folates and the three components. These two markers were mapped with 85 accessions for at least of the traits evaluated in the study, lending credence to the reliability and stability of these two SNPs for MAS.

3.6 GBS-BASED GENOMIC SELECTION IN PULSES

GS, a more advanced version of MAS that makes use of all significant markers associated with a trait across the entire genome to forecast how well the trait will perform, has grown in popularity recently due to the significant decrease in sequencing costs and the availability of several models, as well as its efficiency in plant breeding (Jannink et al., 2010; Varshney et al., 2020). Using molecular markers, MAS identify plants that have genomic areas that are involved in the production of desired traits or QTLs. Linkage or AM studies have frequently employed MAS in the manipulation of large effect alleles associated with markers. In contrast, it has failed to improve quantitative characteristics (Bernardo, 2008; Xu & Crouch, 2008). This inefficiency is attributed to numerous setbacks in MAS (Jannink et al., 2010); therefore, GS offers an efficient and effective means to improve a trait of interest (Annicchiarico et al., 2020; Heffner et al., 2010). The usefulness and accuracy of this strategy are dependent on traits and their heritability, marker density, training, and test populations, LD, population structure, and statistical models (Berro et al., 2019; Bhat et al., 2016; Tayeh et al., 2015). Genetic and phenotypic data from a training population are utilized to train the statistical model, and then the model may be used to predict genomic estimated breeding values (GEBVs) of individuals in a testing population genotyped but not phenotyped (Desta & Ortiz, 2014; Jonas & de Koning, 2013). In the subsequent selection of superior lines, the GEBVs are used rather than the actual phenotypes.

There have been a lot of GS research on pulses such as peas and lentils. GS, MAS, and phenotypic selection (PS) were compared in pea RIL populations for grain yield under severe terminal drought (Annicchiarico et al.,

2020). No substantial difference in yield prediction was found between GS and MAS due to a variety of factors, including the genotype by environment interaction (GE). The selection effectiveness of GS was 18% larger than that of MAS because it utilized drought escape (DE) and inherent DT. The GS and MAS populations outperformed the PS in this study. GS had a 2.8 lower cost per analyzed line than PS. Using GBS-based GS, Annicchiarico et al. (2019b) examined a number of pea attributes and discovered extremely high predictive accuracy (rAC) between predicted values across two locations. In addition to a high grain yield (rAC = 0.59), there was an extremely high farmer acceptance score (rAC = 0.77). Genomic selection of peas in an Italian environment was also studied by Annicchiarico et al. (2019a) and they found that the grain yield was 30%, the flowering time was 66%, and the seed weight was 64%. The lodging susceptibility was 28%. Grain yield, flowering start, seed weight, and lodging susceptibility were all lowered within inter-populations to values of 0.19, 0.40, 0.28, and 0.22, respectively. Genetic gains per unit time were larger for characteristics with GS compared to PS. At least 80% and 20%, respectively, of the predictions made by GS over PS were accurate within the population. If efficiency is measured by the yield, the GS strain was three times more efficient than the PS strain, and at least two times more efficient for features with intrapopulation variation. As a result, inter-population yield was 2.4 times greater, and the other attributes were also more efficient. Pea grain yield and flowering time were predicted by the GS with an average of 0.72 and 0.70, respectively, across three RIL populations under severe terminal drought conditions, as reported by Annicchiarico et al. (2017). Pea Ascochyta blight resistance was studied by another study group using 215 lines from the Plant and Food Research Limited (PFR) breeding program, and the GBLUP model was found to be 0.56 more accurate than phenotypic data (Carpenter et al., 2018).

Natural populations of 324 accessions from gene banks of the International Center for Agricultural Research in the Dry Areas (ICARDA), the USDA, Plant Gene Resources of Canada (PGRC), and cultivars developed at the Crop Development Center (CDC) were used to conduct the first GS application in the lentil breeding program. A total of 110 RILs and 120 RILs were found in LR-01 and LR-11, respectively, from ILL 8006-BM4 CDC Milestone (Haile et al., 2020). With respect to DTF, vegetative phase, days to swollen pods, reproductive period, and other traits, the natural population panel was used to conduct GWAS for all relevant markers, while the other RIL populations were utilized for GWAS + GS (REP).

According to the data for all features and models in the LDP, LR-01 and LDP + LR-11 populations: rAC ranged from 0.65–0.85, 0 36–0.74, 0.40–0.79, and 0.30–0.59. BayesB was the best model, with the highest prediction for traits regulated by few QTLs with relatively large effects; (ii) incorporation of GE improved prediction accuracy by 66%; (iii) GS can be applied in lentil to make within-population and across-environment predictions; (iv) across-population genomic prediction should not be considered in lentil when the population size is small. Pulses such as peas can benefit greatly from the use of genetic selection (GS) to shorten the breeding cycle and increase genetic gains of polygenic characteristics.

3.7 CONCLUSION AND FUTURE OUTLOOK

Arable land is anticipated to shrink as the global population rises to 10 billion by 2050. The increasing spate of climate change, abiotic and biotic stresses threaten food and nutritional security. It requires a more ambitious and ecologically sound approach to increase pulse production while mitigating climate change and other stressors. GBS, linkage mapping, GWAS, and GS applications in pulses hold prospects for meeting the dietary requirements of the teeming global population while preserving the environment for posterity. Environments-specific QTLs mapping has been performed in several pulses via linkage and GWAS mappings. The development of artificial intelligence, machine learning and neural network applications will accelerate the identification of SNPs associated with key QTLs against stressors. Marker density and population-related issues have been recently resolved with the wide application of NGS techniques at a reduced cost of sequencing. High throughput GBS-based SNPs have been identified for linkage mapping, GWAS, and GS application in many pulses. Applications of these advanced breeding approach in pulses usually yield high-quality SNPs for map-based breeding against abiotic and biotic stresses. Further studies could be conducted to integrate GBS-based linkage and AM to meta-QTL analysis to confirm and validate the loci reported in this review. Also, the number of candidate genes reported in the reviewed studies need to be functionally validated by gene overexpression and genome editing (GE) strategies. Lastly not least, the stable markers could be used for practical plant breeding via MAS.

KEYWORDS

- **candidate genes**
- **climate-smart agriculture**
- **food and nutritional security**
- **genomic-assisted breeding**
- **hidden hunger**
- **yield improvement**

REFERENCES

Aldemir, S., Ateş, D., Temel, H. Y., Yağmur, B., Alsaleh, A., Kahriman, A., Özkan, H., et al., (2017). QTLs for iron concentration in seeds of the cultivated lentil (*Lens culinaris* Medic.) via genotyping by sequencing. *Turkish Journal of Agriculture and Forestry, 41*(4), 243–255. doi: 10.3906/tar-1610-33.

Ali, J., Aslam, U. M., Tariq, R., Murugaiyan, V., Schnable, P. S., Li, D., Marfori-Nazarea, C. M., et al., (2018). Exploiting the genomic diversity of rice (*Oryza sativa* L.): SNP-typing in 11 early-backcross introgression-breeding populations. *Frontiers in Plant Science, 9*(849). doi: 10.3389/fpls.2018.00849.

Amelin, A. V., & Parakhin, N. V., (2003). *Scientific Grounds for Pea Selection for Fodder Production* (Vol. 2, pp. 20–25). Kormoproizvodstvo.

Annicchiarico, P., Nazzicari, N., Laouar, M., Thami-Alami, I., Romani, M., & Pecetti, L., (2020). Development and proof-of-concept application of genome-enabled selection for pea grain yield under severe terminal drought. *International Journal of Molecular Sciences, 21*(7), 2414.

Annicchiarico, P., Nazzicari, N., Pecetti, L., Romani, M., & Russi, L., (2019a). Pea genomic selection for Italian environments. *BMC Genomics, 20*(1), 603. doi: 10.1186/s12864-019-5920-x.

Annicchiarico, P., Nazzicari, N., Pecetti, L., Romani, M., Ferrari, B., Wei, Y., & Brummer, E. C., (2017). GBS-based genomic selection for pea grain yield under severe terminal drought. *The Plant Genome, 10*(2). doi: 10.3835/plantgenome2016.07.0072.

Annicchiarico, P., Russi, L., Romani, M., Pecetti, L., & Nazzicari, N., (2019b). Farmer-participatory vs. conventional market-oriented breeding of inbred crops using phenotypic and genome-enabled approaches: A pea case study. *Field Crops Research, 232*, 30–39. doi: https://doi.org/10.1016/j.fcr.2018.11.001.

Ariani, A., Berny, M. Y. T. J. C., & Gepts, P., (2016). Genome-wide identification of SNPs and copy number variation in common bean (*Phaseolus vulgaris* L.) using genotyping-by-sequencing (GBS). *Molecular Breeding, 36*(7), 87. doi: 10.1007/s11032-016-0512-9.

Ates, D., Sever, T., Aldemir, S., Yagmur, B., Temel, H. Y., Kaya, H. B., Alsaleh, A., et al., (2016). Identification QTLs controlling genes for Se uptake in lentil seeds. *PLoS one, 11*(3), e0149210. doi: 10.1371/journal.pone.0149210.

Bernardo, R., (2008). Molecular markers and selection for complex traits in plants: Learning from the last 20 years. *Crop Science, 48*(5), 1649–1664. doi: 10.2135/cropsci2008.03.0131.

Berny, M. Y. T. J. C., Konzen, E. R., Medina, V., Palkovic, A., Ariani, A., Tsai, S. M., Gilbert, M. E., & Gepts, P., (2019). Root and shoot variation in relation to potential intermittent drought adaptation of Mesoamerican wild common bean (*Phaseolus vulgaris* L.). *Annals of Botany, 124*(6), 917–932. doi: 10.1093/aob/mcy221.

Berro, I., Lado, B., Nalin, R. S., Quincke, M., & Gutiérrez, L., (2019). Training population optimization for genomic selection. *The Plant Genome, 12*(3), 190028. doi: 10.3835/plantgenome2019.04.0028.

Bhadauria, V., Ramsay, L., Bett, K. E., & Banniza, S., (2017). QTL mapping reveals genetic determinants of fungal disease resistance in the wild lentil species *Lens ervoides*. *Scientific Reports, 7*(1), 3231. doi: 10.1038/s41598-017-03463-9.

Bhakta, M. S., Jones, V. A., & Vallejos, C. E., (2015). Punctuated distribution of recombination hotspots and demarcation of pericentromeric regions in *Phaseolus vulgaris* L. *PLoS ONE, 10*(1), e0116822. doi: 10.1371/journal.pone.0116822.

Bhat, J. A., Ali, S., Salgotra, R. K., Mir, Z. A., Dutta, S., Jadon, V., Tyagi, A., et al., (2016). Genomic selection in the era of next generation sequencing for complex traits in plant breeding. *Frontiers in Genetics, 7*, 221.

Blair, M. W., Iriarte, G., & Beebe, S., (2006). QTL analysis of yield traits in an advanced backcross population derived from a cultivated Andean × wild common bean (*Phaseolus vulgaris* L.) cross. *Theoretical and Applied Genetics, 112*(6), 1149–1163. doi: 10.1007/s00122-006-0217-2.

Bohra, A., Jha, U. C., Godwin, I. D., & Varshney, R. K., (2020). Genomic interventions for sustainable agriculture. *Plant Biotechnology Journal*. doi: 10.1111/pbi.13472.

Boutet, G., Alves, C. S., Falque, M., Peterlongo, P., Lhuillier, E., Bouchez, O., Lavaud, C., et al., (2016). SNP discovery and genetic mapping using genotyping by sequencing of whole genome genomic DNA from a pea RIL population. *BMC Genomics, 17*(1), 121. doi: 10.1186/s12864-016-2447-2.

Breria, C. M., Hsieh, C. H., Yen, J. Y., Nair, R., Lin, C. Y., Huang, S. M., Noble, T. J., & Schafleitner, R., (2019). Population structure of the world vegetable center mungbean mini core collection and genome-wide association mapping of loci associated with variation of seed coat luster. *Tropical Plant Biology, 13*(1), 1–12. doi: 10.1007/s12042-019-09236-0.

Breria, C. M., Hsieh, C. H., Yen, T. B., Yen, J. Y., Noble, T. J., & Schafleitner, R. (2020). A SNP-Based genome-wide association study to mine genetic loci associated to salinity tolerance in mungbean (*Vigna radiata* L.). *Genes, 11*(7), 759.

Campa, A., Murube, E., & Ferreira, J. J., (2018). Genetic diversity, population structure, and linkage disequilibrium in a Spanish common bean diversity panel revealed through genotyping-by-sequencing. *Genes, 9*(11), 518. doi: 10.3390/genes9110518.

Carpenter, M. A., Goulden, D. S., Woods, C. J., Thomson, S. J., Kenel, F., Frew, T. J., Cooper, R. D., & Timmerman-Vaughan, G. M., (2018). Genomic selection for *Ascochyta* blight resistance in pea. *Frontiers in Plant Science, 9*, 1878.

Chaudhary, J., Shivaraj, S., Khatri, P., Ye, H., Zhou, L., Klepadlo, M., Dhakate, P., et al., (2019). Approaches, applicability, and challenges for development of climate-smart soybean. In: *Genomic Designing of Climate-Smart Oilseed Crops* (pp. 1–74). Springer.

Chu, J., Zhao, Y., Beier, S., Schulthess, A. W., Stein, N., Philipp, N., Röder, M. S., & Reif, J. C., (2020). Suitability of single-nucleotide polymorphism arrays versus genotyping-by-sequencing for GeneBank genomics in wheat. *Frontiers in Plant Science, 11*(42). doi: 10.3389/fpls.2020.00042.

Cortés, J. A., & Blair, M. W., (2018). Genotyping by sequencing and genome–environment associations in wild common bean predict widespread divergent adaptation to drought. *Frontiers in Plant Science, 9*, 128.

Da Silva, L. C., De Souza, T. L. P. O., Cruz, C. D., Carneiro, P. C. S., E Silva, F. F., De Barros, E. G., Vianello, R. P., et al., (2018). Linkage fine-mapping and QTLs affecting morpho-agronomic traits of a Mesoamerican×Andean RIL common bean population. *Euphytica, 214*(12), 221. doi: 10.1007/s10681-018-2299-8.

Desta, A, Z., & Ortiz, R., (2014). Genomic selection: Genome-wide prediction in plant improvement. *Trends in Plant Science, 19*(9), 592–601. doi: https://doi.org/10.1016/j.tplants.2014.05.006.

Diaz, S., Ariza-Suarez, D., Izquierdo, P., Lobaton, J. D., De La Hoz, J. F., Acevedo, F., Duitama, J., et al., (2020). Genetic mapping for agronomic traits in a MAGIC population of common bean (*Phaseolus vulgaris* L.) under drought conditions. *BMC Genomics, 21*(1), 799. doi: 10.1186/s12864-020-07213-6.

Dramadri, I. O., Amongi, W., Kelly, J. D., & Mugisham M. C., (2020). Genome-wide association analysis of resistance to *Pythium ultimum* in common bean (*Phaseolus vulgaris*). *Plant Breeding.* doi: 10.1111/pbr.12855.

Elshire, R. J., Glaubitz, J. C., Sun, Q., Poland, J. A., Kawamoto, K., Buckler, E. S., & Mitchell, S. E., (2011). A robust, simple genotyping-by-sequencing (GBS) approach for high diversity species. *PLoS ONE, 6*(5), e19379. doi: 10.1371/journal.pone.0019379.

Félix-Gastélum, R., Maldonado-Mendoza, I. E., Navarrete-Maya, R., Olivas-Peraza, N. G., Brito-Vega, H., & Acosta-Gallegos, J. A., (2016). Identification of *Pseudomonas syringae* pv. phaseolicola as the causal agent of halo blight in yellow beans in northern Sinaloa, Mexico. *Phytoparasitica, 44*(3), 369–378. doi: 10.1007/s12600-016-0530-5.

Ferreira, J. J., Murube, E., & Campa, A., (2017). Introgressed genomic regions in a set of near-isogenic lines of common bean revealed by genotyping-by-sequencing. *The Plant Genome, 10*(1). doi: 10.3835/plantgenome2016.08.0081.

Foyer, C. H., Lam, H. M., Nguyen, H. T., Siddique, K. H., Varshney, R. K., Colmer, T. D., Cowling, W., et al., (2016). Neglecting legumes has compromised human health and sustainable food production. *Nature Plants, 2*(8), 1–10. doi: 10.1038/NPLANTS.2016.112.

Gali, K. K., Liu, Y., Sindhu, A., Diapari, M., Shunmugam, A. S. K., Arganosa, G., Daba, K., et al., (2018). Construction of high-density linkage maps for mapping quantitative trait loci for multiple traits in field pea (*Pisum sativum* L.). *BMC Plant Biology, 18*(1), 172. doi: 10.1186/s12870-018-1368-4.

Gali, K. K., Sackville, A., Tafesse, E. G., Lachagari, V. B. R., McPhee, K., Hybl, M., Mikić, A., et al., (2019). Genome-wide association mapping for agronomic and seed quality traits of field pea (*Pisum sativum* L.). *Frontiers in Plant Science, 10*(1538). doi: 10.3389/fpls.2019.01538.

Gao, X., Starmer, J., & Martin, E. R., (2008). A multiple testing correction method for genetic association studies using correlated single nucleotide polymorphisms. *Genetic Epidemiology, 32*(4), 361–369. doi: 10.1002/gepi.20310.

Gaulin, E., Jacquet, C., Bottin, A., & Dumas, B., (2007). Root rot disease of legumes caused by *Aphanomyces euteiches*. *Molecular Plant Pathology, 8*(5), 539–548. doi: 10.1111/j.1364-3703.2007.00413.x.

Geleta, M., Gustafsson, C., Glaubitz, J. C., & Ortiz, R., (2020). High-density genetic linkage mapping of lepidium based on genotyping-by-sequencing SNPs and segregating contig tag haplotypes. *Frontiers in Plant Science, 11*(448). doi: 10.3389/fpls.2020.00448.

Glaubitz, J. C., Casstevens, T. M., Lu, F., Harriman, J., Elshire, R. J., Sun, Q., & Buckler, E S., (2014). TASSEL-GBS: A high-capacity genotyping by sequencing analysis pipeline. *PLoS ONE, 9*(2), e90346. doi: 10.1371/journal.pone.0090346.

Gödecke, T., Stein, A. J., & Qaim, M., (2018). The global burden of chronic and hidden hunger: Trends and determinants. *Global Food Security, 17*, 21–29. doi: 10.1016/j.gfs.2018.03.004.

Goel, V., Singla, L. D., & Choudhury, D., (2020). Cuminaldehyde induces oxidative stress-mediated physical damage and death of *Haemonchus contortus*. *Biomedicine & Pharmacotherapy, 130*, 110411. doi: https://doi.org/10.1016/j.biopha.2020.110411.

Gong, H., Rehman, F., Yang, T., Li, Z., Zeng, S., Pan, L., Li, Y., & Wang, Y., (2019). Construction of the first high-density genetic map and QTL mapping for photosynthetic traits in *Lycium barbarum* L. *Molecular Breeding, 39*(7), 106. doi: 10.1007/s11032-019-1000-9.

González, A. M., Yuste-Lisbona, F. J., Saburido, S., Bretones, S., De Ron, A. M., Lozano, R., & Santalla, M., (2016). Major contribution of flowering time and vegetative growth to plant production in common bean as deduced from a comparative genetic mapping. *Frontiers in Plant Science, 7,* 1940–1940. doi: 10.3389/fpls.2016.01940.

Gupta, D., Dadu, R. H., Sambasivam, P., Bar, I., Azad, M., Beera, N., Ford, R., & Biju, S., (2020). Conventional and biotechnological approaches for targeted trait improvement in lentil. In: Gosal, S. S., & Wani, S. H., (eds.), *Accelerated Plant Breeding* (Vol. 3, pp. 67–107). Springer, Cham.

Gupta, P. K., Rustgi, S., & Kulwal, P. L., (2005). Linkage disequilibrium and association studies in higher plants: Present status and future prospects. *Plant Molecular Biology, 57*(4), 461–485. doi: 10.1007/s11103-005-0257-z.

Gupta, R. K., Gangoliya, S. S., & Singh, N. K., (2015). Reduction of phytic acid and enhancement of bioavailable micronutrients in food grains. *J. Food Sci. Technol., 52*(2), 676–684. doi: 10.1007/s13197-013-0978-y.

Haile, T. A., Heidecker, T., Wright, D., Neupane, S., Ramsay, L., Vandenberg, A., & Bett, K. E., (2020). Genomic selection for lentil breeding: Empirical evidence. *The Plant Genome, 13*(1), e20002. doi: 10.1002/tpg2.20002.

Hall, C., Hillen, C., & Garden, R. J., (2017). Composition, nutritional value, and health benefits of pulses. *Cereal Chemistry, 94*(1), 11–31. doi: https://doi.org/10.1094/CCHEM-0-16-0069-FI.

Hart, P. J., & Griffiths, P. D., (2015). Genotyping-by-sequencing enabled mapping and marker development for the by-2 potyvirus resistance allele in common bean. *The Plant Genome, 8*(1). doi: 10.3835/plantgenome2014.09.0058.

He, J., Zhao, X., Laroche, A., Lu, Z. X., Liu, H., & Li, Z., (2014). Genotyping-by-sequencing (GBS), an ultimate marker-assisted selection (MAS) tool to accelerate plant breeding. *Frontiers in Plant Science, 5*, 484. doi: 10.3389/fpls.2014.00484.

Heffner, E., Lorenz, A., Jannink, J. L., & Sorrells, M., (2010). Plant breeding with genomic selection: Gain per unit time and cost. *Crop Science, 50*(5), 1681–1690. doi: 10.2135/cropsci2009.11.0662.

Hickey, L. T., Hafeez, A. N., Robinson, H., Jackson, S. A., Leal-Bertioli S. C. M., Tester, M., Gao, C., et al., (2019). Breeding crops to feed 10 billion. *Nature Biotechnology, 37*(7), 744–754. doi: 10.1038/s41587-019-0152-9.

Hu, H., Scheben, A., & Edwards, D., (2018). Advances in integrating genomics and bioinformatics in the plant breeding pipeline. *Agriculture, 8*(6), 75.

Hwang, W. J., (2015). *GBS-Based Map Construction and QTL Identification for Major Agronomic Traits in Mung Bean.* A Dissertation for the Degree of Doctor of Philosophy in Crop Science and Biotechnology, Department of Plant Science, the Graduate School of Seoul National University.

Ibrahim, A. K., Zhang, L., Niyitanga, S., Afzal, M. Z., Xu, Y., Zhang, L., Zhang, L., & Qi, J., (2020). Principles and approaches of association mapping in plant breeding. *Tropical Plant Biology, 13*(3), 212–224. doi: 10.1007/s12042-020-09261-4.

Jannink, J. L., Lorenz, A. J., & Iwata, H., (2010). Genomic selection in plant breeding: From theory to practice. *Briefings in Functional Genomics, 9*(2), 166–177. doi: 10.1093/bfgp/elq001.

Jha, A. B., Gali, K. K., Tar'an, B., & Warkentin, T. D., (2017). Fine mapping of QTLs for *Ascochyta* blight resistance in pea using heterogeneous inbred families. *Frontiers in Plant Science, 8*, 765. doi: 10.3389/fpls.2017.00765.

Jha, A. B., Gali, K. K., Zhang, H., Purves, R. W., Tar'an, B., Vandenberg, A., & Warkentin, T. D., (2020). Folate profile diversity and associated SNPs using genome wide association study in pea. *Euphytica, 216*(2), 18. doi: 10.1007/s10681-020-2553-8.

Jha, B. A., & Warkentin, T. D., (2020). Biofortification of pulse crops: Status and future perspectives. *Plants (Basel, Switzerland) 9*(1). doi: 10.3390/plants9010073.

Jonas, E., & De Koning, D. J., (2013). Does genomic selection have a future in plant breeding? *Trends in Biotechnology, 31*(9), 497–504. doi: https://doi.org/10.1016/j.tibtech.2013.06.003.

Josephs, E. B., Stinchcombe, J. R., & Wright, S. I., (2017). What can genome-wide association studies tell us about the evolutionary forces maintaining genetic variation for quantitative traits? *New Phytologist, 214*(1), 21–33. doi: 10.1111/nph.14410.

Kang, Y. J., Kim, S. K., Kim, M. Y., Lestari, P., Kim, K. H., Ha, B. K., Jun, T. H., et al., (2014). Genome sequence of mungbean and insights into evolution within *Vigna* species. *Nature Communications, 5*(1), 5443. doi: 10.1038/ncomms6443.

Kang, Y. J., Satyawan, D., Shim, S., Lee, T., Lee, J., Hwang, W. J., Kim, S. K., et al., (2015). Draft genome sequence of adzuki bean, *Vigna angularis*. *Scientific Reports, 5*(1), 8069. doi: 10.1038/srep08069.

Karikari, B., Li, S., Bhat, J. A., Cao, Y., Kong, J., Yang, J., Gai, J., & Zhao, T., (2019). Genome-wide detection of major and epistatic effect QTLs for seed protein and oil content in soybean under multiple environments using high-density bin map. *International Journal of Molecular Sciences, 20*(4), 979.

Kaur, S., Kumari, A., Singh, P., Kaur, L., Sharma, N., & Garg, M., (2020). Biofortification in pulses. In: Sharma, T. R., Deshmukh, R., & S. H., (eds.), *Advances in Agri-Food Biotechnology* (pp. 85–103). Springer, Singapore.

Khan, S. A., Chen, H., Deng, Y., Chen, Y., Zhang, C., Cai, T., Ali, N., et al., (2020). High-density SNP map facilitates fine mapping of QTLs and candidate genes discovery for *Aspergillus flavus* resistance in peanut (*Arachis hypogaea*). *Theoretical and Applied Genetics, 133*(7), 2239–2257. doi: 10.1007/s00122-020-03594-0.

Kumar, A., Sharma, D., Tiwari, A., Jaiswal, J. P., Singh, N. K., & Sood, S., (2016a). Genotyping-by-sequencing analysis for determining population structure of finger millet germplasm of diverse origins. *Plant Genome, 9*(2), plantgenome2015.07.0058. doi: 10.3835/plantgenome2015.07.0058.

Kumar, J., Gupta, D. S., Kumar, S., Gupta, S., & Singh, N. P., (2016b). Current knowledge on genetic biofortification in lentil. *Journal of Agricultural and Food Chemistry, 64*(33), 6383–6396. doi: 10.1021/acs.jafc.6b02171.

Kumar, S., & Pandey, G., (2020). Biofortification of pulses and legumes to enhance nutrition. *Heliyon, 6*, e03682.

Lenaerts, B., Collard, B. C. Y., & Demont, M., (2019). Review: Improving global food security through accelerated plant breeding. *Plant Science, 287*, 110207. doi: https://doi.org/10.1016/j.plantsci.2019.110207.

Li, S., Yang, G., Yang, S., Just, J., Yan, H., Zhou, N., Jian, H., et al., (2019). The development of a high-density genetic map significantly improves the quality of reference genome assemblies for rose. *Scientific Reports, 9*(1), 5985. doi: 10.1038/s41598-019-42428-y.

Lioi, L., Zuluaga, D. L., Pavan, S., & Sonnante, G., (2019). Genotyping-by-sequencing reveals molecular genetic diversity in Italian common bean landraces. *Diversity, 11*(9), 154.

Ma, Y., Marzougui, A., Coyne, C. J., Sankaran, S., Main, D., Porter, L. D., Mugabe, D., et al., (2020). Dissecting the genetic architecture of *Aphanomyces* root rot resistance in lentil by QTL mapping and genome-wide association study. *International Journal of Molecular Sciences, 21*(6), 2129. doi: 10.3390/ijms21062129.

Malvick, D. K., Grau, C. R., & Percich, J. A., (1998). Characterization of *Aphanomyces euteiches* strains based on pathogenicity tests and random amplified polymorphic DNA analyses. *Mycological Research, 102*(4), 465–475. doi: https://doi.org/10.1017/S0953756297005029.

McCrory, M. A., Hamaker, B. R., Lovejoy, J. C., & Pejai Eiche sdoerfer, N., (2010). Pulse consumption, satiety, and weight management. *Advances in Nutrition, 1*(1), 17–30. doi: 10.3945/an.110.1006.

Mhora, T., Ernest, E., Wisser, R., Evans, T., Patzoldt, M., Gregory, N., Westhafer, S., et al., (2016). Genotyping-by-sequencing to predict resistance to lima bean downy mildew in a diversity panel. *Phytopathology, 106*(10), 1152–1158. doi: 10.1094/PHYTO-02-16-0087-FI.

Mogali, S. C. & Hegde, G. M. (2020). Recent Advances in Mungbean Breeding: A Perspective. In: Gosal S.S., & Wani S.H. (eds). *Accelerated Plant Breeding, Volume 3*. Springer, Cham. .235–282. doi:10.1007/978-3-030-47306-8_9.

Moghaddam, S. M., Mamidi, S., Osorno, J. M., Lee, R., Brick, M., Kelly, J., Miklas, P., et al., (2016). Genome-wide association study identifies candidate loci underlying agronomic traits in a middle American diversity panel of common bean. *The Plant Genome, 9*(3), plantgenome2016.02.0012. doi: 10.3835/plantgenome2016.02.0012.

Murube, E., Campa, A., Song, Q., McClean, P., & Ferreira, J. J., (2020). Toward validation of QTLs associated with pod and seed size in common bean using two nested recombinant inbred line populations. *Molecular Breeding, 40*(1), 7. doi: 10.1007/s11032-019-1085-1.

Noble, T. J., Tao, Y., Mace, E. S., Williams, B., Jordan, D. R., Douglas, C. A., & Mundree, S. G., (2018). Characterization of linkage disequilibrium and population structure in a mungbean diversity panel. *Frontiers in Plant Science, 8*(2102). doi: 10.3389/fpls.2017.02102.

Oladzad, A., Porch, T., Rosas, J. C., Moghaddam, S. M., Beaver, J., Beebe, S. E., Burridge, J., et al., (2019a). Single and multi-trait GWAS identify genetic factors associated with production traits in common bean under abiotic stress environments. *G3: Genes|Genomes|Genetics, 9*(6), 1881–1892. doi: 10.1534/g3.119.400072.

Oladzad, A., Zitnick-Anderson, K., Jain, S., Simons, K., Osorno, J. M., McClean, P. E., & Pasche, J. S., (2019b). Genotypes and genomic regions associated with *Rhizoctonia solani* resistance in common bean. *Frontiers in Plant Science, 10*(956). doi: 10.3389/fpls.2019.00956.

Pavan, S., Bardaro, N., Fanelli, V., Marcotrigiano, A. R., Mangini, G., Taranto, F., Catalano, D., et al., (2019). Genotyping by sequencing of cultivated lentil (*Lens culinaris* Medik.) highlights population structure in the Mediterranean gene pool associated with geographic patterns and phenotypic variables. *Frontiers in Genetics, 10*(872). doi: 10.3389/fgene.2019.00872.

Pérez-Vega, E., Pañeda, A., Rodríguez-Suárez, C., Campa, A., Giraldez, R., & Ferreira, J. J., (2010). Mapping of QTLs for morpho-agronomic and seed quality traits in a RIL population of common bean (*Phaseolus vulgaris* L.). *Theoretical and Applied Genetics, 120*(7), 1367–1380. doi: 10.1007/s00122-010-1261-5.

Poland, A. J., & Rife, T. W., (2012). Genotyping-by-sequencing for plant breeding and genetics. *The Plant Genome, 5*(3). doi: https://doi.org/10.3835/plantgenome2012.05.0005.

Poland, J. A., Brown, P. J., Sorrells, M. E., & Jannink, J. L., (2012). Development of high-density genetic maps for barley and wheat using a novel two-enzyme genotyping-by-sequencing approach. *PLoS ONE, 7*(2), e32253. doi: 10.1371/journal.pone.0032253.

Ramdath, D., Renwick, S., & Duncan, A. M., (2016). The role of pulses in the dietary management of diabetes. *Canadian Journal of Diabetes, 40*(4), 355–363. doi: 10.1016/j.jcjd.2016.05.015.

Ray, D. K., Mueller, N. D., West, P. C., & Foley, J. A., (2013). Yield trends are insufficient to double global crop production by 2050. *PLoS One, 8*(6), e66428. doi: 10.1371/journal.pone.0066428.

Ray, S., & Satya, P., (2014). Next generation sequencing technologies for next generation plant breeding. *Frontiers in Plant Science, 5*, 367.

Reddy, V. R. P., Das, S., Dikshit, H. K., Mishra, G. P., Aski, M., Meena, S. K., Singh, A., et al., (2020). Genome-wide association analysis for phosphorus use efficiency traits in mungbean (*Vigna radiata* L. Wilczek) using genotyping by sequencing approach. *Frontiers in Plant Science, 11*, 1546.

Sandhu, K. S., You, F. M., Conner, R. L., Balasubramanian, P. M., & Hou, A., (2018). Genetic analysis and QTL mapping of the seed hardness trait in a black common bean (*Phaseolus vulgaris*) recombinant inbred line (RIL) population. *Molecular Breeding, 38*(3), 34. doi: 10.1007/s11032-018-0789-y.

Schafleitner, R., Huang, S. M., Chu, S. H., Yen, J. Y., Lin, C. Y., Yan, M. R., Krishnan, B., et al., (2016). Identification of single nucleotide polymorphism markers associated with resistance to bruchids (*Callosobruchus spp.*) in wild mungbean (*Vigna radiata* var. sublobata) and cultivated V. radiata through genotyping by sequencing and quantitative trait locus analysis. *BMC Plant Biology, 16*(1), 159. doi: 10.1186/s12870-016-0847-8.

Schröder, S., Mamidi, S., Lee, R., McKain, M. R., McClean, P. E., & Osorno, J. M., (2016). Optimization of genotyping by sequencing (GBS) data in common bean (*Phaseolus vulgaris* L.). *Molecular Breeding, 36*(1), 6. doi: 10.1007/s11032-015-0431-1.

Schwartz, F. H., & Singh, S. P., (2013). Breeding common bean for resistance to white mold: A review. *Crop Science, 53*(5), 1832–1844. doi: 10.2135/cropsci2013.02.0081.

Sekar, S., & Nalini, R., (2017a). Screening of mungbean genotypes against pulse beetle *Callosobruchus chinensis* and evaluating the biochemical basis of resistance. *International Journal of Chemical Studies, 5*, 1296–1301.

Sekar, S., & Nalini, R., (2017b). Varietal screening of mungbean genotypes against whitefly (*Bemisia tabaci* Genn.), mungbean yellow mosaic virus (MYMV) and *Cercospora* leaf spot. *International Journal of Current Microbiology and Applied Sciences, 6*(3), 1278–1285.

Singh, R. K., Prasad, A., Muthamilarasan, M., Parida, S. K., & Prasad, M., (2020). Breeding and biotechnological interventions for trait improvement: Status and prospects. *Planta, 252*(4), 54. doi: 10.1007/s00425-020-03465-4.

Smitchger, J. A., (2017). *Quantitative Trait Loci Associated with Lodging, Stem Strength, Yield, and Other Important Agronomic Traits in Dry Field Peas*. Montana State University-Bozeman, College of Agriculture.

Sokolkova, A., Burlyaeva, M., Valiannikova, T., Vishnyakova, M., Schafleitner, R., Lee, C. R., Ting, C. T., et al., (2020). Genome-wide association study in accessions of the mini-core collection of mungbean (*Vigna radiata*) from the world vegetable gene bank (Taiwan). *BMC Plant Biology, 20*(1), 363. doi: 10.1186/s12870-020-02579-x.

Soltani, A., MafiMoghaddam, S., Walter, K., Restrepo-Montoya, D., Mamidi, S., Schroder, S., Lee, R., et al., (2017). Genetic architecture of flooding tolerance in the dry bean middle-american diversity panel. *Frontiers in Plant Science, 8*(1183). doi: 10.3389/fpls.2017.01183.

Sonah, H., Bastien, M., Iquira, E., Tardivel, A., Légaré, G., Boyle, B., Normandeau, É., et al., (2013). An improved genotyping by sequencing (GBS) approach offering increased versatility and efficiency of SNP discovery and genotyping. *PLoS ONE, 8*(1), e54603. doi: 10.1371/journal.pone.0054603.

Sonah, H., O'Donoughue, L., Cober, E., Rajcan, I., & Belzile, F., (2015). Identification of loci governing eight agronomic traits using a GBS-GWAS approach and validation by QTL mapping in soya bean. *Plant Biotechnology Journal, 13*(2), 211–221. doi: 10.1111/pbi.12249.

Song, Q., Hyten, D. L., Jia, G., Quigley, C. V., Fickus, E. W., Nelson, R. L., & Cregan, P. B., (2015a). Fingerprinting soybean germplasm and its utility in genomic research. *G3: Genes I Genomes I Genetics, 5*(10), 1999–2006. doi: 10.1534/g3.115.019000 %J G3: Genes|Genomes|Genetics.

Song, Q., Jia, G., Hyten, D. L., Jenkins, J., Hwang, E. Y., Schroeder, S. G., Osorno, J. M., et al., (2015b). SNP assay development for linkage map construction, anchoring whole-genome sequence, and other genetic and genomic applications in common bean. *G3: Genes|Genomes|Genetics, 5*(11), 2285–2290. doi: 10.1534/g3.115.020594.

Song, Q., Jia, G., Hyten, D. L., Jenkins, J., Hwang, E. Y., Schroeder, S. G., Osorno, J. M., et al., (2015c). SNP assay development for linkage map construction, anchoring whole-genome sequence, and other genetic and genomic applications in common bean. *G3: Genes|Genomes|Genetics, 5*(11), 2285–2290. doi: 10.1534/g3.115.020594.

Su, K., Xing, H., Guo, Y., Zhao, F., Liu, Z., Li, K., Li, Y., & Guo, X., (2020). High-density genetic linkage map construction and cane cold hardiness QTL mapping for Vitis based on restriction site-associated DNA sequencing. *BMC Genomics, 21*(1), 419. doi: 10.1186/s12864-020-06836-z.

Sudheesh, S., Kimber, R., Braich, S., Forster, J., Paull, J., & Kaur, S., (2019). Construction of an integrated genetic linkage map and detection of quantitative trait loci for *Ascochyta* blight resistance in faba bean (*Vicia faba* L.). *Euphytica, 215*(3), 42. doi: 10.1007/s10681-019-2365-x.

Tafesse, E. G., Gali, K. K., Lachagari, V., Bueckert, R., & Warkentin, T. D., (2020). Genome-wide association mapping for heat stress responsive traits in field pea. *International Journal of Molecular Sciences, 21*(6), 2043. doi: 10.3390/ijms21062043.

Tayeh, N., Klein, A., Le Paslier, M. C., Jacquin, F., Houtin, H., Rond, C., Chabert-Martinello, M., et al., (2015). Genomic prediction in pea: Effect of marker density and training population size and composition on prediction accuracy. *Frontiers in Plant Science, 6*, 941.

Tester, M., & Langridge, P., (2010). Breeding technologies to increase crop production in a changing world. *Science, 327*(5967), 818–822. doi: 10.1126/science.1183700 %J Science.

Tilman, D., Balzer, C., Hill, J., & Befort, B. L., (2011). Global food demand and the sustainable intensification of agriculture. *Proceedings of the National Academy of Sciences of United States of America, 108*(50), 20260–20264. doi: 10.1073/pnas.1116437108.

Tock, A. J., Fourie, D., Walley, P. G., Holub, E. B., Soler, A., Cichy, K. A., Pastor-Corrales, M. A., et al., (2017). Genome-wide linkage and association mapping of halo blight resistance in common bean to race 6 of the globally important bacterial pathogen. *Frontiers in Plant Science, 8*(1170). doi: 10.3389/fpls.2017.01170.

Toonen, R. J., Puritz, J. B., Forsman, Z. H., Whitney, J. L., Fernandez-Silva, I., Andrews, K. R., & Bird, C. E., (2013). ezRAD: A simplified method for genomic genotyping in non-model organisms. *PeerJ, 1*, e203. doi: 10.7717/peerj.203.

van Treuren, R., Hoekstra, R., & van Hintum, T. J. L. (2017). Inventory and prioritization for the conservation of crop wild relatives in The Netherlands under climate change. *Biological Conservation 216,* 123–139. doi:10.1016/j.biocon.2017.10.003.

Varshney, R. K., Pandey, M. K., Bohra, A., Singh, V. K., Thudi, M., & Saxena, R. K., (2019a). Toward the sequence-based breeding in legumes in the post-genome sequencing era. *Theoretical and Applied Genetics, 132*(3), 797–816. doi: 10.1007/s00122-018-3252-x.

Varshney, R. K., Sinha, P., Singh, V. K., Kumar, A., Zhang, Q., & Bennetzen, J. L., (2020). 5Gs for crop genetic improvement. *Current Opinion in Plant Biology.* doi: https://doi.org/10.1016/j.pbi.2019.12.004.

Varshney, R. K., Thudi, M., Roorkiwal, M., He, W., Upadhyaya, H. D., Yang, W., Bajaj, P., et al., (2019b). Resequencing of 429 chickpea accessions from 45 countries provides insights into genome diversity, domestication and agronomic traits. *Nature Genetics, 51*(5), 857–864. doi: 10.1038/s41588-019-0401-3.

Vasconcellos, R. C. C., Oraguzie, O. B., Soler, A., Arkwazee, H., Myers, J. R., Ferreira, J. J., Song, Q., et al., (2017). Meta-QTL for resistance to white mold in common bean. *PLoS ONE, 12*(2), e0171685. doi: 10.1371/journal.pone.0171685.

Von, G. K., Bernstein, J., Patterson, F., Wiemers, M., Chéilleachair, R. N., Foley, C., Gitter, S., et al., (2019). *Global Hunger Index: The Challenge of Hunger and Climate Change*. Available on https://www.globalhungerindex.org/results.html (accessed on 04 January 2022).

Wang, W., He, J., Chen, S., Peng, P., Zhong, W., Wang, X., Zhang, T., & Li, Y., (2020). Construction of a high-density genetic map and fine mapping of a candidate gene locus for a novel branched-spike mutant in barley. *PLoS One, 15*(1), e0227617. doi: 10.1371/journal.pone.0227617.

Wen, L., Chang, H. X., Brown, P. J., Domier, L. L., & Hartman, G. L., (2019). Genome-wide association and genomic prediction identifies soybean cyst nematode resistance in common bean including a syntenic region to soybean Rhg1 locus. *Horticulture Research, 6*(1), 9. doi: 10.1038/s41438-018-0085-3.

Wen, Y. J., Zhang, H., Ni, Y. L., Huang, B., Zhang, J., Feng, J. Y., Wang, S. B., et al., (2017). Methodological implementation of mixed linear models in multi-locus genome-wide association studies. *Briefings in Bioinformatics, 19*(4), 700–712. doi: 10.1093/bib/bbw145.

Wong, M. M., Gujaria-Verma, N., Ramsay, L., Yuan, H. Y., Caron, C., Diapari, M., Vandenberg, A., & Bett, K. E., (2015). Classification and characterization of species within the genus lens using genotyping-by-sequencing (GBS). *PLoS One, 10*(3). doi: 10.1371/journal.pone.0122025.

Wu, C., Mozzoni, L. A., Moseley, D., Hummer, W., Ye, H., Chen, P., Shannon, G., & Nguyen, H., (2020a). Genome-wide association mapping of flooding tolerance in soybean. *Molecular Breeding, 40*(1), 4. doi: 10.1007/s11032-019-1086-0.

Wu, L., Chang, K. F., Conner, R. L., Strelkov, S., Fredua-Agyeman, R., Hwang, S. F., & Feindel, D., (2018). *Aphanomyces euteiches:* A threat to Canadian field pea production. *Engineering, 4*(4), 542–551. doi: https://doi.org/10.1016/j.eng.2018.07.006.

Wu, X., Islam, A., Limpot, N., Mackasmiel, L., Mierzwa, J., Cortés, A. J., & Blair, M. W., (2020b). Genome-wide SNP identification and association mapping for seed mineral concentration in mung bean (*Vigna radiata* L.). *Frontiers in Genetics, 11*, 656. doi: 10.3389/fgene.2020.00656.

Xu, Y., & Crouch, J. H., (2008). Marker-assisted selection in plant breeding: From publications to practice. *Crop Science, 48*(2), 391–407. doi: 10.2135/cropsci2007.04.0191.

Yadav, P., Saxena, K. B., Hingane, A., Kumar, C. V. S., Kandalkar, V. S., Varshney, R. K., & Saxena, R. K., (2019). An "axiom Cajanus SNP array" based high density genetic map and QTL mapping for high-selfing flower and seed quality traits in pigeonpea. *BMC Genomics, 20*(1), 235. doi: 10.1186/s12864-019-5595-3.

Yuste-Lisbona, F. J., González, A. M., Capel, C., García-Alcázar, M., Capel, J., De Ron, A. M., Santalla, M., & Lozano, R., (2014). Genetic variation underlying pod size and color traits of common bean depends on quantitative trait loci with epistatic effects. *Molecular Breeding, 33*(4), 939–952. doi: 10.1007/s11032-013-0008-9.

Zhang, X. F., Wang, G. Y., Dong, T. T., Chen, B., Du, H. S., Li, C. B., Zhang, F. L., et al., (2019a). High-density genetic map construction and QTL mapping of first flower node in pepper (*Capsicum annuum* L.). *BMC Plant Biology, 19*(1), 167. doi: 10.1186/s12870-019-1753-7.

Zhang, Y. M., Jia, Z., & Dunwell, J. M., (2019b). The applications of new multi-locus GWAS methodologies in the genetic dissection of complex traits. *Frontiers in Plant Science, 10*(100). doi: 10.3389/fpls.2019.00100.

Ziegler, G., Nelson, R., Granada, S., Krishnan, H. B., Gillman, J. D., & Baxter, I., (2018). Genome wide association study of ionomic traits on diverse soybean populations from germplasm collections. *Plant Direct, 2*(1), 1–14. doi: 10.1002/pld3.33.

Zitnick-Anderson, K., Oladzadabbasabadi, A., Jain, S., Modderman, C., Osorno, J. M., McClean, P. E., & Pasche, J. S., (2020). Sources of resistance to *Fusarium solani* and associated genomic regions in common bean diversity panels. *Frontiers in Genetics, 11*, 475.

CHAPTER 4

Harnessing Genome Editing Approaches in Crop Improvement to Reshape Modern Agriculture

HARSHA SOMASHEKAR

Plant Cytogenetics Laboratory, National Institute of Genetics (NIG), The Graduate University for Advanced Studies (SOKENDAI), Mishima, Shizuoka Prefecture, Japan

ABSTRACT

Recombination involves mixing up of the parental DNA and introduction of genetic variation that is the basis for all crop improvement programs. Most of the conventional crop improvement approaches exploit the national variation already existing in the crop plants and involve mating between two plant genotypes with contrasting phenotypes, followed by selection of an elite genotype in the subsequent generations. These approaches are mostly intractable with respect to the carryover of unwanted/obnoxious genes or traits to offsprings, and takes several generations to have a desired genomic constitution.

Modern genome editing technologies make it possible to insert a desired genetic change with higher precision and efficiency, thus reinforcing the crop improvement programs. There are a wide array of genome editing tools which use either of two major molecular scissors called site-specific nucleases and site-specific recombinases. Both systems involve a small nucleotide sequence that can recognize the target sequence of the genome where they induce a cut. The SSNs include 5 major types of enzymes - meganucleases/homing meganucleases, zinc-finger nucleases, transcription activator-like effector nucleases, RNA-guided nuclease systems and DNA-guided nuclease

Smart Breeding: Molecular Interventions and Advancements for Crop Improvement. Kailash Chandra, Subhash Chand, Ravi Prakash Saini, and Rohit Sharma (Eds.)

system. These enzymes can recognize the target sequence and induce double-strand breaks that triggers an endogenous DNA repair pathways that are often prone to errors. Improvement of crops via genome editing offers several advantages over the traditional methods as such tools are highly efficient and precise in inducing the mutation with minimal off target effects. In this chapter, we will discuss about different genome editing tools currently available for crop editing and highlight their importance in the context of crop improvement.

4.1 INTRODUCTION

About 11,000 years ago, wild plant species of natural habitats were brought under human management and humans started domesticating various crop species like rice, wheat, barley, rye, legumes, cotton, and several fruits and vegetable crops suitable for various needs of the human society of the human society (Ladizinsky, 1998; Zeder, 2011). Since then, agricultural crops have been subjected to selection pressures and breeding procedures with an objective of improving various desirable crop traits/characters suitable for food, medicine, timber, ornaments, shelter, etc. (McCouch, 2004; Tang et al., 2010; Wieczorek & Wright, 2012). Most of the crop plants and the unique traits they bear today are a result of such continuous selection and breeding efforts carried out for centuries. Ever since the start of agriculture, amidst numerous challenges encountered in agriculture farming, there was always a demand for further refinement of crops. As the world population continues to burgeon each year, the demand for food is also growing exponentially. While the population is expected to hit 9.7 billion by the year 2050, it is estimated that the demand for food will have increased by 60–70% by 2050 (Hunter et al., 2017; Tilman et al., 2011). Besides, fluctuating environmental conditions, declining cultivable areas, drastic changes in agriculture climate, and several plant stress conditions, etc., jeopardize the future of agriculture. Further, the problem of malnourishment in the human population is still prevalent across the globe (Bastin et al., 2019; Development Initiatives, 2018; Wieczorek & Wright, 2012; Yu & Li, 2021). All these collectively make contemporary agriculture and also future of agriculture even more challenging, and demands innovation in agricultural production. The practice of robust crop improvement techniques is crucial for sustainable agriculture and to ensure food security. Agriculture crop improvements techniques/methods that plant geneticists and breeders employ today are mostly conventional and the use of modern genetic engineering/genome editing (GE) techniques is very much limited. Resolving the future food crisis through the traditional crop improvement techniques could be difficult to materialize. Thus there is an urgent need for exploiting the potentials of modern GE tools and inculcating them alongside with the traditional methods in crop improvement programs for rapid and efficient crop improvement (Foley et al., 2011; Qaim, 2020).

Recombination involves shuffling of genetic composition of an organism. It is key for generating new genotypes with improved traits in the segregating populations. Most conventional breeding methods involve hybridization between two genotypes with contrasting phenotypes to generate segregating population in the subsequent generations (Goulet et al., 2017; Soltis & Soltis, 2019). These populations are then subjected to screening and selection with an aim to develop an elite variety with improved traits or performance. Although these methods can yield improved varieties/hybrids, the procedures involved in such methods are arduous and time consuming. Though transgenics technology touted as a promising technique for rapid generation of improved crops to ensure food security, its use however is very much limited to few countries such as South Africa and Sudan (Huang et al., 2016; Key et al., 2008; Turnbull et al., 2021). Many other countries are still in debate about its use and regulatory frameworks owing to several environmental concerns and strict regulations concerning transgenic crops usage for public (Huang et al., 2016; Turnbull et al., 2021). As a result, the production of genetically modified (GM) crops is confined to a small number of cultivated crops and regions.

Recent advancements in genetic engineering have come up with very efficient modern GE tools/techniques which allow scientists to precisely and efficiently target and modify the genome of an organism (Ainley et al., 2013; Chen et al., 2019). The availability of these modern gene editing techniques and the genome sequence information of several crop species has made it possible to induce a desirable change in any crop genome and improve the performance of any given trait in terms of growth, yield, and adaptation (Ainley et al., 2013; Chen et al., 2019, 2020; Gurbachan, 2017; Hsu et al., 2014) Besides exploiting these advanced gene editing techniques to sought novel insights into functional genomics of an organism, the use of such modern GE tools for crop improvement programs has received attention in the last several decades. In this chapter, we are going to explore the recent advancements made in the field of GE and discuss about the potentials of such tools for the betterment of crops. Further, we also discuss on the implications of these tools for crop improvement programs in various agricultural and horticultural important plants.

GE technique comprise of all advanced genetic toolkits which enable the modification of specific site of an organism's genome with higher specificity, efficiency, and precision. With the advent of modern GE technologies, it is possible to bring about a genetic variation, and reinforce the modern crop improvement programs. Hence GE techniques have emerged as the most promising tools for crop protection and for ensuring global food security. There are a wide array of GE tools available today and most of these tools use either of two major molecular scissors called site-specific nucleases (SSNs) and SSRs to edit any given organism's DNA (Arora & Narula, 2017; Gao et al., 2010; Nasti & Voytas, 2021). Both systems involve a specialized endonucleases that can recognize the target sequence of genome and induce a genetic change via double-strand breaks (DSBs) (Symington & Gautier, 2011). The endonucleases employed in GE are grouped into four major classes – meganucleases (MNs); zinc-finger nucleases (ZFNs); transcription activator-like effector nucleases (TALENs); and CRISPR (clustered regularly interspersed palindromic repeats). All these classes of enzymes are artificially synthesized through protein engineering to enable them recognize the target DNA sequence in the genome and induce double-strand breaks (DSBs) with higher precision and efficiency (Char et al., 2015; Chen et al., 2019; Ku & Ha, 2020). Following the DSBs, the cell triggers the endogenous DNA repair pathways, and as the cell tries to repair the breaks by non-homologous end joining (NHEJ) and/or homologous recombination (HR), it is prone to induce various types of mutations like insertions, deletions, gene replacements/corrections in the nucleotide sequence causing gene knockouts/gene alterations (Shrivastav et al., 2008; Symington & Gautier, 2011). In this way, the site-specific mutations of nucleotide sequences are generated. The GE strategies that various enzymes exploit to induce mutations at the target site is shown in Figure 4.1. The four major classes of GE tools implicated in crop improvement programs are described in further sections.

4.2 MEGANUCLEASES (MNS)

Meganucleases (MNs) are also called homing meganucleases (HNs) are a family of endonucleases that are present in eukaryotes, bacteria, and several archaea bacteria which can recognize DNA sequences of over 12 bp long. These enzymes can execute self-splicing activity of exon-exon junction that has intron regions between. The well-studied family of MNs are the enzymes containing LAGLIDADG motif that has been conserved across multiple organisms. LAGLIDADG motif can organize onto PI-Sec1 domain motif that has a splicing activity. Because of their self-splicing role of intron which they reside in, they function as RNA maturases and these family of endonucleases can able to target and cut large sequences of DNA (Figure 4.2(a)). Because these nucleases can able to recognize large sequences of target DNA, they are good at achieving high target specificity compared to other genome engineering tools such as ZFNs. Further, MNs are less toxic to the injected cells compared to other endonucleases like ZFNs. Over the years, several modified versions of MNs were created to achieve the cuts at user-defined locations on the genome while retaining the specificity of the enzymes. One of the disadvantages of MNs is that protein engineering is costly and very time-consuming, making this system the most challenging SSN system to work with. The use of MNs is very much limited in crop plants due to constraints in isolating their DNA binding and cleavage properties. MNs ability of target mutagenesis have been demonstrated in liguleless locus (liguleless1) in maize. In *Nicotiana tabacum* (tobacco), it was possible to reduce the toxicity levels of nicotine substance successfully by MNs (Gao et al., 2010; Silva et al., 2011; Zhang et al., 2018).

4.3 ZINC-FINGER NUCLEASES (ZFNS)

ZFNs were the first discovered GE tool demonstrated for their ability to specifically recognize DNA sequences in 1996. Zing finger motifs are a class of genetically engineered nucleases containing zinger-finger domain that can be able to bind to DNA, and are coupled with DNA cutting domain like type II restriction endonucleases Fokl that cleaves DNA at the bound site and induce mutations by exploiting endogenous DNA repair mechanisms (Figure 4.2(b)). The zinc-finger DNA binding domains can vary between 3 and 6 repeats that allows this enzyme to efficiently bind to the DNA of interest. Different protein engineering techniques have been utilized to improve the specificity of DNA binding of Zinc finger domains as well as improving the efficiency of cleavage by Fokl domain. ZFNs can disrupt the gene function rapidly and are able to integrate into any given genomic loci, hence, ZFNs are suitable tools for gene knock-outs and functional studies in animals and plants. ZFNs have been employed to create genetic variation and improve several desirable traits in various crop plants like rice, wheat, maize, soybean, brassica, tobacco, etc., and also in several fruits and vegetables crops (*refer* Table 4.2). However, ZFNs are highly toxic for the cells, limiting their usage for crop improvement programs (Cantos et al., 2014; Shukla et al., 2009; Zhang et al., 2018).

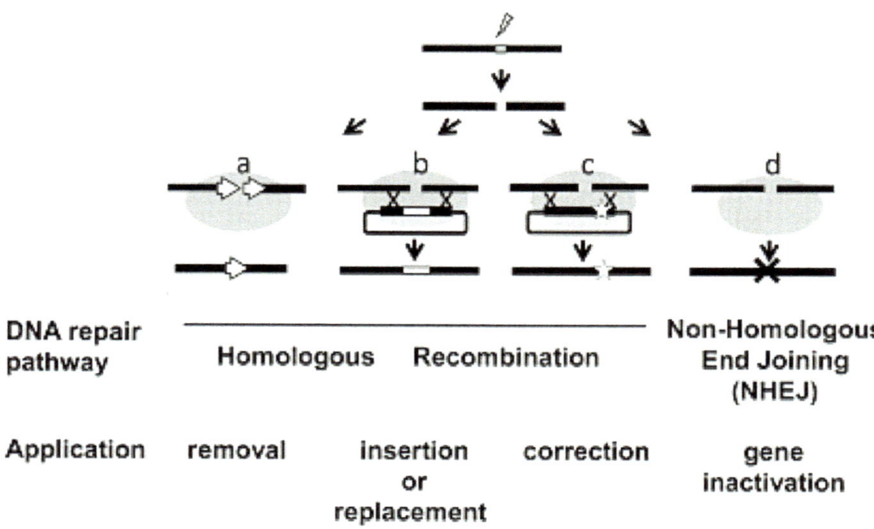

FIGURE 4.1 Gene targeting approaches of endonucleases.

Note: Cleavage of DNA by endonucleases triggers endogenous repair system that results in one of several outcomes. When a double-strand break occurs, it causes (a) homologous recombination that can result in the deletion. Gene insertion (b) or correction (c) can be achieved by the introduction of a DNA repair matrix containing sequences homologous to the endogenous sequence surrounding the DNA break. As the cell tries to correct mutations either at the site of break or distal to the break, with the frequency of correction decreases with increasing distance. The DNA mis-repair happens frequently by non-homologous end joining (d) which is highly error-prone and can result in insertions or deletions of various sizes, leading to gene inactivation.

Source: Reprinted from Silva et al. (2011). © 2011 Bentham Science Publishers Ltd. https://creativecommons.org/licenses/by/2.5/

4.4 TRANSCRIPTION ACTIVATOR-LIKE EFFECTOR NUCLEASES (TALENS)

Similar to ZFNs, TALENs are artificial restriction endonucleases produced by combining the catalytic domain of restriction endonuclease FokI with

DNA-binding domain to construct that can target any given genome site (Figure 4.2(c)). With improved efficiency, specificity, and cleavage ability, TALENs were developed from the transcription activator-like effectors (TALES) derived from genus *Xanthomonas* which is a plant pathogenic bacterium in 2011. In nature, Xanthomonas bacteria secretes the TAL effector protein into its host genome which binds within ~60 bp from the start codon and activates the transcription of that gene. The DNA binding domain of TAL effector was discovered by Boch and colleagues, which then lead to the discovery of TALENs with a fusion of Fok1 with a DNA binding repeat as an efficient GE tool. Compared to ZFNs, TALENs offer some advantages in terms of cell toxicity and feasibility (Boch et al., 2009; Char et al., 2015; Jung & Alpeter, 2016; Li et al., 2012; Malzahn et al., 2017).

TALENs have been used to improve a variety of crop traits and disease tolerance/resistance in several crop species. In rice, TALENs tool was employed to target TALE effector proteins of *Xanthomonas oryzae* and confer resistance to bacterial blight disease in rice. FAD2 gene in soybean was targeted for improved oil content by TALENs. Improved storage capacities of rice seeds such as storage tolerance, seed fragrance, better cold storage and processing traits in potato, powdery mildew (PM) resistance in wheat are some other traits successfully improved by TALENs GE tool (*refer* Table 4.2).

4.5 CRISPR (CLUSTERED REGULARLY INTERSPERSED PALINDROMIC REPEATS)

Following the discovery of TALENs, a new tool called CRISPR has emerged as the most promising GE tool for crop improvement till date. CRISPR is the widely adopted GE tool today due to its ease of engineering, simple, and quick way of creating genetic variation (Chen et al., 2019; Malzahn et al., 2017). The adaption of CRISPR technology traces back to its role in adaptive immunity in prokaryotes, which is a natural defense mechanism in the domains of single-cell microorganisms such as bacteria and archaebacteria. When a virus invades a bacteria, the virus DNA fragments called 'spacers' are integrated into the bacterial genome in a 'repeat-spacer-repeat-spacer' fashion, this organization of repeats and spacers is termed 'CRISPR.' Bacteria containing such genetic elements use them as 'genomic memory' of invading pathogens and help bacteria to detect the subsequent infection by similar viruses. When the virus infects bacteria containing CRISPR elements, the CRISPR locus gets activated and produce pre-crRNA and tracerRNA (Jinker et al., 2012). These two RNAs are then stabilized by Cas proteins of bacteria which forms crRNA:tracrRNA:Cas complex which then cuts the foreign viral DNA and renders bacteria resistant to virus infection. There are 6 types of CRISPR-Cas systems, of these Type-II system is the most widely exploited system as it requires only one cutting enzyme 'Cas9' to mediate DSBs. CRISPR-Cas9 mediated DSBs involves formation of single guide RNA (sgRNA) and Cas9 complex (sgRNA:Cas9) which interrogates DNA in the cell searching first for the appropriate protospacer adjacent motif (PAM) sequence, which is a 3 nucleotide motif sequence for recognition of sgRNA:Cas protein complex. PAM sequence in the case of Cas9 is 5′-NGG-3′. Upon recognition of PAM, sgRNA:Cas complex binds to site adjacent to the target sequence and unwinds the DNA from the first 10–12 nucleotides following the PAM sequence, this is called "seed region" (Figure 4.2(d)). If the interrogated DNA sequence matches the sgRNA target sequence, the HNH nuclease domain of the Cas9 nuclease lobe cleaves the target strand, while the RuVC-like nuclease domain of Cas9, i.e., α-helical lobe cleaves the non-targeted strand (other strands) and induces a successful double-strand break at the target site. Single or often multiple mismatches will be tolerated by Cas9-RNA complex with mismatches being more tolerated in the downstream regions of the seed sequence region adjacent to the PAM (Atkins & Voytas, 2020; Jinek et al., 2012; Wang et al., 2019; Xu et al., 2019).

The exploitation of the concept of CRISPR as a promising GE technique was developed by Emmanuelle Charpentier and Jennifer Doudna in 2011 for which they were awarded Nobel Prize in Chemistry in 2020. CRISPR/Cas9 technology is a valuable genetic tool in the plant science research and has benefitted agriculture in many ways. Due to its simple yet powerful target specific DNA editing ability, CRISPR/Cas9 system has been used to improve several different crop traits ranging from plant architecture, yield-related traits, and also improving performance of the crops to biotic and abiotic stresses. Furthermore, CRISPR is very powerful in precise editing of genomic DNA and hence it is widely used in reverse genetics and/or functional characterization of genes. Owing to the advancements in CRISPR research and protein and DNA engineering, various modified versions of CRISPR are available, which include: CRISPR-interference (CRISPRi), CRISPR-activation (CRISPRa), etc., CRISPRi involves a vector construct containing sgRNA and Cas9 protein encoding sequence with inactivation of cleavage site of Cas9. As a result CRISPRi technique

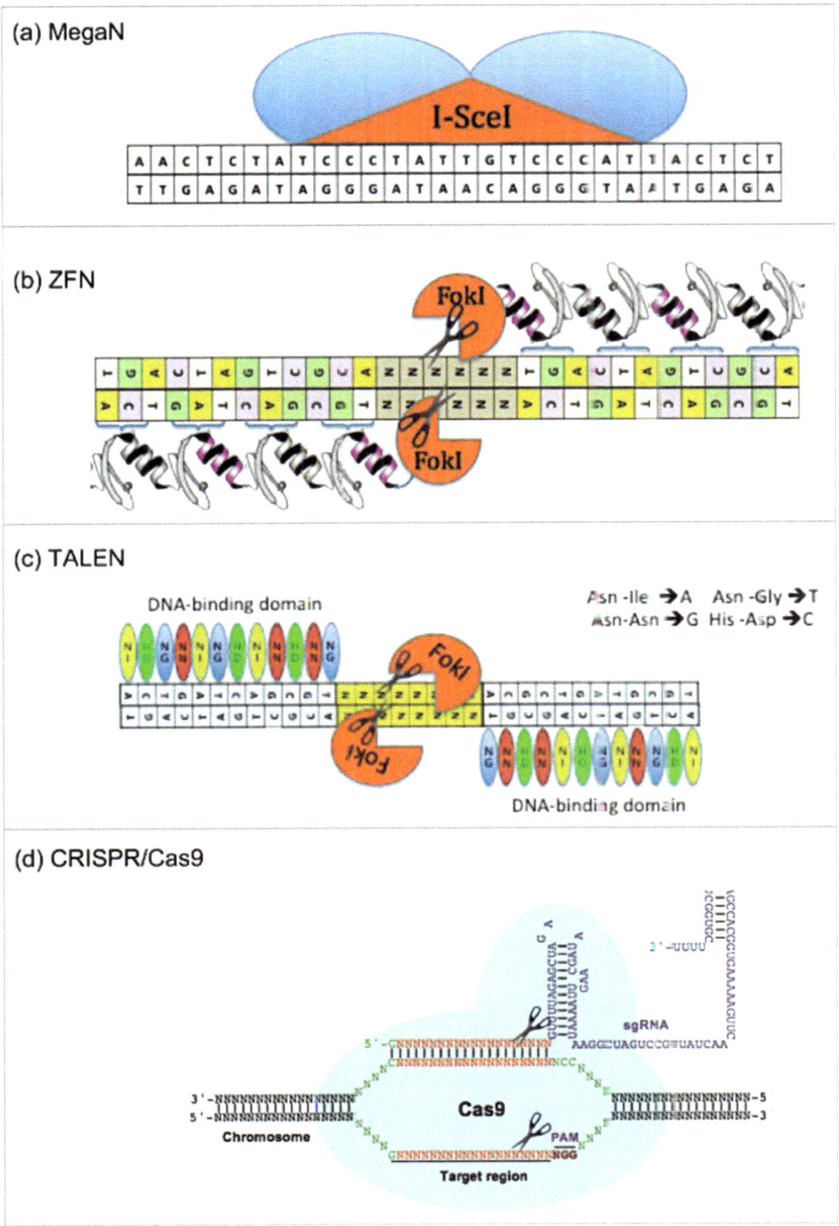

FIGURE 4.2 The structure of genome editing of different site-specific nucleases. (a) Meganucleases; (b) zing finger nucleases (ZFNs); (c) transcription activator-like effector nucleases (TALENs); and (d) CRISPR (clustered regularly interspersed palindromic repeats).

Source: Reprinted from Naglaa et al. (2015). © 2015 Taylor & Francis Group, Open access.

interferes in expression of a particular gene and is useful for studying the effects of gene suppression in functional genomics (Ainley et al., 2013; Jinek et al., 2012; Lu et al., 2020; Menz et al., 2020). CRISPRa is a novel technique where gRNA sequence is fused to transcription factors which can bind to target gene and activate the expression of a particular gene. Thus, CRISPRa confers advantages in studying the effect of overexpression of a given gene on a defined phenotype. CRISPR tool can also be used in epigenetics studies as well by fusing the construct with acetylaes/methyltranferases, etc., to promote target gene/protein acetylation/methylation. CRISPR-Cas9 technology is the most widely employed GE tool for crop improvement and is revolutionizing the field of crop breeding ever since its first application in the year 2013. This system is utilized to improve a variety of characters across many crop plants such as rice, wheat, maize, soybean, sorghum, barley, sugarcane, flax, sunflower, and so on. In case of horticultural crops, GE tools are mostly in vegetables, followed by fruits, flower crops and medicinal and aromatic plants. CRISPR has been utilized in several fruit crops like orange, strawberry, grapefruits, and vegetable crops like tomato, potato, brassica, cucumber, cassava, etc., CRISPR-Cas9 system is also used to improve browning character in mushroom. Besides agronomic traits, CRISPR-Cas9 system has been adopted to confer resistance to various pathogen attacks, insect infestations and herbicide tolerance (Gao et al., 2010, 2015; Haun et al., 2014; Hsu et al., 2014; Ku & Ha, 2020; Merker et al., 2020; Naglaa et al., 2015; Osakabe et al., 2018; Oz et al., 2021; Zaidi et al., 2016, 2020). A comparison between different genome edit tools is presented in Table 4.1.

4.6 CONCLUSION

For centuries the conventional breeding methods, which mostly rely on the natural variation, have contributed greatly to the agricultural and horticultural crop improvement. Conventional breeding offers several benefits over modern plant breeding techniques, such as increasing the genetic resources via introgression of multiple desirable traits, and also they are useful to improve traits like yield, plant height and architecture, resistance to diseases and pest, grain number etc., which are quantitative in nature. Although traditional plant breeding practices have their own advantages in crop improvement programs, these procedures are often tedious and may take several generations to develop a commercial crop

TABLE 4.1 Comparison between Meganucleases (MNs), Zinc-Finger Nucleases (ZFNs), Transcription Activator-Like Effector Nucleases (TALENs), and CRISPR (Clustered Regularly Interspersed Palindromic Repeats)

Sl. No.	Parameter	Meganucleases	ZFNs	TALENs	CRISPR/Cas
1.	Efficiency	High	Normal	Normal	Very high
2.	Specificity	Highest	Low	High	Very High
3.	Ease of engineering	Very difficult	difficult	difficult	Easy
4.	Design	Complex	Moderately complex	Slightly complex	Simple
5.	Off-target	Less-off-target effects	High off target than TALENs	Less off target	High off-targets than ZFNs and TALENs
6.	Nuclease	Fok1	Fok1	Fok1	Cas9
7.	Repair mechanism	HR	NHEJ	HR	NHEJ
8.	Applicability in plants	Very limited	Moderate	Moderate	Widely used
9.	Toxicity	Less toxic than ZFNs	Toxic	Less toxic	Relatively toxic
10.	Cost-effective	No	No	Moderate	Yes

Note: NHEJ: non-homologous end joining; HR: homologous recombination.

TABLE 4.2 Genetically Modified Crops, Genes Edited, Traits Improved, Repair Mechanism Involved by Using Various Genome Editing Tools

Crop	Tool	Gene Targeted	Repair Mechanism	Trait Improved
Rice	ZFNs	OsQQR	HDR	Trait stacking
Rice	TALENs	OsSWEET14	NHEJ	Bacterial blight resistance
Rice	TALENs	OsBADH2	NHEJ	Fragrance of grain
Rice	CRISPR/Cas9	Pi-ta	–	Resistance to rice blast
Rice	CRISPR/Cas9	EPSPS	NHEJ	Resistance to herbicide
Rice	CRISPR/Cas9	OsMATL	NHEJ	Haploids production
Rice	CRISPR/Cas9	OsSWEET13	NHEJ	Resistance to bacterial blight
Rice	CRISPR/Cas9	LAZY1	NHEJ	Tiller-spreading
Rice	CRISPR/Cas9	Gn1a, GS3, DEP1	NHEJ	Grain number, size, and dense erect panicle
Rice	CRISPR/Cas9	SBEIIb	NHEJ	High amylose content
Wheat	TALENs	TaMLO	NHEJ	Resistance to powdery mildew

TABLE 4.2 *(Continued)*

Crop	Tool	Gene Targeted	Repair Mechanism	Trait Improved
Wheat	CRISPR/Cas9	GW2	NHEJ	Grain weight, size, and grain protein
Wheat	CRISPR/Cas9	EDR1	NHEJ	Resistance to powdery mildew
Maize	Meganucleases	LG1	NHEJ	Liguleless
Maize	ZFNs	IPK1	HDR	Herbicide tolerance and reduced phytate content
Maize	ZFNs	TLP	HDR	Trait stacking
Maize	TALENs	GL2	NHEJ	Reduced epicuticular wax in leaves
Maize	TALENs	MTL	NHEJ	Haploids production
Maize	CRISPR/Cas9	Wx1	NHEJ	High amylopectin content
Maize	CRISPR/Cas9	ARGOS8	HDR	Drought tolerance
Maize	CRISPR/Cas9	ALS	HDR	Resistance to herbicide
Maize	CRISPR/Cas9	TMS5	NHEJ	Thermosensitive male-sterility
Sugarcane	TALENs	SoCOMT	NHEJ	Reduced lignin
Sugarcane	TALENs	SoCOMT	NHEJ	Saccharification efficiency
Sugarcane	CRISPR/Cas9	ALS	HDR	Tolerance to herbicide
Tobacco	Meganucleases	BBL	NHEJ	Reduce nicotine content
Tobacco	CRISPR/Cas9	NtPDS, NtPDR6	–	Etiolation of leaves, enhanced branches
Tobacco	CRISPR/Cas9	NtAna and NtAnb	–	Enhanced lipid in seeds
Tomato	TALENs	ANT1	HDR	High anthocyanin
Tomato	CRISPR/Cas9	SlMLO1	NHEJ	Powdery mildew resistance
Tomato	CRISPR/Cas9	SlJAZ2	NHEJ	Bacterial speck resistance
Tomato	CRISPR/Cas9	SP5G	NHEJ	Earlier harvest time
Tomato	CRISPR/Cas9	SlAGL6	NHEJ	Parthenocarpy
Tomato	CRISPR/Cas9	SP, SP5G, CLV3, WUS, GGP1	NHEJ	Tomato domestication
Soybean	TALENs	FAD2-1A, FAD2-1B	NHEJ	High unsaturated fatty acid (oleic acid)
Soybean	TALENs	FAD2-1A, FAD2-1B, FAD3A	NHEJ	High unsaturated fatty acid (oleic acid), low linoleic acid
Soybean	CRISPR/Cas9	ALS	HDR	Resistance to herbicide
Potato	TALENs	StVInv	NHEJ	Cold storage
Potato	TALENs	StSSR2	–	Reduced glycoalkaloids
Potato	CRISPR/Cas9	Wx1	NHEJ	High amylopectin content

TABLE 4.2 *(Continued)*

Crop	Tool	Gene Targeted	Repair Mechanism	Trait Improved
Potato	CRISPR/Cas9	StALS1	HDR	Resistance to herbicide
Brassica oleracea	TALENs	FRIGIDA	NHEJ	Early flowering time
Brassica oleracea	CRISPR/Cas9	BoPDS, BoSRK, BoMS1	–	Self-incompatibility, Albino phenotype, male sterility
Camelina sativa	CRISPR/Cas9	FAD2	NHEJ	Reduced polyunsaturated fatty acids
Apple	CRISPR/Cas9	DIPM1-4	–	Fire blight resistance
Apple	CRISPR/Cas9	MdPDS	–	Albino lacking green tissue
Grapefruit	CRISPR/Cas9	CsLOB1 promoter	NHEJ	Alleviated citrus canker
Orange	CRISPR/Cas9	CsLOB1 promoter	NHEJ	Citrus canker resistance
Strawberry	CRISPR/Cas9	TAA1, FveARFB	–	Auxin biosynthesis and response
Grapefruit	CRISPR/Cas9	CsLOB1	NHEJ	Citrus canker resistance
Cucumber	CRISPR/Cas9	eIF4E	NHEJ	Viral resistance
Mushroom	CRISPR/Cas9	PPO	NHEJ	Anti-browning property
Flax	CRISPR/Cas9	EPSPS	HDR	Resistance to herbicide
Cassava	CRISPR/Cas9	EPSPS	HDR	Resistance to herbicide
Cassava	CRISPR/Cas9	CYP79D1, CYP79D2	–	Reduced cyanide levels

Note: CRISPR: clustered regularly interspaced short palindromic repeats; *HDR*: homology-directed recombination; *NHEJ:* non-homologous end joining; *TALEN*: transcription activator-like effector nuclease; and *ZFN:* zinc-finger nuclease.

variety or hybrid. Furthermore, undergoing standard regulatory procedures will increase cost and time in bringing a promising hybrid/variety/GM crops to the fields. In addition the degree of increment in yield and other traits through conventional breeding methods is insufficient to cope up with global food requirement. Hence, these methods are reaching close to their limit in feeding the ever-increasing global population. With the advent of genetic engineering, molecular biology and genomics, a major technical advancements in crop improvement programs have occured. GM technology can potentially be a good substitute for conventional programs in creating genetic variation and crop improvement. Although, the GM crops have the potential to increase yields, confer resistance to disease pests, reduced the cost of production and so on, however, the availability of GM crops for public use often rises concerns regarding human health and environmental safety. Moreover, stringent regulations concerning the release of GM crops makes it hard for GM crops to reach to the public. It is thus becoming necessary now for breeders and scientists to switch to a safe yet an efficient, rapid, and advanced crop improvement tools. Genome editing is one of such promising tool which can reach beyond the limits of the current plant breeding programs. These tools help in creating genetic variation in a given plant population which would otherwise takes several years/generations to bring out the similar change in the same population by hybridization and recombination. Genome editing is very powerful in terms of precision, specificity, simplicity, speed and efficiency over traditional and modern plant breeding methods. As discussed above, GE tools allow us to further refine the crops by eliminating the negative undesirable traits, besides being a useful laboratory toolkits for performing target gene mutation and gene functional analyses studies. Although, GE tools offer several advantages over current breeding methods, yet there are some bottlenecks, such as the ease of engineering the nucleases, low HDR-efficiency that limits use of system for gene replacement/insertions, technical difficulties in delivery of the genome editing constructs. In addition, unavailability of the reference sequence information makes genome editing tools inaccessible for some economically important crops. Thus, further advancement in protein biology, genetic engineering techniques, transformation techniques is a prerequisite to address the current hurdles involved in genome editing tools. Large scale sequencing of minor crops and other lesser known economically important crops is essential to fully exploit the potentials of genetic variation in such crops, however, given the technical difficulties, this might take some time to happen. Nonetheless, genome editing tools provide a rapid way for generating the genetic variation and improving the desirable characters in most crop plants today, thus making it possible to ease the performance of various agricultural and horticultural crops at a greater speed. Given the hurdles that agriculture and horticulture production encountering today and constraints in the present crop breeding protocols, coupled with an ever-growing global population and malnutrition crisis, it is high time to integrate the modern genome editing tools in plant breeding programs to meet the food demands of the future generations and safeguard the global food security.

KEYWORDS

- **agriculture breeding**
- **bacterial resistance**
- **CRISPR**
- **genome**
- **genome editing**
- **molecular scissors**
- **nucleases**
- **transcription factors**

REFERENCES

Ainley, W. M., Sastry-Dent, L., Welter, M. E., Murray, M. G., Zeitler, B., Amora, R., Corbin, D. R., et al., (2013). Trait stacking via targeted genome editing. *Plant Biotechnology Journal, 11*(9), 1126–1134.

Arora, L., & Narula, A., (2017). Gene editing and crop improvement using CRISPR-Cas9 system. *Frontiers in Plant Science, 8*, 1932.

Atkins, P. A., & Voytas, D. F., (2020). Overcoming bottlenecks in plant gene editing. *Current Opinion in Plant Biology, 54*, 79–84.

Bastin, J. F., Clark, E., Elliott, T., Hart, S., Van, D. H. J., Hordijk, I., Ma, H., et al., (2019). Understanding climate change from a global analysis of city analogues. *PloS One, 14*(7), e0217592.

Boch, J., Scholze, H., Schornack, S., Landgraf, A., Hahn, S., Kay, S., Lahaye, T., et al., (2009). Breaking the code of DNA binding specificity of TAL-type III effectors. *Science, 326*(5959), 1509–1512. New York, N.Y.

Cantos, C., Francisco, P., Trijatmiko, K. R., Slamet-Loedin, I., & Chadha-Mohanty, P. K., (2014). Identification of "safe harbor" loci in indica rice genome by harnessing the property of zinc-finger nucleases to induce DNA damage and repair. *Frontiers in Plant Science, 5*, 302.

Char, S. N., Unger-Wallace, E., Frame, B., Briggs, S. A., Main, M., Spalding, M. H., Vollbrecht, E., et al., (2015). Heritable site-specific mutagenesis using TALENs in maize. *Plant Biotechnology Journal, 13*(7), 1002–1010.

Chen, K., & Gao, C., (2020). Genome-edited crops: How to move them from laboratory to market. *Front. Agric. Sci. Eng., 7*, 181–187.

Chen, K., Wang, Y., Zhang, R., Zhang, H., & Gao, C., (2019). CRISPR/Cas genome editing and precision plant breeding in agriculture. *Annual Review of Plant Biology, 70*, 667–697.

Development Initiatives, (2018). *Global Nutrition Report 2018*. Bristol: Development Initiatives.

Foley, J. A., Ramankutty, N., Brauman, K. A., Cassidy, E. S., Gerber, J. S., Johnston, M., Mueller, N. D., et al., (2011). Solutions for a cultivated planet. *Nature, 478*(7369), 337–342.

Gao, H., Smith, J., Yang, M., Jones, S., Djukanovic, V., Nicholson, M. G., West, A., et al., (2010). Heritable targeted mutagenesis in maize using a designed endonuclease. *The Plant Journal: For Cell and Molecular Biology, 61*(1), 176–187.

Gao, J., Wang, G., Ma, S., Xie, X., Wu, X., Zhang, X., Wu, Y., et al., (2015). CRISPR/Cas9-mediated targeted mutagenesis in *Nicotiana tabacum*. *Plant Molecular Biology, 87*(1, 2), 99–110.

Goulet, B. E., Roda, F., & Hopkins, R., (2017). Hybridization in plants: Old ideas, new techniques. *Plant Physiology, 173*(1), 65–78.

Gurbachan, S. M., (2017). Genome editing in crop improvement: Present scenario and future prospects. *Journal of Crop Improvement, 31*(4), 453559.

Haun, W., Coffman, A., Clasen, B. M., Demorest, Z. L., Lowy, A., Ray, E., Retterath, A., et al., (2014). Improved soybean oil quality by targeted mutagenesis of the fatty acid desaturase 2 gene family. *Plant Biotechnology Journal, 12*(7), 934–940.

Hsu, P. D., Lander, E. S., & Zhang, F., (2014). Development and applications of CRISPR-Cas9 for genome engineering. *Cell, 157*(6), 1262–1278.

Huang, S., Weigel, D., Beachy, R. N., & Li, J., (2016). A proposed regulatory framework for genome-edited crops. *Nature Genetics, 48*(2), 109–111. 109–111.

Hunter, M. C., Smith, R. G., Schipanski, M. E., Atwood, L. W., & Mortensen, D. A., (2017). Agriculture in 2050: Recalibrating targets for sustainable intensification. *BioScience, 67*(4), 386–391.

Jinek, M., Chylinski, K., Fonfara, I., Hauer, M., Doudna, J. A., & Charpentier, E., (2012). A programmable dual-RNA-guided DNA endonuclease in adaptive bacterial immunity. *Science (New York, N.Y.), 337*(6096), 816–821.

Jung, J. H., & Altpeter, F., (2016). TALEN mediated targeted mutagenesis of the caffeic acid O-methyltransferase in highly polyploid sugarcane improves cell wall composition for production of bioethanol. *Plant Molecular Biology, 92*(1, 2), 131–142.

Key, S., Ma, J. K., & Drake, P. M., (2008). Genetically modified plants and human health. *Journal of the Royal Society of Medicine, 101*(6), 290–298.

Ku, H. K., & Ha, S. H., (2020). Improving nutritional and functional quality by genome editing of crops: Status and perspectives. *Frontiers in Plant Science, 11*, 577313.

Ladizinsky, G., (1998). *Plant Evolution Under Domestication*. Springer Netherlands.

Li, T., Liu, B., Spalding, M. H., Weeks, D. P., & Yang, B., (2012). High-efficiency TALEN-based gene editing produces disease-resistant rice. *Nature Biotechnology, 30*(5), 390–392.

Lu, Y., Tian, Y., Shen, R., Yao, Q., Wang, M., Chen, M., Dong, J., et al., (2020). Targeted, efficient sequence insertion and replacement in rice. *Nature Biotechnology, 38*(12), 1402–1407.

Malzahn, A., Lowder, L., & Qi, Y., (2017). Plant genome editing with TALEN and CRISPR. *Cell & Bioscience, 7*, 21.

McCouch, S., (2004). Diversifying selection in plant breeding. *PLoS Biology, 2*(10), e347.

Menz, J., Modrzejewski, D., Hartung, F., Wilhelm, R., & Sprink, T., (2020). Genome edited crops touch the market: A view on the global development and regulatory environment. *Frontiers in Plant Science, 11*, 586027.

Merker, L., Schindele, P., Huang, T. K., Wolter, F., & Puchta, H., (2020). Enhancing in planta gene targeting efficiencies in *Arabidopsis* using temperature-tolerant CRISPR/LbCas12a. *Plant Biotechnology Journal, 18*(12), 2382–2384.

Naglaa, A. A., Channapatna, S. P., & McHughen, A. G., (2015). Genome editing for crop improvement: Challenges and opportunities. *GM Crops & Food, 6*(4), 183–205.

Nasti, R. A., & Voytas, D. F., (2021). Attaining the promise of plant gene editing at scale. *Proceedings of the National Academy of Sciences of the United States of America, 118*(22), e2004846117.

Osakabe, Y., Liang, Z., Ren, C., Nishitani, C., Osakabe, K., Wada, M., Komori, S., et al., (2018). CRISPR-Cas9-mediated genome editing in apple and grapevine. *Nature Protocols 13*(12), 2844–2863.

Oz, M., Altpeter, A., Karan, R., Merotto, A., & Altpeter, F., (2021). CRISPR/Cas9-mediated multi-allelic gene targeting in sugarcane confers herbicide tolerance. *Frontiers in Genome Editing.*

Qaim, M., (2020). Role of new plant breeding technologies for food security and sustainable agricultural development. *Applied Economic Perspectives and Policy, 42*(2), 129–150.

Shrivastav, M., De Haro, L. P., & Nickoloff, J. A., (2008). Regulation of DNA double-strand break repair pathway choice. *Cell Research, 18*(1), 134–147.

Shukla, V. K., Doyon, Y., Miller, J. C., DeKelver, R. C., Moehle, E. A., Worden, S. E., Mitchell, J. C., et al., (2009). Precise genome modification in the crop species Zea mays using zinc-finger nucleases. *Nature, 459*(7245), 437–441.

Silva, G., Poirot, L., Galetto, R., Smith, J., Montoya, G., Duchateau, P., & Pâques, F., (2011). Meganucleases and other tools for targeted genome engineering: Perspectives and challenges for gene therapy. *Current Gene Therapy, 11*(1), 11–27.

Soltis, P. S., & Soltis, D. E., (2009). The role of hybridization in plant speciation. *Annual Review of Plant Biology, 60,* 561–588.

Symington, L. S., & Gautier, J., (2011). Double-strand break end resection and repair pathway choice. *Annual Review of Genetics, 45*, 247–271.

Tang, H., Sezen, U., & Paterson, A. H., (2010). Domestication and plant genomes. *Current Opinion in Plant Biology, 13*(2), 160–166.

Tilman, D., Balzer, C., Hill, J., & Befort, B. L., (2011). Global food demand and the sustainable intensification of agriculture. *Proceedings of the National Academy of Sciences of the United States of America, 108*(50), 20260–20264.

Turnbull, C., Lillemo, M., & Hvoslef-Eide, T., (2021). Global regulation of genetically modified crops amid the gene edited crop boom: A review. *Frontiers in Plant Science, 12,* 630396.

Wang, T., Zhang, H., & Zhu, H., (2019). CRISPR technology is revolutionizing the improvement of tomato and other fruit crops. *Horticulture Research, 6*, 77.

Wieczorek, A. M., & Wright, M. G., (2012). History of agricultural biotechnology: How crop development has evolved. *Nature Education Knowledge, 3*(10), 9.

Xu, J., Hua, K., & Lang, Z., (2019). Genome editing for horticultural crop improvement. *Horticulture Research, 6*, 113.

Yu, H., & Li, J., (2021). Short- and long-term challenges in crop breeding. *National Science Review, 8*(2), nwab002.

Zaidi, S. S., Mahas, A., Vanderschuren, H., & Mahfouz, M. M., (2020). Engineering crops of the future: CRISPR approaches to develop climate-resilient and disease-resistant plants. *Genome Biology, 21*(1), 289.

Zaidi, S. S., Tashkandi, M., Mansoor, S., & Mahfouz, M. M., (2016). Engineering plant immunity: Using CRISPR/Cas9 to generate virus resistance. *Frontiers in Plant Science, 7*, 1673.

Zeder, M. A., (2011). The origins of agriculture in the near east. *Current Anthropology, 52*(S4), S221–S235.

Zhang, Y., Massel, K., Godwin, I. D., & Gao, C., (2018). Applications and potential of genome editing in crop improvement. *Genome Biology, 19*(1), 210.

CHAPTER 5

Towards Smart Plant Breeding: Genome Editing Tools for Efficient Manipulation of Economically Important Traits of Crop Species

AALOK SHIV,[1] KAJAL SAMANTARA,[2] K. T. RAVIKIRAN,[3] SUBHASH CHAND,[4] AMAR KANT KUSHWAHA,[5] LAKSHYA GOYAL,[6] INDU,[4] RAKESH CHAUDHARY,[7] and SANJEEV KUMAR[1]

[1]*Division of Crop Improvement, ICAR–Indian Institute of Sugarcane Research, Lucknow, Uttar Pradesh, India*

[2]*Institute of Technology, University of Tartu, Estonia*

[3]*Regional Research Station, ICAR–Central Soil Salinity Research Institute, Lucknow, Uttar Pradesh, India*

[4]*Division of Crop Improvement, ICAR–Indian Grassland and Fodder Research Institute, Jhansi, Uttar Pradesh, India*

[5]*Division of Crop Improvement and Biotechnology, ICAR–Central Institute of Subtropical Horticulture, Lucknow, Uttar Pradesh, India*

[6]*Department of Plant Breeding and Genetics, Punjab Agricultural University, Ludhiana, Punjab, India*

[7]*Department of Genetics and Plant Breeding, Rani Lakshmi Bai Central Agricultural University, Jhansi, Uttar Pradesh, India*

Smart Breeding: Molecular Interventions and Advancements for Crop Improvement. Kailash Chandra, Subhash Chand, Ravi Prakash Saini, and Rohit Sharma (Eds.)

ABSTRACT

Breeding crop varieties tolerant to multiple stresses coupled with high yield and nutritive potential is the need of the hour in the contexts of climate change and burgeoning human population. Over the past several decades, natural variability exploited through various traditional plant breeding methods paid huge dividends in agriculture. However, biotechnological advances have significantly hastened the conventional process of plant breeding and helped in achieving breeding targets in lesser time. Genome editing (GE) is one such efficient and reliable technique for making targeted changes in the genome precisely and also paves the way to generate novel variability. Since the inception of zinc finger nucleases (ZFNs) to CRISPR/Cas9, GE has been successfully employed in crop plants for methodical manipulation of myriad of traits and has ample scope for further applications. Moreover, CRISPR/Cas9 has been proven to be an efficient transgene-free technique. The present chapter elaborates on how GE technologies evolved over time and contributed towards the improvement of various economically important crop attributes. The limitations of these technologies and their future applications were also discussed.

5.1 INTRODUCTION

Homo sapiens has solved uncountable mysteries present on this earth and is on the verge of controlling everything present in this universe, whether it is about inhabiting a new planet like Mars, generation of new crops like triticale, or finding solutions to incurable genetic disorders like sickle cell anemia. Elucidation of the DNA structure by Watson and Crick in 1953 accelerated development of biological sciences. After knowing the structure of DNA, its manipulation and tailoring in the desired way are some of the most sought-after goals in biological sciences.

Genome editing (GE) comprises insertion, deletion, and replacement of a particular genome sequence at the desired target site. Different types of genomes editing techniques are available with their own limitations. Zinc finger nucleases (ZFNs) and transcription activator-like effector nucleases (TALENs) were the two main techniques used for GE until CRISPR/Cas9 became the poster child for GE technology. Both ZFNs and TALENs are chimeric proteins possessing two domains, i.e., DNA binding and cleaving domain. They both have Fok1 nuclease as cleaving domain but differ in their

DNA binding domain; ZFNs have g zinc finger residues while TALENS have TALEs for the purpose of DNA binding. Despite being old, ZFNs, and TALENs are still not widely used technologies due to some limitations. Designing of ZFNs and TALENs is quite cumbersome (Gaj et al., 2016) along with poor target specificity and lower precision. Moreover, the large structure of ZFNs and TALENs also hinders their efficient delivery into the cell.

The CRISPR/Cas9 is an adaptive immune system in bacteria (Horvath & Barrangou, 2010) and has been evolved in same to protect them against phage infection and plasmids. It includes RNA guided DNA cleavage of foreign agents by employing Cas9. Broadly, there are two classes (Class I and Class II) of CRISPR systems and Class II has been customized and used on a large scale for GE purposes (Doudna & Charpentier, 2014). Briefly, a typical CRISPR locus of CLASS II systems include an array of palindromic repeat sequences separated by segments of unique sequences known as spacers, along with a group of CRISPR associated genes (Cas genes) near this array (Jiang et al., 2017). Preceding this array, a gene for trans-activating CRISPR RNA (tracrRNA) is also present, which is a noncoding RNA. Interestingly, spacers are sequences derived from phages and plasmids which have attacked bacteria in the past. During immune response, these sequences get transcribed to form CRISPR RNA or CrRNA which makes a complex with tracrRNA later on this complex is further processed with the help of different enzymes to make a single guide RNA (SgRNA) of approximately 20 nucleotides. Now, this SgRNA along with the Cas9 enzyme cleaves the foreign DNA and generates double-stranded breaks (DSBs), based on its complementarity and availability of protospacer adjacent motif (PAM) sequence. For the Cas9 enzyme, the PAM requirement is 5′-NGG-3′ where N can be any of four nucleotides. Before cutting any sequence Cas9 will search for specific PAM sites and then it checks for suitable complementarity for efficient cleavage. The significance of PAM sites is very crucial as it prevents the bacteria from its own DNA cleavage by Cas9 as there is no PAM site is present in the bacteria's own genome.

After elucidation of the complete mechanism of CRISPR/Cas9 in bacteria, it has been shown that this mechanism can be customized to generate DSBs at desired sites *in-vitro* (Jinek et al., 2012). The identification of the target sites and the creation of DSBs are done in the same way as described above. The induction of DSBs activates the inner cellular repair machinery of the cell to repair the DSBs generated. There are two pathways to repair these DSBs, non-homologous end joining (NHEJ) and homology directed repair

(HDR). The NHEJ generates random insertions or deletions (InDels) at the target site leading to gene knock out whereas HDR results in the integration of suitable donor fragment at the target site provided that the donor segment shares homology with the targeted segment and is provided externally. But HDR pathway occurs at very lower frequency in a cell as compared to NHEJ and is limited by inefficient delivery of donor fragments in a cell. The other problem with HDR is its dependency on a particular stage of cell division as it cannot work in a non-dividing cell. Therefore, introducing desirable point mutations or integrating particular gene sequence using this mechanism is a big challenge, but recent techniques like base editing (Nishida et al., 2016; Gaudelli et al., 2017) and prime editing (Anzalone et al., 2019) has been proved very efficient techniques to overcome these challenges.

5.2 EVOLUTION OF GENOME EDITING (GE) METHODS

5.2.1 ZINC FINGER NUCLEASE (ZFN)

Over a decade ago, ZFN (zinc finger nucleases) conferred the possibility of targeting particular genes of the genome (Chen et al., 2019). The ZFNs are sequence specific nucleases, and their construction includes the amalgamation of DNA-binding zinc-finger motifs with the nonspecific cleavage domain of the FokI endonuclease (Carroll, 2011; Voytas, 2013). One Zn finger comprises of nearly 30 amino acids which fold to form a ββα configuration and gets stabilized with the help of a zinc ion (Chen & Gao, 2014). A single zinc finger is able to recognize three base pairs of DNA at its main groove through the insertion of a helix and one complete zinc-finger unit comprises of around 3–4 binding modules (Pabo et al., 2001). We require two Zn finger nucleases to introduce DSB at target sites as dimerization is a pre-requisite for FokI endonuclease domain to act (Kumar & Jain, 2015). These two ZFN monomers bind to a unique 18 to 24 bp sequence along with a 5–6 bp gap. As ZFNs functions through protein-DNA binding, each editing site requires new ZFNs. There are several crop plants for example, Arabidopsis (Osakabe et al., 2010; Zhang et al., 2010), tobacco (Townsend et al., 2009), soybean (Curtin et al., 2011), and corn (Shukla et al., 2009) where ZFNs have been employed for gene targeting. However, it has its own limitations such as lack of targeting sites in the genome, complicacy of designing process, high yet variable off-target rate, expensive, and high skill requirement (Chen et al., 2019) which hinders its application in breeding program.

5.2.2 TRANSCRIPTION ACTIVATOR-LIKE EFFECTOR NUCLEASE (TALEN)

Transcription activator-like effectors (TALE) are found to be present in plant pathogen *Xanthomonas*, from which TALENS (Transcription activator-like effector nucleases) have been derived (Carroll, 2011; Weeks et al., 2016). These are as similar to the ZFNs in their basic structure and are synthetically made up of custom-built DNA binding TALE domain to the FokI cleavage domain. TALE domain contains secretion and translocation signals at its N-terminal, nuclear localization signals at its C-terminal (NLS) and an acidic transcription-activation domain (AD), and a central DNA binding domain (DBD) (Boch & Bonas, 2010; Bogdanove et al., 2010). The central DBD usually comprises of 14–20 tandem arrays of extremely conserved 34 amino acid repeats. The amino acid sequences of the repeats are almost similar excluding the di-residues at positions 12 and 13 which are specified as "repeat variable di-residues" (RVDs) (Chen & Gao, 2013; Graham et al., 2020). In every repeat, the repeat-variable di-residues establish pairing with one base of the target DNA sequence in a one-RVD-to-one-nucleotide fashion (Streubel et al., 2012). Usually, the 2 monomers of TALENs fuse to nearly a 50 to 60 bp long target sequence within 14 to 18 bp spacer, which is essential for its functioning. As compared to ZFNs, which only identify a restricted target sites in the genome, TALENs can readily be engineered to modify any recognition site providing a thymine (T) before the first nucleotide of the target site. This TALEN approach has been utilized in various crop species viz., tobacco (Zhang et al., 2013), rice (Li et al., 2012; Shan et al., 2013), barley (Wendt et al., 2013), and Arabidopsis (Cermak et al., 2011; Christian et al., 2010, 2013). But on account of the high repeat numbers of RVDs, the design of TALENs continues to be a difficulty (Chen et al., 2019).

5.2.3 CLUSTERED REGULARLY INTERSPACED SHORT PALINDROMIC REPEATS (CRISPR)-ASSOCIATED ENDONUCLEASE CAS9 (CRISPR/CAS 9)

As compared to the previous tools of GE, CRISPR/Cas9 tool achieved major attention owing to its efficacy and ease of use (Mishra et al., 2018) as mentioned earlier.

5.2.4 BASE EDITING

The CRISPR-Cas9 and CRISPR-Cas12a tools of gene editing need DNA breakage in the form of DSBs and are then repaired through HDR (for insertions or replacement) or NHEJ (to knock out a gene), carrying out the modification at the targeted sites (Zetsche et al., 2015). The application of base editors enables precise editing to be taken place at the needed site without the requirement for DSBs or donor template to enhance its efficiency as compared to other tools for generating desirable mutations at the target site (Yang et al., 2017). The fusion of Cas9 nickase and cytidine deaminase enzyme promotes the modifications of cytidine (C) to uridine (U), allowing C → T or G → A substitutions to made available at requires site in the genome (Komor et al., 2016).

5.2.5 PRIME EDITING

As mentioned earlier, the application of Homology-directed repair (HDR) is limited due to its lower efficacy (Ali et al., 2020). However, a recent approach named prime editing provides the higher potential for doing precise editing in mammalian cells with better efficacy (Anzalone et al., 2019). In this approach, two components are included viz., an RNA-programmable nickase (Cas9-H840A) fused to reverse transcriptase (RT), and a prime editing guide RNA (pegRNA) that determines the genomic target site and encodes the required edit (Zafar et al., 2020). In contrast to the HDR approach, this does not need a donor repair template, but the required edits are integrated into the pegRNA sequence. With the help of RT, the pegRNA is used as a template to copy the required edits into the genome (Lin et al., 2020). This has been used successfully in rice to obtain varieties tolerance to herbicide (Xu et al., 2020).

5.3 APPLICATION OF GE FOR YIELD IMPROVEMENT

Higher yield has always been the main objective of plant breeding and molecular breeding techniques have further accelerated the pace of plant breeding to achieve this everlasting goal. There are several ways of utilizing gene editing to gain the higher yield, mainly by knocking out the genes which negatively control the yield and yield contributing traits (Ma et al., 2016; Song et al., 2016). For instance, knockout mutant generated through CRISPR/Cas9 for *Gn1a*, *GS3*, and *DEP1* genes in rice showed higher grain number, larger grain size and dense-erect panicle, respectively, which ultimately resulted in higher yield per plant (Li et al., 2016). Similarly, Xu et al. (2016) demonstrated an increase of 1,000 grain weight (GW) by knocking out *GW2*, *GW5*, and *TGW6* genes in rice. Recently, Usman et al. (2021) observed higher grain yield in rice by editing *OsSPL16* gene using CRISPR/Cas9. Similar results were also reported in wheat where targeting of three homoallele of a gene *GASR7* resulted in increased 1,000 kernel weight (Zhang et al., 2016) and targeting three homoeologous copies of a gene *TaGW2* resulted in increased seed size and 1000 GW (Wang et al., 2018). Yield improvement through gene editing is not restricted to just wheat and rice but has also been demonstrated and explored in various crops like maize (Shi et al., 2017), tomato (Rodriguez et al., 2017), soybean (Nguyen et al., 2021), etc. This shows that gene editing could be successfully employed to obtain higher yield in other crops as well by identifying and eliminating the negative regulator of the yield and related traits.

Heterosis has been exploited for higher yield in many crops and gene editing could assist in the exploitation of heterosis. For example, gene editing could be used for the development of male sterile (MS) line, which is one of the major components of economical hybrid seed production. In this way, Singh et al. (2018) used the CRISPR/Cas9 to generate MS line in wheat by modifying the three homeologs of *Ms45* gene. Similarly, Du et al. (2020) generated MS tomato line by mutagenizing the *S1STR1* gene using CRISPR/Cas9. Recently, editing of *GmMs1* by CRISPR/Cas9 in soybean resulted in male sterile phenotype (Nadeem et al., 2021). Several such examples in different crops have been demonstrated by various researchers and a detailed review was done by Chen et al. (2020). Several such genes could be targeted for the generation of MS line and efficient hybrid seed production system in different crops using gene editing.

For instance, lodging and seed shattering cause great loss in yield. Therefore, development of varieties with semi-dwarf nature and reduced seed shattering could ultimately results in higher yield. Zhai et al. (2019) demonstrated the essential role of *INDEHISCENT* (*IND*) and *ALCATRAZ* (*ALC*) gene in pod shattering resistance in *Brassica napus L.* using CRISPR/Cas9. Similarly, Mamidi et al. (2020) validated the role of *Less Shattering1* (*SvLes1*) gene in setaria species which was found to be involved in controlling seed shattering. Jiang et al. (2020) knock out the several *SPL* genes in rice and found variation for plant height, panicle size and grain length. Guo et al. (2021) generated *OsTIR1* and *OsAFB2-5* mutants through CRISPR/Cas9 and demonstrated the role of *OsTIR1/AFB* family members in grain yield, tillering, plant height, root system, germination, and herbicide resistance. Further, fine tuning flowering pattern and flowering time could be used to improve the yield of crop plants. For example, Liu et al. (2020) used

CRISPR/Cas9 to alter spike morphogenesis and grain threshability in wheat. Liu et al. (2021), also successfully used the CRISPR/Cas9 to delay the rice heading date by editing the *Hd2* uORFs. Very recently, Dai et al. (2021) used CRISPR/Cas9 to edit flowering related genes in *Hevea brasiliensis.*

These examples showed that gene editing methods could be successfully employed to improve the yield. Further, speed breeding and high throughput phenotyping could be used to identify other similar traits which contribute to yield and ways could be explored to harness these genes for higher yield through GE.

5.4 APPLICATION OF GE FOR BIOTIC STRESS RESISTANCE IN PLANTS

There are significant yield penalties in agriculturally essential crops due to various biotic and abiotic stresses globally. Yield losses can be minimized or avoided by deploying stress tolerance varieties, and attainable goals can be achieved by employing GE tools.

5.4.1 DISEASES

There is a wide range of biotic stresses, and plants are mainly affected by various pathogenic species like fungi, bacteria, plant viruses, nematodes, plant parasites, insects, etc. Globally, plant viruses alone can reduce crop production by about 10–15% (Vats et al., 2019; Mores et al., 2021). The CRISPR/Cas9 system has been utilized in various crop plants to build resistance against biotic stresses. For example, in cassava for brown steak and mosaic virus, spinach for powdery mildew (PM), apple for blight, orange for greening, maize for lethal sphacelus, grapevine for mildew, tomato for yellow leaf curl virus. In cotton, multiple disease-resistant plants were developed against begomoviruses through targeting reserved mono-nucleotide sequences (TYLCV, CLCuKoV, TYLCSV, BCTV-Worland, MeMV, and BCTV-logan) of cotton leaf curl Kokhran virus (CLCuKoV) (Kamburova et al., 2017). Similarly, a resistant tomato plant was developed against begomovirus by targeting a single nucleotide change in coat protein (Mushtaq et al., 2020; Rai et al., 2021). It has been reported that *MLO* (S gene) suppresses the defense mechanism of plants against PM. TALEN technology was used in hexaploid bread wheat (*T. aestivum*) to introduce targeted mutations in the three homoalleles encoding mildew-resistant locus

(MLO) proteins. The selected mutated plants for three *TaMLO* homeologs were resistant to PM (Wang et al., 2014; Arazoe, 2021). Initiation factors like *eIF4E* and *eIF(iso)4E* have already been established in *Arabidopsis*, which are responsible for Turnip mosaic virus (TuMV) infection (Kamburova et al., 2017). In cucumber (*Cucumis sativus* L.) initiation factor-*eIF4E* was targeted and selected mutated plants were more tolerant to potyviruses (papaya ringspot mosaic virus-W and Zucchini yellow mosaic virus) and ipomovirus (cucumber vein yellowing virus) groups (Mushtaq et al., 2020). CRISPR/Cas9 system was used for GE for its accuracy and rapidity in *phytophthora sojae*, an oomycete soybean pathogen. Fang & Tyler (2016) validated the contribution of the RXLR effector gene *Avr4/6* of *P. sojae* to recognize the resistance *R* gene of soybean, i.e., *Rps4* and *Rps6*. GE tools are more accurate and precise and could be used to study functional genomics in virulent genes of pathogens and resistant loci of plants.

5.4.2 INSECTS

Amongst biotic stresses, insect pests harm plant growth and development and are considered one of the major threats to maintaining and sustaining crop productivity globally. GE tools have helped substantially in reducing insect pests' adverse effects and would improve varietal stability without yield penalty. GE can be used at insect and plant levels to modify and mitigate insect pest load (Tyagi et al., 2020). Cadherin act as a receptor of *Bt Cry1A* toxins in cotton bollworm (*Helicoverpa armigera*). Cadherin gene (*HaCad*) disruption is linked to *Bt* toxin *Cry1Ac* resistance in *H. armigera HaCad* gene of *Cry1Ac* susceptible *H. armigera* was knocked out by CRISPR/Cas9 system, and selected mutant revealed many fold resistance to *Cry1Ac* than control, however, no significant difference for *Cry2Ab* (Wang et al., 2016). Knockdown of *CYP6AB14* and *CYP9A98* genes in beet armyworm (*Spodoptera exigua*) employing RNAi and showed more sensitive response to delta methrin insecticide-commonly used in the cotton field (Hafeez et al., 2019). These genes can be targeted to avoid the development of insecticide resistance in insects.

GE in plants for insect pest management has been meager subjugated. Insects recognize the host plants employing plants own derived volatile compounds, physical appearance, oviposition spots and their interactions as they are coevolved with each other. Host plants can be saved from the attack of insect pests after changing their volatile blends (Beale et al., 2006; Tyagi et al., 2020). Aphid attack leads to the secretion of sesquiterpene hydrocarbon (E)-β-farnesene (EβF) in crop plants and attract parasitic wasp that controls the aphid attack on host plants as observed in transgenic studies (Beale et al., 2006; Tyagi et al., 2020). Plants' visual appearance has a significant role in recognizing and attacking the host plants by the insect pests. For example, transgenic tobacco, i.e., red leaf tobacco, was developed by changing the anthocyanin biosynthesis pathway. Overproduction of anthocyanin pigment leads to the red coloration of tobacco leaves and acts as a deterrent to the *S. litura* and *H. armigera* in transgenic tobacco plants (Malone et al., 2009). In rice (*O. sativa*), serotonin secretion is induced by the insect attack, and its suppression leads to deliberate resistance to stem borer and plant hoppers (Lu et al., 2018). In this context, the cytochrome P450 gene (*CYP71A1*) encodes the last product serotonin via different biochemical pathways. However, knocked-out *CYP71A1* plants were reported high salicylic acid and no serotonin production; and mutated plants were more insect tolerant than wild-type rice (Lu et al., 2018).

5.4.3 WEEDS

Weeds adversely affect crop yield since their domestication and their contribution in yield loss are more than diseases and insects that need to be controlled by different approaches. Long-term application of herbicides to control weeds leads to the development of herbicide-resistance weeds and adversely affects natural flora and fauna (Hussain et al., 2021). GE technology comes up with possibilities to sustain crop productivity by controlling weed menace to meet burgeoning global food demand. In maize (*Zea mays* L.), CRISPR/Cas9 system targeted acetolactate synthase gene (*ALS2*), and selected plants were reported to be resistant against chlorsulfuron herbicide (Svitashev et al., 2015). In soybean (*Glycine max* L.), *DD20* and *DD43* genomic sites were targeted employing CRISPR/Cas9 with mutation frequency of 59% and 76%, respectively, and selected plants were resistant to chlorsulfuron (Li et al., 2015). In rice, the *ALS* gene was targeted using CRISPR/Cas9 and selected homozygous mutated plants were resistant to chlorsulfuron and bispyribac sodium herbicides (Sun et al., 2016). In rice, endogenous gene 5-endolpyruvylshikimate-3-phosphate synthase (*EPSPS*) was targeted employing CRISPR/Cas9 and the selected plants with the *OsEPSPS* gene were resistant to glyphosate (Li et al., 2016).

Thus, developing resistant crop varieties to combat biotic stress employing gene editing tools could sustain the global yield production to meet the global food and nutritional demand.

5.5 APPLICATION OF GE FOR ABIOTIC STRESS TOLERANCE

With respect to abiotic stresses under changing climatic scenario, drought, salinity, heat, and cold are the major stresses, affecting crops invariably across the globe. Breeding varieties for environmental stress tolerance have not been as fruitful due to complex mechanisms and limited variability for the tolerance in the crops. To address these challenges exploration of genes from eco-geographical niches and understanding their mechanism is of prime importance. With the advent of new genomics tools, the long process can be shortened, and with confidence we can create new variabilities. To date, CRISPR CAS system have been mostly utilized in the confirmation and validation of genes involved under abiotic stresses, and in some cases, it has created novel mutants for providing tolerance against abiotic stresses. The subsections will elaborate on the uses and achievements of CRISPR technology in various abiotic stress tolerance in crop plants.

5.5.1 SALINITY

Salinity describes the prominence of salts (sodium chloride, calcium sulfates, magnesium sulfates, magnesium bicarbonates and calcium bicarbonates) in soil and affect plant growth and productivity. Salinity damage crops by inducing osmotic stress and cellular toxicity. Salinity tolerance is achieved via three mechanisms viz., lowering the osmotic stress, promotion of root and leaf growth under water stress and low salt accumulation, i.e., salt exclusion. Before the discovery of molecular biology tools like RNA interference and CRISPR systems, fewer studies had been done to show the role of genes in salinity stress tolerance (Munns, 2005).

Many salinity resistant genes (nearly 51) have been identified alone in rice (Khan et al., 2021). One of these genes, *OsRR2* has a role in salt tolerance through cytokinin signal transduction and its metabolism. The tolerance is the result of loss of function of this gene (Takagi et al., 2015). Engineering this gene using CRISPR/Cas9 has shown increased tolerance against salinity in rice at seedling stage as compared to wild type counterpart (Zhang et al., 2019). CRISPR-Cas9 modification of *OsGT*γ*-2*, a transcription factor has indicated its role in salinity adaption through regulation of expression of ion transporters (Liu et al., 2020). Identification of *NAC* transcription factor encoding gene *OsNAC041* and knocking it down using CRISPR-Cas9 have shown its role in regulation of expression of salinity stress related genes as mutant was more tolerant to saline conditions than that of wild types (Boet et al., 2019). CRISPR-Cas9 mediated modification of *dst* gene led to generation of loss of function mutant in rice with improved tolerance against osmotic and salt stress (Kumar et al., 2020). CRISPR mediated modification of *ARF4* gene in tomato led to tolerance against salt and osmotic stress (Bouzroud et al., 2020).

Though CRISPR-Cas9 system have been successfully utilized in rice in improving salinity tolerance since it is a model crop in terms of simplicity and availability of genomic resource but there is a need to focus on other crops as well.

5.5.2 DROUGHT

Drought is the condition where soil moisture availability to crop has reduced to a point where it harmfully affects the crop yield (Manochhi et al., 2004). Under extreme drought condition complete crop loss is a surety. The genetic variation among crop plants for drought tolerance (DT) exits, and the inheritance is complex in nature. Plant responds to drought by altering molecular, biochemical, cellular, physiological, and morphological processes (Illyas et al., 2021). Plants sustain its productivity under drought via drought avoidance (DA), DT, drought escape (DE) and drought recovery (DR) (Fang & Xiong, 2015). Targeting gene/gene complexes involved in these mechanisms can lead to the generation of genetic material that can cope up with the stress. Though the classical breeding program across the globe resulted in a number of drought tolerant varieties there is potential of modern tools like CRISPR in designing crop to combat drought. Most of the studies involving CRISPR-Cas9 have been done to understand the molecular mechanism of drought imparting component traits but they also resulted in potential breeding material.

The role of leaf rolling in DT has been reported by many workers. Liao et al. (2019) has used the CRISPR-Cas9 to generate mutant of SLR1 and SLR2. The mutants led to the identification of genes involved in the modulation of expression proteins affecting different biological processes, cellular components and molecular function pathways. These mutants can be incorporated in breeding program for the development of drought tolerant varieties. NPR1 (non expressor of pathogenesis-related gene-1) positively regulates the resistance against pathogen it is well characterized and its defense pathway have been well elucidated (Zhong et al., 2015). Recently the role of NPR1 have shown to be involved in tolerance against cold, salt, and oxidative stress (Olate et al., 2018; Jayakannan et al., 2015; Srinivasan et al., 2019). Quill et al. (2008) through transgene overexpression of the NPR1 from Arabidopsis in rice showed its involvement in providing tolerance against salinity and drought as this made the plant hypersensitive to the above stresses. CRISPR-Cas9 mediated mutation in *SLNPR1 gene* lead to production of tomato mutant slnpr1 and shown its role in DT as it was expressed in all plant tissue of the mutant under drought condition (Li et al., 2019). slnpr1 mutants so generated had lower-level DT accompanied with lower level of expression of drought stress-related genes viz., *SIGST*, *SIDHN*, and *SIDREB*. Morpho-physiological and biochemical level were also as that of sensitive plants when compared with wild type SLNPR1.

Phytohormones are key players in DT in plants viz., abscisic acid (ABA), auxin, and brassinosteroids (Gupta et al., 2020). Among these ABA is a major regulator (Fujita et al., 2013). Several genes involved in signaling of ABA have been subjected to tailor the DT in plants (Umezawa et al., 2006). CRISPR-Cas9 induced mutation (frame-shift mutation) in gene OsERA1 led to the production of OsERA1 mutant in rice, with enhanced root growth and response, thus proving OsERA1 as a negative regulator for drought related traits (Ogata et al., 2020). The role of various anti-oxidant enzymes in DT in tomato has been shown by creating CRISPR mediated mutation in MAPK signaling molecule encoding gene, i.e., SIMAPK3. The mutant SIMAPK3 has shown more activity of H_2O_2 under drought stress reveals SIMAPK3 reduces the oxidative stress by scavenging reactive oxygen species (Wang et al., 2017). In tomato, a gene LBD40 belonging to the LBD gene family (Lateral organ boundaries domain) was found to have a role in drought stress tolerance through the modulation of the jasmonic acid signaling pathway. The mutant of this gene created through CRISPR-Cas9, i.e., SILBD40 was sensitive to jasmonic acid and negatively regulated the DT (Liu et al., 2020).

5.5.3 HEAT STRESS

As such CRISPR-Cas9 has not resulted in any mutation that led to identification of heat related gene(s) or heat tolerant mutants but it has been shown to induce mutation effectively in heat tolerant crops. CRISPR/Cas12b is a heat-induced system which requires a temperature ranging between 40°C and 55°C for effective cleavage which normal plant cells cannot tolerate. Tetraploid cotton undergoes such temperature during its growth and development hence has potential to induce targeted mutagenesis (Wang et al., 2020). In rice, the efficiency of CRISPR-Cas9 has been improved under heat stress by using heat shock promoter from the fact that rate of mutagenesis has been increased from a mere 16% to 63% (Nandy et al., 2019). An evolutionary conserved molecular chaperone, i.e., heat shock protein 70 (Hsp70), has been shown to have a role in many cellular processes like protein folding and transport under both heat and non-heat stress condition. It has been of interest among researchers working with Hsp70 to identify client protein which binds to them. Identification of target protein is one of the important aspects for understanding heat tolerance. CRISPR-Cas9 has been utilized to understand how these Hsp70 interacts with other proteins under stress. CRISPR have advantage in detection of such interaction over mass spectrometry (MS) technique which suffers from biasness due to the over-expression of interfering the protein-protein complex stoichiometry (Nikita & Truman, 2017).

5.5.4 COLD STRESS

The species of lower altitude regions viz., tropics or subtropics as compared to temperate flora show more injury or even die under low temperature and manifestation of this is in terms of chlorosis, necrosis, or growth retardation. Changes in gene expression pattern and interaction are primary responses of plants under low temperatures, which have an impact on survival and yield. Cold tolerant genes have been classified into four classes, viz., transcription factors, protein kinases, microRNAs (miRNAs) and others (Shen et al., 2017). There are very limited reports on the utilization of CRISPR-Cas9 in inducing mutagenesis for improved cold tolerance. In rice, an annexin gene *OsAnn3* was targeted to induce knockout mutation to identify its role in cold stress. The mutant showed higher relative electrical conductivity and lower survival under cold stress compared to the wild type suggesting OsAnn3 has a potential role in cold stress (Shen et al., 2017). Simultaneously targeting three genes viz., *OsPIN5b* (panicle length), *GS3* (grain size) and *OsMYB30* (cold tolerance) using CRISPR-Cas9 system led to production of mutants having cold tolerance along with better yield under stress condition (Zeng et al., 2020).

Though variation for abiotic stress tolerance in crops exists, there is an urgent need to incorporate them into breeding program. To utilize them effectively, it is important to know the factors involved in the tolerance and create more allelic variants of them.

5.6 APPLICATION OF GE FOR IMPROVEMENT OF NUTRITIONAL QUALITY

5.6.1 GRAIN QUALITY

Grain quality of food crops is an important factor for the consumers/miller as well as for farmers. The quality of end product dictates consumer preferences. Grain quality, particularly of cereals, is an amalgamation of myriad of factors. For instance, rice grain quality includes several attributes such as hulling percentage, milling percentage, head rice recovery, aroma, grain dimensions, cooking quality, amylose content (AC) and grain micronutrient profile. The improvement of grain quality through conventional breeding methods is highly laborious. GE provides a fast, efficient, sustainable, and cost-effective alternative to efficiently manipulate the genes and/or QTLs governing various quality traits in crops. The succeeding sections summarize the recent achievements in this direction.

5.6.1.1 GRAIN APPEARANCE AND DIMENSIONS

Grain shape is one of the key grain quality attributes of consumer preference. Several genes were reported for grain size and dimensions in cereals and genes encoding these traits are the frequent targets of editing in rice. For instance, *GS3* (*GRAIN SIZE 3*) was the first identified QTL governing grain length which has been successfully edited in five japonica rice varieties. T1 lines of different genetic backgrounds showed increased grain length relative to wild type (Shen et al., 2018; Yuyu et al., 2020). Manipulating grain shape will have repercussions for GW (Xu et al., 2016). By disrupting *GW2*, *GW5*, and *GW6*, the GW negative regulators, rice GW has been significantly improved. Similarly, by editing (knockout) of *TaGW7* and *TaGW2*, the role of the gene in increasing GW and width in wheat has been established (Xu et al., 2016; Zhang et al., 2018). Similarly in wheat, disrupting three *TaGASR7* homeologs increased the test weight significantly (Zhang et al., 2016). The gene (1,3;1,4)-beta-glucan synthases of barley were targeted for GE to lower its levels to make it favorable for brewing and distilling. This has resulted in increased test weight, grain length, width, and surface area (Garcia-Gimenez et al., 2020).

5.6.1.2 GRAIN COMPOSITION AND QUALITY

Amylose content (AC) in rice grain is the deciding factor for cooking quality. Different ACs are preferred in different parts of the world. For instance, very low amylose (waxy) with medium bold grain type is preferred in Lao People's Democratic Republic, whereas low AC with medium grain type is preferred in Cambodia, Thailand, China, and Taiwan and moderate AC is favored in India. Hence, modulation of grain AC received much attention for GE studies. *OsWaxy* which regulates amylose synthesis is routinely targeted for decreasing AC of the varieties (Huang et al., 2020; Xu et al., 2020; Zeng et al., 2020). Similarly for increasing AC, genes encoding starch branching enzymes, *SBEI* and *SBEIIb* and debranching enzyme, *IAS*, were successfully modified (Sun et al., 2017; Chao et al., 2019). Gluten sensitivity is quite common food related disorder across the world. This is due to the gluten proteins of wheat encoded by gliadin genes which trigger celiac disease in genetically sensitive individuals. Low-gluten wheat was efficiently produced through CRISPR Cas meditated editing of conserved region in α-gliadin genes which reduced immunoreactivity by 85% (Sanchez-Leon et al., 2018). Moreover, targeted editing of *TaSBEIIa* by CRISPR/Cas9 successfully enhanced the amylose and resistant starch contents in wheat (Li et al., 2020). Similarly, four grain quality-related genes *viz. pinb*, *waxy*, *ppo*, and *psy*, governing grain hardness, starch quality and dough color, respectively, were targeted for GE and wheat lines with hard dough, low AC and improved dough color have been obtained (Zhang et al., 2021). Similarly in maize, to reduce the antinutritional zein protein and phytic acid contents, a MADS box protein coding gene *MADS4* and inositol phosphate encoding gene, *IPK1* was effectively edited out, respectively (Qi et al., 2016). In oilseed crops, enhancing the fatty acid profile is the primary target of plant breeders. An attempt via GE was also done in rape seed where a Fatty acid desaturase 2 gene (*FAD2_Aa*) was edited and mutated lines with enhanced oleic acid content were recovered (Okuzaki et al., 2018).

5.6.2 FRUIT QUALITY IN HORTICULTURAL CROPS

Horticultural crops, comprising fruits, vegetables, and ornamental plants are an important component of the global agriculture production systems and play a vital role in sustaining human nutrition. These crop species have also been the targets of GE in parallel to the field crops. However, the primary focus of researchers in case of fruits and vegetables is to enhance the shelf life. Accordingly, the genes encoding pectin degradation such as *pectate lyase*, *RIN* (*ripening inhibitor*) are the frequent targets of GE. For instance, in tomato, CRISPR/CAS9 mediated editing of *Pectate lyase*, *Polygalacturonase*, *2a&β-galactanase* which are responsible for pectin degradation during fruit ripening has significantly increased the shelf life of tomatoes (Uluisik et al., 2016; Wang et al., 2019). Similarly, modification of RIN locus, which encodes a MADS-Box transcription factor controlling fruit ripening has significantly delayed ripening in tomato (Ito et al., 2015; Jung et al., 2018). Amelioration of fruit color in tomato is also one of the popular targets of GE. *anthochyanin 1* (*ANT1*), *phytoene synthase* (*PSY1*), *LCY-E*, *Blc*, *LCY-B2*, and *CycB* are some of the genes encoding various carotenoids (such as lycopene) which were modified through GE to improve the fruit color in tomatoes (Čermák et al., 2015; Filler Hayut et al., 2017; Li et al., 2018d; Zsögön et al., 2018). Polyphenol oxidases (PPOs) are responsible for the browning of tissues in vegetables or fruits. Knocking down of these genes will significantly prolong the shelf life. For instance, *StPPO* in potato and *SmelPPO4*, *SmelPPO5*, and *SmelPPO6* in eggplant were successfully edited to reduce browning in tubers and fruits, respectively (Gonzalez et al., 2019; Maioli et al., 2020).

In addition to the above studies, GE was utilized for fortifying crops. Enhanced β-carotene content in rice grains were achieved by modifying *SSU-crtI* and *ZmPsy* involved in carotenoid biosynthesis (Dong et al., 2020). Similarly, cadmium accretion in the rice grains was significantly curtailed by editing Cd transporter OsNramp5 (Tang et al., 2017). By editing *MaLCYepsilon* gene, β-carotene content was increased in banana (Kaur et al., 2020). Targeted modification of *GGP1* in tomato increased the vitamin C content in fruits (Li et al., 2018c). Thus, extensive GE studies were reported primarily in vegetables, particularly in tomato, compared to fruits, and there is an absolute need to standardize the protocols for other horticultural crops as well. While the progress in field crops seems to be appreciable, albeit, like horticultural crops, it is restricted to a few major crops such as rice, wheat, maize, and barley and needs to be expanded to others such as millets. No GE studies were reported in pulses which needs to be addressed in future studies (Table 5.1).

TABLE 5.1 List of Genes Targeted for Improving Grain and Fruit Quality in Various Field and Horticultural Crops

Crop	Gene Targeted	Editing Tool	Gene Function	Altered Phenotype	References
Field Crops					
Rice	*OsWaxy*	CRISPR/Cas9	Amylose synthesis	Decreased amylose content	Ma et al. (2015); Zhang et al. (2018); Huang et al. (2020); Xu et al. (2020); Zeng et al. (2020)
	OsBADH2	TALENS	Aroma	Increased 2AP content or fragrance	Shan et al. (2014); Usman et al. (2020)
	OsBEI and *OsBEIIb*	CRISPR/Cas9	Starch branching enzyme	Increased amylose content	Chao et al. (2019)
	OsAPL2	CRISPR/Cas9	Starch biosynthesis	Starch deterioration in leaves	Perez et al. (2018)
	IAS	CRISPR/Cas9	Starch debranching enzyme	Decreased amylose content and increased TSS	Chao et al. (2019)
	SSU-crtI and *ZmPsy*	CRISPR/Cas9	Carotenoid biosynthesis	Increased β-carotene content	Dong et al. (2020)
	OsNramp5	CRISPR/Cas9	Cd accumulation	Low Cd accretion in grains	Tang et al. (2017)
	SBEI and *SBEIIb*	CRISPR/Cas9	Amylopectin synthesis	High amylose	Sun et al. (2017)
	GS3 and *Gn1a*	CRISPR/Cas9	Grain size and grain number, respectively	Differential effect based on genetic background	Shen et al. (2018)
	GW2, *GW5*, and *TGW6*	CRISPR/Cas9	Grain width and test weight	Increased grain length, width, and test weight	Xu et al. (2016)
	OsGS3, *OsGW2*, and *OsGn1a*	CRISPR/Cas9	Grain size, width, and weight, and number	Increase in respective traits with a significant increase in yield	Zhou et al. (2019)
Wheat	*Taα-gliadin*	CRISPR/Cas9	Gluten synthesis	Low gluten content in grains	Sanchez-Leon et al. (2018)
	TaGW7	CRISPR/Cas9	TONNEAU1-recruiting motif (TRM) protein	Increased grain width and weight, but decreased grain length	Wang et al. (2019a)
	Pinb, *waxy*, *ppo*, and *psy*	CRISPR/Cas9	Grain hardness, starch quality and dough color	Hard dough with improved color and low amylose content	Zhang et al. (2021)

TABLE 5.1 *(Continued)*

Crop	Gene Targeted	Editing Tool	Gene Function	Altered Phenotype	References
	TaLOX2	CRISPR/Cas9	Lipoxygenase-storability of wheat grains	Improved wheat quality	Shan et al. (2014); Zhang et al. (2016); Zong et al. (2017)
	TaSBEIIa	CRISPR/Cas9	Amylose synthesis	High amylose	Li et al. (2020)
Maize	*IPK1*	ZNF/TALEN/ CRISPR-Cas9	Inositol phosphate	Reduced phytic acid content	Shukla et al. (2009); Liang et al. (2014)
	MADS4	CRISPR/Cas9	MADS-box protein ZmMADS47	Decreased zein protein	Qi et al. (2016)
	Wx1	CRISPR/Cas9	Waxy protein/Granule-bound starch synthase	High amylopectin starch	Waltz (2016)
Barley	*Hv d-Hordein*	CRISPR/Cas9	D-hordein proteins	Increased contents of starch, amylose, and β-glucan	Yang et al. (2020)
	HvHPT and *HvHGGT*	CRISPR/Cas9	Vitamin E biosynthesis	Reduction of tocopherols and tocotrienols	Zeng et al. (2020)
Rapeseed	*FAD2_Aa*	CRISPR/Cas9	Fatty acid desaturase 2	Increased oleic acid content	Okuzaki et al. (2018)
	BnITPK	CRISPR/Cas9	Inositol tetrakisphosphate kinase of PA synthesis pathway	Low phytic acid and high protein level in oil	Sashidhar et al. (2020)
	BnSFAR4 and *BnSFAR5*	CRISPR/Cas9	Lipases responsible for low FA composition of seeds	Enhanced seed oil content	Karunarathna et al. (2020)
Horticultural Crops					
Banana	*MaLCYepsilon*	CRISPR/Cas9	β-carotene synthesis	Fortification of banana fruits with β-carotene	Kaur et al. (2020)
Tomato	*SlCLV3* and *SlWUS*	CRISPR/Cas9	Locule number in fruits	Increased fruit size	Xu et al. (2015); Rodriguez-Leal et al. (2017)

TABLE 5.1 *(Continued)*

Crop	Gene Targeted	Editing Tool	Gene Function	Altered Phenotype	References
	Pectate lyase, *Polygalacturonase 2a*, and *β-galactanase*	CRISPR/Cas9	Pectin degradation during fruit ripening	Delayed shelf life	Wang et al. (2019a)
	SlFUL1 and *FUL2*	CRISPR/Cas9	Master regulators of fruit ripening	Delayed fruit ripening	Wang et al. (2019b)
	SlALC	CRISPR/Cas9	Fruit ripening	Longer shelf life	Yu et al. (2017)
	SlENO	CRISPR/Cas9	Stem cell proliferation in floral meristem	Increased fruit number and size	Yuste-Lisbona et al. (2020)
	Slagamous-like 6 (SlAGL6)	CRISPR/Cas9	Parthenocarpy	Better red fruit yields and higher Brix grade under heat stress	Klap et al. (2017)
	Ripening inhibitor (RIN)	CRISPR/Cas9	MADS-box transcription factor regulating fruit ripening	Fruits never turn red, altered firmness	Ito et al. (2015)
	Pectate lyase	CRISPR/Cas9	Pectate hydrolysis and fruit softening	Fruit softening inhibited and shelf life enhanced	Uluisik et al. (2016)
	Leafy-cotyledon-1-like 4 (L1L4)	ZFN	Biosynthesis for seed storage proteins and fatty acids	Greater content of soluble solids, fiber, phenol, and β-carotene	Gago et al. (2017)
	Anthochyanin 1 (ANT1)	CRISPR/ Cas9TALEN	Anthocyanin production	Constitutive accumulation of anthocyanin	Čermák et al. (2015)
	Phytoene synthase (PSY1)	CRISPR/Cas9	Carotenoid synthesis	Yellow fruits with red sectors	Filler Hayut et al. (2017)
	ALC	CRISPR/Cas9	Alcobaca (ALC)/ NAC domain protein (NAC-NOR)	Delayed fruit ripening causing long-shelf life	Yu et al. (2017)

TABLE 5.1 *(Continued)*

Crop	Gene Targeted	Editing Tool	Gene Function	Altered Phenotype	References
	GAD2, *GAD3*	CRISPR/Cas9	Glutamate decarboxylase 2 and 3	Higher γ-aminobutyric acid content	Nonaka et al. (2017)
	GABA-TP1, *GABA-TP3*, *CAT9*, *SSADH*	CRISPR/Cas9	Pyruvate-dependent GABA transaminases 1 and 3, cationic amino acid transporter, succinate semialdehyde dehydrogenase	Higher γ-aminobutyric acid content	Li et al. (2018b)
	lncRNA1459	CRISPR/Cas9	Ripening-related long non-coding RNA1459	Repressed fruit ripening, ethylene production and lycopene accumulation	Li et al. (2018a)
	L1L4/NFYB6	ZFN	Leafy-cotyledon-1-like-4/ nuclear transcription factor Y (NF-Y) gene	Reduced contents of the oxalic acid	Gago et al. (2017)
	MYB12	CRISPR/Cas9	R2R3-MYB transcription factor	Pink color trait by absence of yellow-colored flavonoid naringenin chalcone in tomato peels	Deng et al. (2018)
	RIN/LeMADS	CRISPR/Cas9	MADS-box transcription factor/ripening inhibitor	Delayed fruit ripening and reduced ethylene production	Jung et al. (2018)
	LCY-E, *Blc*, *LCY-B2*	CRISPR/Cas9	Lycopene b-cyclase, beta-lycopene cyclase, lycopene β-cyclase 2	Increased contents of lycopene and beta-carotenoids	Li et al. (2018d)
	CycB, *SP*, *O*, *MULT*, *FAS*, *FW2.2*	CRISPR/Cas9	Lycopene b-cyclase, self-pruning, ovate, multiflora, fasciated, fruit weight 2.2	Increased contents of lycopene (5x) as nutritional value with alteration in the size (3x) and number (10x) of the fruits	Zsögön et al. (2018)

TABLE 5.1 *(Continued)*

Crop	Gene Targeted	Editing Tool	Gene Function	Altered Phenotype	References
	GGP1, *SP*, *SP5*, *CLV3*, *WUS*	CRISPR/Cas9	GGP1 encoding a vitamin C-biosynthetic enzyme, Self-pruning, Self-pruning 5 g, CLV3, homeobox-encoding gene WUS	Increased contents of vitamin C	Li et al. (2018c)
Potato	*StPPO*	CRISPR/Cas9	Polyphenol oxidase	Prevention of enzymatic browning of tubers	Gonzalez et al. (2019)
	Vacuolar invertase (VInv)	TALEN	Catalysis of sucrose to glucose and fructose	Undetectable level of reducing sugar in tubers	Clasen et al. (2016)
	Granule-bound starch synthase (GBSS)	CRISPR/Cas9	Amylose synthesis	Modified amylose: amylopectin ratio in tubers	Andersson et al. (2017, 2018); Kusano et al. (2018)
	16DOX	CRISPR/Cas9	Steroid 16α-hydroxylase	Absence of antinutritional factors steroidal glycoalkaloids	Nakayasu et al. (2018)
	SSR2	TALEN	Sterol side chain reductase 2	Reduced accumulation of steroidal glycoalkaloids	Yasumoto et al. (2019)
Eggplant	*SmelPPO4*, *SmelPPO5*, and *SmelPPO6*	CRISPR/Cas9	Polyphenol oxidases-browning of fruit pulp	Low flesh browning	Maioli et al. (2020)

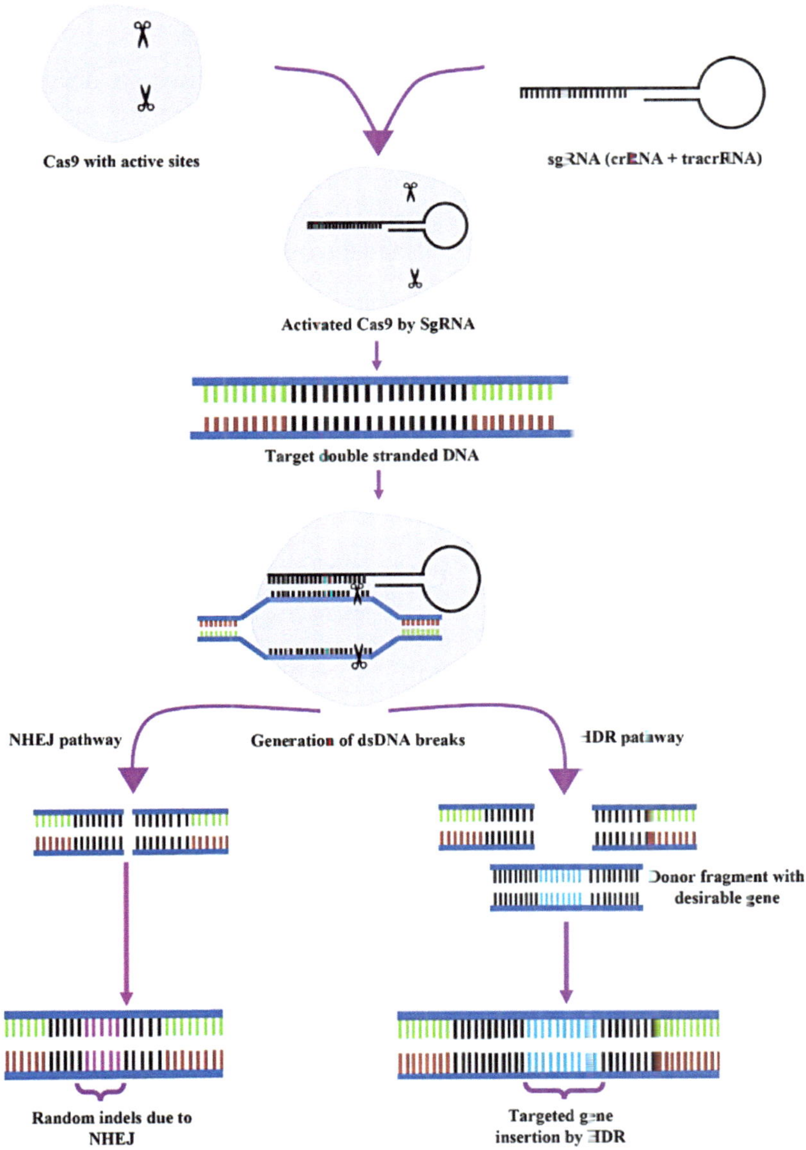

FIGURE 5.1 Basic mechanism of CRISPR/Cas9 based genome editing technique for inducing dsDNA breaks at targeted sites.

Note: dsDNA breaks can be repaired by either error prone repair pathway, i.e., NHEJ for inducing indels or can follow HDR pathway for insertion of desired sequence.

5.7 CONCLUSION AND FUTURE THRUST

GE methods particularly CRISPR/Cas9 has been used to develop high yielding and nutritionally rich genotypes (Figure 5.1). In the changing climate scenario, GE has the potential to play a significant role in smart plant breeding by developing abiotic and biotic stress resistant genotypes. GE has enormous potential, and with time many applications are yet to come. Overcoming the limitations of GE and its amalgamation with speed breeding and artificial intelligence (AI) will make it a perfect tool for smart plant breeding to tackle future challenges of crop improvement.

KEYWORDS

- **CRISPR/Cas9**
- **crop improvement**
- **genome editing**
- **homology-directed repair**
- **non-homologous end joining**
- **single guide RNA**
- **smart plant breeding**
- **trans-activating CRISPR RNA**

REFERENCES

Alfatih, A., Wu, J., Jan, S. U., Zhang, Z. S., Xia, J. Q., & Xiang, C. B., (2020). Loss of rice PARAQUAT TOLERANCE 3 confers enhanced resistance to abiotic stresses and increases grain yield in the field. *Plant Cell Environ., 43*(11), 2743–2754.

Ali, Z., Shami, A., Sedeek, K., Kamel, R., Alhabsi, A., Tehseen, M., et al., (2020). Fusion of the Cas9 endonuclease and the VirD2 relaxase facilitates homology-directed repair for precise genome engineering in rice. *Biol. Commun., 3*, 1–13.

Anzalone, A. V., Randolph, P. B., Davis, J. R., Sousa, A. A., Koblan, L. W., Levy, J. M., Chen, P. J., et al., (2019). Search-and-replace genome editing without double-strand breaks or donor DNA. *Nature, 576*, 149–157.

Beale, M. H., Birkett, M. A., Bruce, T. J., Chamberlain, K., Field, L. M., Huttly, A. K., Martin, J. L., et al., (2006). Aphid alarm pheromone produced by transgenic plants affects aphid and parasitoid behavior. *Proceedings of the National Academy of Sciences, 103*(27), 10509–10513.

Bo, W. A. N. G., Zhaohui, Z. H. O. N. G., Huanhuan, Z. H. A. N. G., Xia, W. A. N. G., Binglin, L. I. U., Lijia, Y. A. N. G., Xiangyan, H. A. N., et al., (2019). Targeted mutagenesis of NAC transcription factor gene, OsNAC041, leading to salt sensitivity in rice. *Rice Science, 26*(2), 98–108.
Boch, J., & Bonas, U., (2010). Xanthomonas AvrBs3 family-type III effectors: Discovery and function. *Annu. Rev. Phytopathol., 48*, 419–436.
Bogdanove, A. J., Schornack, S., & Lahaye, T., (2010). TAL effectors: Finding plant genes for disease and defense. *Curr. Opin. Plant Biol., 13*, 394–401.
Bolotin, A., Quinquis, B., Sorokin, A., & Ehrlich, S. D., (2005) Clustered regularly interspaced short palindrome repeats (CRISPRs) have spacers of extrachromosomal origin. *Microbiology (Reading, England), 151*, 2551–2561.
Bouzroud, S., Gasparini, K., Hu, G., Barbosa, M. A. M., Rosa, B. L., Fahr, M., Bendaou, N., et al., (2020). Down regulation and loss of auxin response factor 4 function using CRISPR/Cas9 alters plant growth, stomatal function and improves tomato tolerance to salinity and osmotic stress. *Genes, 11*(3), 272.
Carroll, D., (2011). Genome engineering with zinc-finger nucleases. *Genetics, 188*, 773–782.
Cermak, T., Doyle, E. L., Christian, M., Wang, L., Zhang, Y., Schmidt, C., Baller, J. A., et al., (2011). Efficient design and assembly of custom TALEN and other TAL effector-based constructs for DNA targeting. *Nucleic Acids Res., 39*, e82.
Chen, G., Zhou, Y., Kishchenko, O., Stepanenko, A., Jatayev, S., Zhang, D., & Borisjuk, N., (2021). Gene editing to facilitate hybrid crop production. *Biotechnology Advances*. Elsevier Inc.
Chen, K., & Gao, C., (2014). Targeted genome modification technologies and their applications in crop improvements. *Plant Cell Reports, 33*(4), 575–583.
Chen, K., Wang, Y., Zhang, R., Zhang, H., & Gao, C., (2019). CRISPR/Cas genome editing and precision plant breeding in agriculture. *Annual Review of Plant Biology, 70*, 667–697.
Christian, M., Cermak, T., Doyle, E. L., Schmidt, C., Zhang, F., Hummel, A., Bogdanove, A. J., & Voytas, D. F., (2010). Targeting DNA double-strand breaks with TAL effector nucleases. *Genetics, 186*, 757–761.
Christian, M., Qi, Y., Zhang, Y., & Voytas, D. F., (2013). *Targeted Mutagenesis of Arabidopsis Thaliana Using Engineered TAL Effector Nucleases (TALENs)*. G3 (Bethesda).
Cong, L., Ran, F. A., Cox, D., Lin, S., Barretto, R., Habib, N., et al., (2013). Multiplex genome engineering using CRISPR/Cas systems. *Science, 339*, 819–823.
Curtin, S. J., Zhang, F., Sander, J. D., et al., (2011). Targeted mutagenesis of duplicated genes in soybean with zinc finger nucleases. Plant *Physiology, 156*, 466–473
Dai, X., Yang, X., Wang, C., Fan, Y., Xin, S., Hua, Y., Wang, K., & Huang, H., (2021). CRISPR/Cas9-mediated genome editing in *Hevea brasiliensis*. *Ind. Crops Prod., 164*, 113418.
Doudna, J. A., & Charpentier, E., (2014). The new frontier of genome engineering with CRISPR-Cas9. *Science*, 346.
Du, M., Zhou, K., Liu, Y., Deng, L., Zhang, X., Lin, L., Zhou, M., et al., (2020). A biotechnology-based male-sterility system for hybrid seed production in tomato. *Plant J., 102*(5), 1090–1100.
Fang, Y., & Tyler, B. M., (2016). Efficient disruption and replacement of an effector gene in the oomycete *Phytophthora sojae* using CRISPR/C as9. *Molecular Plant Pathology, 17*(1), 127–139.

Fang, Y., & Xiong, L., (2015). General mechanisms of drought response and their application in drought resistance improvement in plants. *Cellular and Molecular Life Sciences, 72*(4), 673–689.

Fujita, Y., Nakashima, K., Yoshida, T., Fujita, M., Shinozaki, K., & Yamaguchi-Shinozaki, K., (2013). Role of abscisic acid signaling in drought tolerance and preharvest sprouting under climate change. *Climate Change and Plant Abiotic Stress Tolerance*, 521–554.

Gaj, T., Sirk, S. J., Shui, S. L., & Liu, J., (2016). Genome-editing technologies: principles and applications. *Cold Spring Harb. Perspect. Biol., 8*, a023754.

Gaudelli, N. M., Komor, A. C., Rees, H. A., Packer, M. S., Badran, A. H., Bryson, D. I., & Liu, D. R., (2017). Programmable base editing of A• T to G• C in genomic DNA without DNA cleavage. *Nature, 551*, 464–471.

Gil-Humanes, J., Wang, Y., Liang, Z., Shan, Q., Ozuna, C. V., Sánchez-León, S., et al., (2017). High-efficiency gene targeting in hexaploid wheat using DNA replicons and CRISPR/Cas9. *Plant J., 89*, 1251–1262.

Graham, N., Patil, G. B., Bubeck, D. M., Dobert, R. C., Glenn, K. C., Gutsche, A. T., & Morrell, P. L., (2020). Plant genome editing and the relevance of off-target changes. *Plant Physiology, 183*(4), 1453–1471.

Guo, F., Huang, Y., Qi, P., Lian, G., Hu, X., Han, N., Wang, J., et al., (2021). Functional analysis of auxin receptor OsTIR1/OsAFB family members in rice grain yield, tillering, plant height, root system, germination, and auxinic herbicide resistance. *New Phytol., 229*(5), 2676–2692.

Gupta, A., Sinha, R., Fernandes, J. L., Abdelrahman, M., Burritt, D. J., & Tran, L. S. P., (2020). Phytohormones regulate convergent and divergent responses between individual and combined drought and pathogen infection. *Critical Reviews in Biotechnology, 40*(3), 320–340.

Hafeez, M., Liu, S., Jan, S., Shi, L., Fernández-Grandon, G. M., Gulzar, A., Ali, B., et al., (2019). Knock-down of gossypol-inducing cytochrome P450 genes reduced deltamethrin sensitivity in *Spodoptera exigua* (Hübner). *International Journal of Molecular Sciences, 20*(9), 2248.

Horvath, P., & Barrangou, R., (2010). CRISPR/Cas, the immune system of bacteria and archaea. *Science, 327*, 167–170.

Hussain, A., Ding, X., Alariqi, M., Manghwar, H., Hui, F., Li, Y., Cheng, J., et al., (2021). Herbicide resistance: Another hot agronomic trait for plant genome editing. *Plants, 10*(4), 621.

Ilyas, M., Nisar, M., Khan, N., Hazrat, A., Khan, A. H., Hayat, K., Fahad, S., et al., (2021). Drought tolerance strategies in plants: A mechanistic approach. *Journal of Plant Growth Regulation, 40*(3), 926–944.

Ishino, Y., Shinagawa, H., Makino, K., Amemura, M., & Nakata, A., (1987). Nucleotide sequence of the IAP gene, responsible for alkaline phosphatase isozyme conversion in *Escherichia coli*, and identification of the gene product. *J. Bacteriol., 169*, 5429–5433.

Jayakannan, M., Bose, J., Babourina, O., Shabala, S., Massart, A., Poschenrieder, C., & Rengel, Z., (2015). The NPR1-dependent salicylic acid signaling pathway is pivotal for enhanced salt and oxidative stress tolerance in *Arabidopsis*. *Journal of Experimental Botany, 66*(7), 1865–1875.

Jiang, F., & Doudna, J. A., (2017). CRISPR–Cas9 structures and mechanisms. *Annu. Rev. Biophys., 46*, 505–529.

Jiang, M., He, Y., Chen, X., Zhang, X., Guo, Y., Yang, S., Huang, J., & Traw, M. B., (2020). CRISPR-based assessment of genomic structure in the conserved SQUAMOSA promoter-binding-like gene clusters in rice. *Plant J., 104*(5), 1301–1314.

Jinek, M., Chylinski, K., Fonfara, I., Hauer, M., Doudna, J. A., & Charpentier, E., (2012). A programmable dual-RNA–guided DNA endonuclease in adaptive bacterial immunity. *Science, 337*, 816–821.

Jinek, M., Jiang, F., Taylor, D. W., Sternberg, S. H., Kaya, E., Ma, E., et al., (2014). Structures of Cas9 endonucleases reveal RNA-mediated conformational activation. *Science, 343*, 1247997.

Kantor, A., McClements, M. E., & MacLaren, R. E., (2020). CRISPR-Cas9 DNA base-editing and prime-editing. *Int. J. Mol. Sci., 21*, 6240.

Khan, I., Khan, S., Zhang, Y., Zhou, J., Akhoundian, M., & Jan, S. A., (2021). CRISPR-Cas technology based genome editing for modification of salinity stress tolerance responses in rice (*Oryza sativa* L.). *Molecular Biology Reports,* 1–11.

Komor, A. C., Kim, Y. B., Packer, M. S., Zuris, J. A., & Liu, D. R., (2016). Programmable editing of a target base in genomic DNA without double-stranded DNA cleavage. *Nature 533*, 420–424.

Kumar, V. S., Verma, R. K., Yadav, S. K., Yadav, P., Watts, A., Rao, M. V., & Chinnusamy, V., (2020). CRISPR-Cas9 mediated genome editing of drought and salt tolerance (OsDST) gene in indica mega rice cultivar MTU1010. *Physiology and Molecular Biology of Plants, 26*(6), 1099.

Kumar, V., & Jain, M., (2015). The CRISPR–Cas system for plant genome editing: Advances and opportunities. *Journal of Experimental Botany, 66*(1), 47–57.

Li, J., Meng, X., Zong, Y., Chen, K., Zhang, H., Liu, J., Li, J., & Gao, C., (2016). Gene replacements and insertions in rice by intron targeting using CRISPR–Cas9. *Nature Plants, 2*(10), 1–6.

Li, M., Li, X., Zhou, Z., Wu, P., Fang, M., Pan, X., Lin, Q., et al., (2016). Reassessment of the four yield-related genes Gn1a, DEP1, GS3, and IPA1 in rice using a CRISPR/Cas9 system. *Front. Plant Sci., 7*, 1–13.

Li, R., Liu, C., Zhao, R., Wang, L., Chen, L., Yu, W., Zhang, S., et al., (2019). CRISPR/Cas9-mediated SlNPR1 mutagenesis reduces tomato plant drought tolerance. *BMC Plant Biology, 19*(1), 1–13.

Li, T., Liu, B., Spalding, M. H., Weeks, D. P., & Yang, B., (2012). High-efficiency TALEN-based gene editing produces disease-resistant rice. *Nat. Biotechnol., 30*, 390–392.

Li, Z., Liu, Z. B., Xing, A., Moon, B. P., Koellhoffer, J. P., Huang, L., Ward, R. T., et al., (2015). Cas9-guide RNA directed genome editing in soybean. *Plant Physiology, 169*(2), 960–970.

Liao, S., Qin, X., Luo, L., Han, Y., Wang, X., Usman, B., Nawaz, G., et al., (2019). CRISPR/Cas9-induced mutagenesis of semi-rolled Leaf1, 2 confers curled leaf phenotype and drought tolerance by influencing protein expression patterns and ROS scavenging in rice (*Oryza sativa* L.). *Agronomy, 9*(11), 728.

Lin, Q., Zong, Y., Xue, C., Wang, S., Jin, S., Zhu, Z., Wang, Y., et al., (2020). Prime genome editing in rice and wheat. *Nature Biotechnol., 38*, 582–585.

Liu, H., Wang, K., Tang, H., Gong, Q., Du, L., Pei, X., & Ye, X., (2020). CRISPR/Cas9 editing of wheat TaQ genes alters spike morphogenesis and grain threshability. *J. Genet. Genomics*, *47*(9), 563–575.

Liu, L., Zhang, J., Xu, J., Li, Y., Guo, L., Wang, Z., Zhang, X., et al., (2020). CRISPR/Cas9 targeted mutagenesis of SlLBD40, a lateral organ boundaries domain transcription factor, enhances drought tolerance in tomato. *Plant Science, 301*, 110683.

Liu, X., Liu, H., Zhang, Y., He, M., Li, R., Meng, W., Wang, Z., et al., (2021). Fine-tuning flowering time via genome editing of upstream open reading frames of heading date 2 in rice. *Rice, 14*(1).

Liu, X., Wu, D., Shan, T., Xu, S., Qin, R., Li, H., Negm, M., et al., (2020). The trihelix transcription factor OsGTγ-2 is involved adaption to salt stress in rice. *Plant Molecular Biology, 103*, 545–560.

Ma, X., Zhu, Q., Chen, Y., & Liu, Y. G., (2016). CRISPR/Cas9 platforms for genome editing in plants: Developments and applications. *Mol. Plant, 9*(7), 961–974.

Malone, L. A., Barraclough, E. I., Lin-Wang, K., Stevenson, D. E., Allan, A. C., (2009). Effects of red-leaved transgenic tobacco expressing a MYB transcription factor on two herbivorous insects, *Spodoptera litura* and *Helicoverpa armigera. Entomologia Experimentalis et Applicata, 133*(2), pp.117–127.

Mannocchi, F., Todisco, F., & Vergni, L., (2005). Agricultural drought: Indices, definition and analysis. *Water and Energy Abstracts., 15*(2), 21.

Miladinovic, D., Antunes, D., Yildirim, K., Bakhsh, A., Cvejić, S., Kondić-Špika, A., Marjanovic, J. A., et al., (2021). Targeted plant improvement through genome editing: From laboratory to field. *Plant Cell Reports* (pp. 935–951). Springer Science and Business Media Deutschland GmbH.

Mishra, R., Joshi, R. K., & Zhao, K., (2018). Genome editing in rice: Recent advances, challenges, and future implications. *Frontiers in Plant Science, 9*, 1361.

Munns, R., (2005). Genes and salt tolerance: Bringing them together. *New Phytologist, 167*(3), 645–663.

Nadeem, M., Chen, A., Hong, H., Li, D., Li, J., Zhao, D., Wang, W., et al., (2021). GmMs1 encodes a kinesin-like protein essential for male fertility in soybean (*Glycine max* L.). *J. Integr. Plant Biol., 63*(6), 1054–1064.

Nandy, S., Pathak, B., Zhao, S., & Srivastava, V., (2019). Heat-shock-inducible CRISPR/Cas9 system generates heritable mutations in rice. *Plant Direct, 3*(5), e00145.

Nasti, R. A., & Voytas, D. F., (2021). Attaining the promise of plant gene editing at scale. *Proc. Natl. Acad. Sci. U. S. A., 118*(22).

Nguyen, C. X., Paddock, K. J., Zhang, Z., & Stacey, M. G., (2021). GmKIX8-1 regulates organ size in soybean and is the causative gene for the major seed weight QTL QSw17-1. *New Phytol., 229*(2), 920–934.

Nishida, K., Arazoe, T., Yachie, N., Banno, S., Kakimoto, M., Tabata, M., Mochizuki, M., et al., (2016). Targeted nucleotide editing using hybrid prokaryotic and vertebrate adaptive immune systems. *Science*, 353.

Nitika, & Truman, A. W., (2018). Endogenous epitope tagging of heat shock protein 70 isoform Hsc70 using CRISPR/Cas9. *Cell Stress and Chaperones, 23*, 347–355.

Ogata, T., Ishizaki, T., Fujita, M., & Fujita, Y., (2020). CRISPR/Cas9-targeted mutagenesis of OsERA1 confers enhanced responses to abscisic acid and drought stress and increased primary root growth under nonstresses conditions in rice. *PloS One, 15*(12), e0243376.

Olate, E., Jiménez-Gómez, J. M., Holuigue, L., & Salinas, J., (2018). NPR1 mediates a novel regulatory pathway in cold acclimation by interacting with HSFA1 factors. *Nature Plants, 4*(10), 811–823.

Osakabe, K., Osakabe, Y., & Yoki, S., (2010). Site directed mutagenesis in *Arabidopsis* using custom-designed zinc finger nucleases. *Proceedings of the National Academy of Sciences, USA., 107*, 12034–12039.

Pabo, C. O., Peisach, E., & Grant, R. A., (2001). Design and selection of novel Cys2His2 zinc finger proteins. *Annu. Rev. Biochem., 70*, 313–340.

Quilis, J., Penas, G., Messeguer, J., Brugidou, C., & Segundo, B. S., (2008). The *Arabidopsis AtNPR1* inversely modulates defense responses against fungal, bacterial, or viral pathogens while conferring hypersensitivity to abiotic stresses in transgenic rice. *Mol. Plant Microbe Interact., 21*, 1215–1231.

Ran, F. A., et al., (2013). Genome engineering using the CRISPR-Cas9 system. *Nat. Protoc. 8*, 2281–2308.

Rodríguez-Leal, D., Lemmon, Z. H., Man, J., Bartlett, M. E., & Lippman, Z. B., (2017). Engineering quantitative trait variation for crop improvement by genome editing. *Cell 171*(2), 470–480.e8.

Sedeek, K. E. M., Mahas, A., & Mahfouz, M., (2019). Plant genome engineering for targeted improvement of crop traits. *Frontiers in Plant Science*. Frontiers Media S.A.

Shan, Q., Wang, Y., Chen, K., Liang, Z., Li, J., Zhang, Y., Zhang, K., et al., (2013). Rapid and efficient gene modification in rice and *Brachypodium* using TALENs. *Mol Plant., 6*, 1365–1368.

Shen, C., Que, Z., Xia, Y., Tang, N., Li, D., He, R., & Cao, M., (2017). Knock out of the annexin gene OsAnn3 via CRISPR/Cas9-mediated genome editing decreased cold tolerance in rice. *Journal of Plant Biology, 60*(6), 539–547.

Shi, J., Gao, H., Wang, H., Lafitte, H. R., Archibald, R. L., Yang, M., Hakimi, S. M., et al., (2017). ARGOS8 variants generated by CRISPR-Cas9 improve maize grain yield under field drought stress conditions. *Plant Biotechnol. J., 15*(2), 207–216.

Shukla, V. K., Doyon, Y., Miller, J. C., et al., (2009). Precise genome modification in the crop species *Zea mays* using zinc finger nucleases. *Nature, 459*, 437–443.

Singh, M., Kumar, M., Albertsen, M. C., Young, J. K., & Cigan, A. M., (2018). Concurrent modifications in the three homeologs of Ms45 gene with CRISPR-Cas9 lead to rapid generation of male sterile bread wheat (*Triticum aestivum* L.). *Plant Mol. Biol., 97*(4, 5), 371–383.

Song, G., Jia, M., Chen, K., Kong, X., Khattak, B., Xie, C., Li, A., & Mao, L., (2016). CRISPR/Cas9: A powerful tool for crop genome editing. *Crop J., 4*(2), 75–82.

Srinivasan, T., Kumar, K. R. R., Meur, G., & Kirti, P. B., (2009). Heterologous expression of *Arabidopsis* NPR1 (AtNPR1) enhances oxidative stress tolerance in transgenic tobacco plants. *Biotechnology Letters, 31*(9), 1343–1351.

Streubel, J., Blücher, C., Landgraf, A., & Boch, J., (2012). TAL effector RVD specificities and efficiencies. *Nature Biotechnology, 30*, 593–595.

Sun, Y., Zhang, X., Wu, C., He, Y., Ma, Y., Hou, H., Guo, X., et al., (2016). Engineering herbicide-resistant rice plants through CRISPR/Cas9-mediated homologous recombination of acetolactate synthase. *Molecular Plant, 9*(4), 628–631.

Svitashev, S., Young, J. K., Schwartz, C., Gao, H., Falco, S. C., & Cigan, A. M., (2015). Targeted mutagenesis, precise gene editing, and site-specific gene insertion in maize using Cas9 and guide RNA. *Plant Physiology, 169*(2), 931–945.

Takagi, H., Tamiru, M., Abe, A., Yoshida, K., Uemura, A., Yaegashi, H., Obara, T., et al., (2015). MutMap accelerates the breeding of a salt-tolerant rice cultivar. *Nature Biotechnology, 33*(5), 445–449.

Townsend, J. A., Wright, D. A., Winfrey, R. J., et al., (2020). High frequency modification of plant genes using engineered zinc finger nucleases. *Nature, 459*, 437–443.

Tyagi, S., Kesiraju, K., Saakre, M., Rathinam, M., Raman, V., Pattanayak, D., & Sreevathsa, R., (2020). Genome editing for resistance to insect pests: An emerging tool for crop improvement. *ACS Omega, 5*(33), 20674–20683.

Umezawa, T., Fujita, M., Fujita, Y., Yamaguchi-Shinozaki, K., & Shinozaki, K., (2006). Engineering drought tolerance in plants: Discovering and tailoring genes to unlock the future. *Current Opinion in Biotechnology, 17*(2), 113–122.

Usman, B., Nawaz, G., Zhao, N., Liao, S., Qin, B., Liu, F., Liu, Y., & Li, R., (2021). Programmed editing of rice (*Oryza sativa* l.) Osspl16 gene using Crispr/Cas9 improves grain yield by modulating the expression of pyruvate enzymes and cell cycle proteins. *Int. J. Mol. Sci., 22*(1), 1–19.

Voytas, D. F., (2013). Plant genome engineering with sequence-specific nucleases. *Annual Review of Plant Biology., 64*, 327–350.

Wang, J., Zhang, H., Wang, H., Zhao, S., Zuo, Y., Yang, Y., & Wu, Y., (2016). Functional validation of cadherin as a receptor of Bt toxin Cry1Ac in *Helicoverpa armigera* utilizing the CRISPR/Cas9 system. *Insect Biochemistry and Molecular Biology, 76*, 11–17.

Wang, L., Chen, L., Li, R., Zhao, R., Yang, M., Sheng, J., & Shen, L., (2017). Reduced drought tolerance by CRISPR/Cas9-mediated SlMAPK3 mutagenesis in tomato plants. *Journal of Agricultural and Food Chemistry, 65*(39), 8674–8682.

Wang, Q., Alariqi, M., Wang, F., Li, B., Ding, X., Rui, H., Li, Y., et al., (2020). The application of a heat-inducible CRISPR/Cas12b (C2c1) genome editing system in tetraploid cotton (G. *hirsutum*) plants. *Plant Biotechnology Journal, 18*(12), 2436–2443.

Wang, Y., Cheng, X., Shan, Q., Zhang, Y., Liu, J., Gao, C., & Qiu, J. L., (2014). Simultaneous editing of three homoeoalleles in hexaploid bread wheat confers heritable resistance to powdery mildew. *Nature Biotechnology, 32*(9), 947–951.

Weeks, D. P., Spalding, M. H., & Yang, B., (2016). Use of designer nucleases for targeted gene and genome editing 1197 in plants. *Plant Biotechnol. J., 14*, 483–495.

Wendt, T., Holm, P. B., Starker, C. G., Christian, M., Voytas, D. F., Brinch, P. H., & Holme, I. B., (2013). TAL effector nucleases induce mutations at a pre-selected location in the genome of primary barley transformants. *Plant Mol. Biol., 83*, 279–285.

Xu, W., Zhang, C., Yang, Y., Zhao, S., Kang, G., He, X., et al., (2020). Versatile nucleotides substitution in plant using an improved prime editing system. *Mol. Plant., 13*, 675–678.

Yang, B., Li, X., Lei, L., & Chen, J., (2017). APOBEC: From mutator to editor. *J. Genet. Genomics, 44*, 423–437.

Zafar, K., Sedeek, K. E., Rao, G. S., Khan, M. Z., Amin, I., Kamel, R., & Mahfouz, M. M., (2020). Genome editing technologies for rice improvement: Progress, prospects, and safety concerns. *Frontiers in Genome Editing, 2*, 5.

Zeng, Y., Wen, J., Zhao, W., Wang, Q., & Huang, W., (2020). Rational improvement of rice yield and cold tolerance by editing the three genes OsPIN5b, GS3, and OsMYB30 with the CRISPR–Cas9 system. *Frontiers in Plant Science, 10*, 1663.

Zetsche, B., Gootenberg, J. S., Abudayyeh, O. O., Slaymaker, I. M., Makarova, K. S., Essletzbichler, P., et al., (2015). Cpf1 is a single RNA-guided endonuclease of a class 2 CRISPR-Cas system. *Cell, 163*, 759–771.

Zhai, Y., Cai, S., Hu, L., Yang, Y., Amoo, O., Fan, C., & Zhou, Y., (2019). CRISPR/Cas9-mediated genome editing reveals differences in the contribution of INDEHISCENT homologues to pod shatter resistance in *Brassica napus* L. *Theor. Appl. Genet., 132*(7), 2111–2123.

Zhang, A., Liu, Y., Wang, F., Li, T., Chen, Z., Kong, D., Bi, J., et al., (2019). Enhanced rice salinity tolerance via CRISPR/Cas9-targeted mutagenesis of the *OsRR22* gene. *Molecular Breeding, 39*(3), 1–10.

Zhang, F., Maeder, M. L., Unger-Wallace, E., et al., (2010). High frequency targeted mutagenesis in *Arabidopsis thaliana* using zinc finger nucleases. *Proceedings of the National Academy of Sciences, USA, 107*, 12028–12033.

Zhang, Y., Zhang, F., Li, X., Baller, J. A., Qi, Y., Starker, C. G., Bogdanove, A. J., & Voytas, D. F., (2013). Transcription activator-like effector nucleases enable efficient plant genome engineering. *Plant Physiol., 161*, 20–27.

Zhong, X., Xi, L., Lian, Q., Luo, X., Wu, Z., Seng, S., Yuan, X., & Yi, M., (2015). The NPR1 homolog GhNPR1 plays an important role in the defense response of gladiolus hybridus. *Plant Cell Reports, 34*(6), 1063–1074.

CHAPTER 6

Smart Breeding for Climate-Resilient Agriculture

APARNA,[1,2] DALPAT LAL,[2] CHARUPRIYA CHAUHAN,[1] ANJALI JOSHI,[3] INDRA DEO,[1] and SUMAN SINGH[2]

[1]*Department of Genetics and Plant Breeding, GBPUAT, Pantnagar, Uttarakhand, India*

[2]*Departments of Agriculture, Jagan Nath University, Jaipur, Rajasthan, India*

[3]*Forest Genetics and Tree Improvement Division, Arid Forest Research Institute, Jodhpur, Rajasthan, India*

ABSTRACT

Climate change is caused by small variations in weather patterns over a period of time. After the industrial revolution, human activities triggers an increase in greenhouse gases leading to global warming. Both Climate change and agriculture are global processes. Changes in average temperature, precipitation, extremes heat waves, elevated atmospheric CO_2 and ozone concentrations are all ill effects of global warming Due to climatic fluctuations, agriculture and food production system suffer serious problems caused by drought, floods, and upsurge of insect pest and diseases. Improved varieties produced by plant breeding are one of the most important ways to have stable production in the face of new challenges posed by climate change. Role of plant breeding in human civilization is indispensable as it has performed revolutionary task to feed the fast growing population of the globe. Opening the archive of genetic diversity and intensive use of wild resources are inevitably critical for any plant breeding system. To withstand

Smart Breeding: Molecular Interventions and Advancements for Crop Improvement. Kailash Chandra, Subhash Chand, Ravi Prakash Saini, and Rohit Sharma (Eds.)

the ill effect of climate change, breeding strategies focuses on the generation of varieties with distinct growth habits and time span to avoid unforeseeable occurrences of stress at crucial stage in plant life cycles. Current developments in modern plant breeding and biotechnological approaches such as genomics, proteomics, phenomics, sequencing, cloning, marker assisted breeding (MAB) methodologies and cutting-edge genome-editing technologies along with artificial intelligence (AI), unfold new avenues for speedy climate-resilient smart breeding. Hence, climate smart breeding provides a comprehensive approach to combat climate fluctuations and to produce better-adapted crop varieties.

6.1 INTRODUCTION

Increasing human activities have affected the environment in many ways. Both, climate change and agriculture are global process. Global warming influences agriculture by causing variation in average temperature, concentration of atmospheric carbon dioxide, rainfall, ground-level ozone concentrations, incidence of insect-pests and diseases and deterioration of nutritional quality of food crops (Hoffmann, 2013; Milius, 2017). Further, it is also reported that climate change has a significant impact on agricultural productivity and farming operations of all crops, with a wide range of consequences that are highly dependent on geographical location (Ackerly et al., 2010). Factors that adversely affect agricultural production include the speedy upsurge in mean temperatures, as well as the escalating frequency and severity of extreme weather incidence which leads to heat, drought, flood, and chilling stresses (Figure 6.1) (Gornall et al., 2010). Water scarcity is a global issue that has an impact on plant growth and development. Drought also has a negative effect on yield as well as the quality of the plant. Water scarcity has been proven to affect flower blooming, pollination, and grain filling in most of the crops. However, excess of water availability may also harm plants due to elevated relative humidity that exposed plants to several disease and insect pest outbreaks (Rosenzweig et al., 2001). Nevertheless, it is observed that the severity of the various stresses relies on the plant's phenological condition (Nuruddin et al., 2003). Several studies have been conducted to see the high effects of heat waves in wheat (Ottman et al., 2021, rice (Zhao et al., 2016), maize (Bassu et al., 2014), and soybean (Deryng et al., 2014). Increases in seasonal temperatures of 1°C lead to yield reductions ranging from 3.1% in soybean to 7.4% in maize. CO_2 level in the atmosphere have opposing impacts on crops. Plant response is clearly influenced by the mixture of individual stresses, the degree of each stress, and the developmental stage of the plant. Studies have shown that different, and sometimes contradictory signaling pathways control plant stress and defensive responses, and the plant activates specific signaling cascades and metabolic pathways depending on whether it is stressed or not. Ihsan et al. (2016) reported that high temperatures and drought have a deleterious impact on cereal growth, development, and reproduction, lowering the productivity. In arid or semi-arid regions, monsoon pattern varies a lot from year to year, as the climate becomes more unstable and involves a shift in seasons.

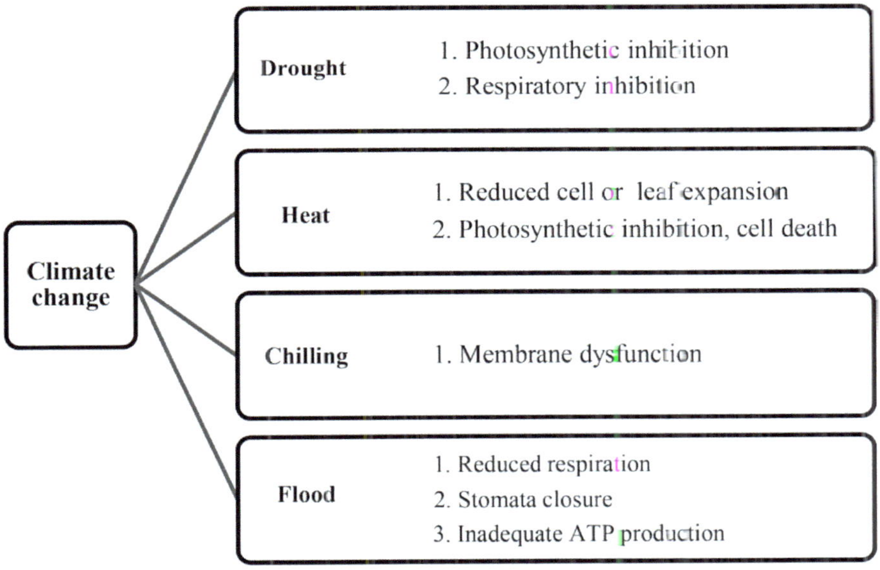

FIGURE 6.1 Abiotic stresses and their effect on crop.

Agriculture is struggling to meet the increasing demand of food in many parts of the world as natural wealth such as soils, water, and biodiversity are in danger and declining day by day. The food security challenges will become more difficult and the world will need to produce more food. Environmental stresses have become more frequent and aggravated by rapid climate change in recent decades, and plant breeding is an essential part of the solution. The world's food security problem will get aggravated, leading to additional load on a food production system. Challenges posed by climate change can be efficiently counteracted by use of improved varieties developed through plant breeding and biotechnological approaches. The basic breeding strategies to cope up with climate change is to develop varieties with varied life span to avoid anticipated occurrences of stress at critical periods in crop life cycles, moving temperature optima for crop growth and re-emphasizing population breeding. In view of this, there is a dire need for climate smart breeding to develop novel varieties which can sustain the adverse effect of climate fluctuation.

6.2 BREEDING TOOLS TO COMBAT CLIMATE CHANGE

Plant breeding techniques range from the basic selection of plants in farmer's fields with desirable qualities to more advanced molecular or biotechnological approaches to generate variations to challenge climate change. At present, several classical and molecular techniques have been identified to develop climate resilient crops. These strategies are selection, germplasm collection, mutation breeding, marker assisted selection (MAS), transgenic technology, evolutionary plant breeding and genetic enhancement (Figure 6.2). All these strategies focus on discovering and utilizing sources

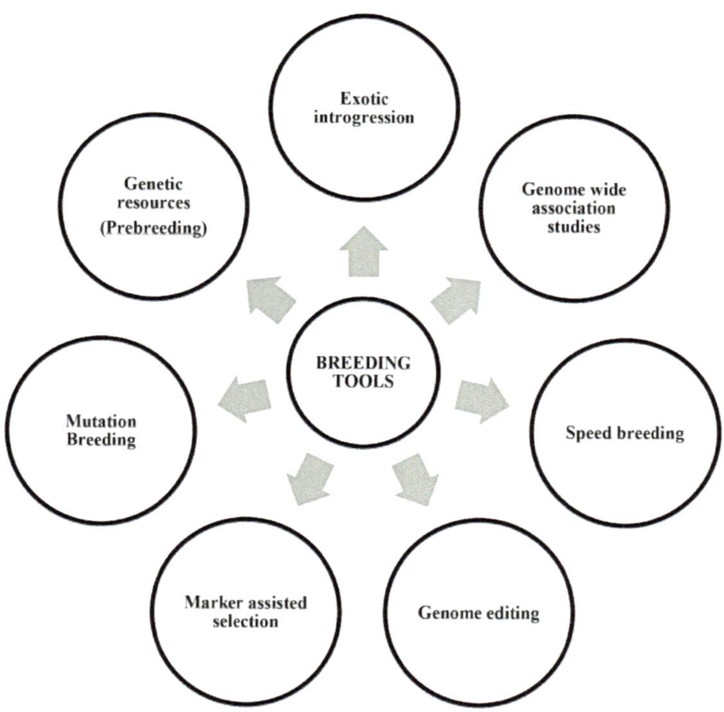

FIGURE 6.2 Plant breeding and molecular tools used for development of climate resilience crop varieties.

of genetic diversity for increased biotic and abiotic stress tolerance. Further, rapid climate change requires varieties to be developed in a short stretch of time to cope with the unforeseeable weather fluctuation The idea behind climate smart breeding is to unify all the breeding tools viz; traditional breeding technologies, advanced molecular and genomic tools in order to develop effective plant breeding system for climate resilient agriculture. To reduce the breeding cycle while offering systematic evaluation approaches for abiotic and biotic stresses, recent breeding technologies like rapid generation advancement through single seed descent method, double haploid (DH), MAS, and genomic selection, etc., have been adopted. Breeders can use data mining to evaluate changing weather susceptibility and tolerance in the present genetic resources. Data mining information can help breeders in finding temperature and photoperiod sensitive germplasm which can be used for further genetic improvement of crop plants. Some successful breeding efforts for climate resilience have been summarized in Table 6.1.

TABLE 6.1 Crop Varieties Developed for Climate Resilience

Sl. No.	Crops	Variety Name	Stress Tolerance	Plant Breeding Methods
1.	Rice	Sahbhagidhan	Drought	Conventional breeding
2.	Rice	Anjali	Submergence	Conventional breeding
3.	Rice	Naveen	Submergence	Conventional breeding
4.	Rice	Swarna sub-1	Submergence	Marker-assisted selection
5.	Wheat	KRL 1-4	Salinity	Pedigree selection
6.	Rice	Damodar (CSR 1)	Salinity	Pureline selection
7.	Rice	Getu (CSR 3)	Salinity	Pureline selection
8.	Maize	RCRMH2	Heat	Heterosis breeding
9.	Potato	Kufri Sheetman	Frost resistant	Hybridization
10.	Rice	TCCP 266-2-49-B-B-3	Salinity	Somaclonal variation
11.	Maize	Drought gard	Drought	Transgenic
12.	Rice	IR64-sub-1	Submergence	Marker-assisted selection
13.	Rice	Samba-Mahsuri sub-1	Submergence	Marker-assisted selection
14.	Rice	Ranjit-sub-1	Submergence	Marker-assisted selection
15.	Rice	DRR Dhan 50	Drought and submergence	Marker-assisted selection
16.	Chickpea	RSG-44	Drought and frost	Conventional method
17.	Chickpea	Ujjwal	Drought	Conventional method

6.2.1 PLANT GENETIC RESOURCES (PGRS) TO COMBAT CLIMATE CHANGE

Germplasm can be utilized for evaluation of character, pre-breeding (including genetic enhancement and base-broadening), diversification of crops, evolution of novel varieties and their commercialization. These activities can help to address a variety of issues related to crop production during recent times of climate change. Well suited landraces and farmer varieties have proven to be a reliable source of adaptive genes for crop improvement (Mba, 2013). Crops are more sensitive to stresses caused by climate change due to increasing homogeneity and uniformity. Characters resilient to biotic and abiotic stresses present in crop wild relatives (CWR) (Maxted et al., 2008), as well as landraces and farmer varieties, may be taken into consideration while developing improved cultivars to minimize genetic susceptibility. Widespread distribution of a flood-tolerant rice variety (Scuba Rice) in flood-prone areas like Bangladesh, India, and the Philippines, is an example of successful breeding of a crop variety that supports climate-smart agriculture. In the areas of extreme climate events such as flooding, drought, climate-ready varieties are being adopted. Many edible plant species such as Bambara groundnut (*Vigna subterranea*), yam bean (*Pachyrhizus erosus*) and Moringa (*Moringa oleifera*) are underappreciated and neglected, but they are robust and well suited to harsh climatic conditions. It would make sense to replace staple crops like maize, rice, wheat with drought-resistance crops like cassava and millets in drought-prone areas. However, it is possible only if peasants are enthusiastic about adoption of these newly developed varieties which will provide viable means to achieve climate-smart agriculture. Plant genetic resources (PGRs) are more important with respect to climate adaptation as compared to climate mitigation. However, a number of measures can help to improve greenhouse gas sequestration and thus support climate change mitigation. Growing 4 carbon compound plants such as maize, sorghum, sugarcane, and millets can increase plant carbon content. The nitrogen-fixing potential of legume crops, such as garden peas, lentils, groundnuts, mung beans, cowpeas, pigeon pea, and chickpea, varies significantly. Greater nitrogen fixation is also linked to increased carbon sequestration, thus cultivating pulses and other legumes would provide extra support for climate change mitigation.

6.2.2 EXOTIC INTROGRESSION INTO ELITE VARIETIES

Wild species harbor valuable genes for various environmental stresses which have contributed considerably to combat abiotic stresses. Distant hybridization is a common method for the introgression of biotic and abiotic stress resistance gene. Comparative pool sequencing of the genome can be used to study the sources of biotic and abiotic stresses in wild species. This technique has been utilized in crops like maize chickpea and barley (Mason & Batley, 2015). Genomic technologies, such as molecular markers, aid in the characterization and creation of populations that consist of gene for introgressions, which may be easily included in breeding programs to address the escalating environmental problems. Polyploidy is another potential means for increasing diversity and sustaining adaptation to various stresses.

6.2.3 TILLING (TARGETING INDUCED LOCAL LESIONS IN GENOMES)

In the mid-19th century, mutation breeding was popularized with the goal of increasing genetic variation. A recently developed reverse genetics method called TILLING (targeting induced local lesions in genomes) has transformed the branch of mutation breeding. TILLING finds uncommon mutations in populations by using chemical-induced mutagenesis with a DNA screening technique (Scarascia-Mugnozza et al., 1990). Traditionally, TILLING processes screened the population and selected mutagenized lines using enzymatic or physical means. Loss-of-function, gain-of-function, and hypomorphic alleles can be discovered and perhaps linked to phenotypes. The discovery of the rare variants in a population could be successfully accelerated by integrating TILLING with multifaceted pooling (Tsai et al., 2011). TILLING by sequencing is being utilized to discover allelic variants accompanying agro morphological traits which are involved in climate change response (Tadele et al., 2010). TILLING was employed in particular to find novel allelic variations in the Hsp26 gene family that were linked to heat tolerance in wheat (Comastri, 2018). By utilizing TILLING process, barley mutants were created to study nucleotide base pair difference in the era1 gene which is associated with drought tolerance (DT). This era 1 gene show differential control under drought condition in soybean and wheat. Eco-TILLING was proposed by Comai et al. (2004) for detecting natural mutations in a population. It serves as a valuable technique for identifying allelic variants involved in crop adaptation to several environmental stresses brought upon by extreme agro-climatic conditions.

6.2.4 MARKER-ASSISTED BREEDING (MAB)

Basically, marker-assisted backcrossing (MABC) and marker-assisted gene pyramiding (MAGP) are most commonly utilized approach of MAS to develop improved varieties for various purposes. Identifying and utilizing QTLs/genes in marker-assisted breeding (MAB) can aid in the development of novel varieties that are resistant to stresses such as cold, heat, salinity, and submergence. In AMDROUT (Asia Maize Drought Tolerance) project, CIMMYT has got significant success in generating early generation yellow drought tolerant maize inbred lines utilizing the MARS approach. In chickpea, germplasm were used to evaluate DT, heat tolerance, and yield traits in order to identify significant marker-trait associations (MTAs) to be used for developing superior varieties with enhanced drought and heat tolerance (Thudi et al., 2014). Furthermore, SNP has been reported in cereals and legumes for climate change response and to develop new resilient crops. Several researches have been reported in sorghum that showed association of SNP with heat stress at the vegetative stage under field conditions (Chen et al., 2017). Climate resilient varieties developed through MAS have been given in Table 6.1.

6.2.5 GWAS AND GENOMIC SELECTION

GWAS technique has been applied for the identification of climate resilient crop varieties for several environmental stresses. GWAS is being considerably supported by genome-wide distributed SNPs in identifying the genomic region linked with phenotypic variants in large germplasm pools. As a result, GWAS has emerged as a crucial strategy for mapping quantitative characteristics and investigating natural variation in recent times. The GWAS approach entails analysis of statistical relationships between DNA polymorphisms and trait variations across distantly related and heterogeneous genotypes from varied collections for traits of interest. GWAS effectively find causal loci/gene underlying natural phenotypic variation by screening vast and diverse collections of germplasm with the help of ample genetic markers. Coupled with improved genome sequencing technology, GWAS enhances the mapping resolution for accurate location of allele/QTL/genes. GWAS uses prior recombination events in association panels to find the genes which reveal phenotypic variables with higher precision as compared to QTL analysis. Crain et al. (2018) reported various Genomic Selection approaches for phenotypic data collected for heat and DT at the CIMMYT.

6.2.6 GENOME EDITING (GE)

Transgenic plants are not much accepted by the public and also have several stringent regulations. Therefore, scientist has discovered the genome editing (GE) tools which are flexible and popular methods for crop improvement program. GE can be achieved by utilizing mega-nucleases (MNs) (homing endonucleases), zinc finger nucleases (ZFNs), transcription activator-like effectors nucleases (TALENs), and clustered regularly interspaced short palindromic repeats (CRISPR) associated protein (CRISPR/Cas). These gene editing methods applied to generate insertion, deletion, substitution, and replacement of genes. Editing of genes involving abiotic and biotic stress responses has been revealed by several workers (Cardi et al., 2017). The modification of promoter area using TALEN in rice for bacterial blight susceptibility gene OsSWEET14 is one of the successful examples of the GE. The effectors binding motif was removed as a result of this modification, resulting in resistance to bacterial blight (Li et al., 2012). The recent advancement of GE technologies, particularly CRISPR/Cas9, has opened new pathways for fast and accurate genome modification, offering speedy translation of knowledge from the laboratory to the field. Insertions/deletions of a new sequence at a specific position in the target genome are possible with GE Several genome altered crop plants which includes drought and salt tolerant soybeans, Camelina with improved oil content have been reported and reached the final phases of commercialization in the United States (Waltz, 2018). CRISPR technology can be used to create durable crops that fix more carbon. Fortunately, CRISPR, a precise GE technique, has the potential to reduce agricultural emissions and better prepare crops for climate change. For example, utilizing speed marker-assisted backcrossing (MABC), a transformant carrying a Cas9 foreign gene is being utilized as a donor parent to establish a pool of superior inbred lines. CRISPR may be used to design heat tolerant, drought tolerant and salt tolerant plants. Further, several drought tolerant crops has been develops from CRISPER, for examples, *SlLBD*40 knockout mutant in tomato (Lin et al., 2020a); *OsERA*1 rice mutant (Ogata et al., 2020); *ZmSRL*5 mutant in maize (Pan et al., 2020). Integration of these procedures eliminates the use of foreign DNA and thus results in GMO label free end products.

6.2.7 SPEED BREEDING

Speed breeding is a strategy for reducing generation time and speeding up breeding and research activities. If Climate change will continue, it will result in insufficient production of wheat, rice, and maize to fulfill the growing future demands of food (Ray et al., 2012). The longer generation time of the crops, which typically allow one or two generations per year, is a major restrictive element in breeding improved varieties. Breeders are able to harvest up to 6 generations per year by cutting generation time by more than half with the help of "speed breeding" techniques that uses prolonged photoperiods and regulated temperatures (Pfeiffer, 1926). The single seed descent method is an efficient method for quick generation advancement which aids in reducing the generation time. In one year, 4–6 generations of spring wheat, durum wheat, barley, chickpea can be achieved instead of two–three under standard glass house conditions, by employing light-emitting diode (LED) (Watson et al., 2018; Hickey et al., 2017; Collard et al., 2017). Speed breeding can be used with modern techniques such as GE to accurately modify plant genes for greater adaptation to biotic and abiotic challenges in the face of looming climate change. According to Riaz et al. (2017), CRISPR-Cas9 technique has been utilized to produce single or multiple alternations in various crops. It is possible to construct genotypes that are 'CRISPR ready' and contain a heterologous Cas9 gene.

6.3 CONCLUSION

Agriculture is facing unprecedented challenges due to increasing climate change threats. Indeed, climate change is affecting ecosystem dynamics through changing the behavior of numerous species. To feed the growing population, there is a dire need to boost global food supply with shrinking arable land. Although, the breeder's skill is still evolving, but it has expanded enough to allow us to start looking at the plant breeding with a broad horizon. Technological developments in genotypic and phenotypic analysis, along with the biotechnological and digital transformation, preservation, and maintenance of genetic assets will result in a cost-effective climate-smart agriculture. In fact, breeding for climate-resilient agriculture is benefiting from novel revolutionary technologies based on genomics and new generation sequencing, which is predicted to yield successful results in the future. Furthermore, an integrated approach that includes both conventional and advanced breeding methods will help in uncovering the resistance mechanism to various stresses generated by climate change and to develop the climate resilient crop varieties.

KEYWORDS

- **genomics**
- **global warming**
- **marker-assisted backcrossing**
- **marker-assisted gene pyramiding**
- **phenomics**
- **plant breeding**
- **targeting induced local lesions in genomes**
- **transcription activator-like effectors nucleases**

REFERENCES

Ackerly, D. D., Loarie, S. R., Cornwell, W. K., Weiss, S. B., Hamilton, H., Branciforte, R., & Kraft, N. J. B., (2010). The geography of climate change: Implications for conservation biogeography. *Diversity and Distributions, 16*(3), 476–487.

Bassu, S., Brisson, N., Durand, J. L., Boote, K., Lizaso, J., Jones, J. W. & Waha, K., (2014). How do various maize crop models vary in their responses to climate change factors? *Global Change Biology, 20*(7), 2301–2320.

Cardi, T., Batelli, G., & Nicolia, A., (2017). Opportunities for genome editing in vegetable crops. *Emerging Topics in Life Sciences, 1*(2), 193–207.

Chen, J., Chopra, R., Hayes, C., Morris, G., Marla, S., Burke, J., & Burow, G., (2017). Genome-wide association study of developing leaves' heat tolerance during vegetative growth stages in a sorghum association panel. *The Plant Genome, 10*(2), 1–9.

Collard, B. C., Beredo, J. C., Lenaerts, B., Mendoza, R., Santelices, R., Lopena, V., & Islam, M. R., (2017). Revisiting rice breeding methods–evaluating the use of rapid generation advance (RGA) for routine rice breeding. *Plant Production Science, 20*(4), 337–352.

Comai, L., Young, K., Till, B. J., Reynolds, S. H., Greene, E. A., Codomo, C. A., & Henikoff, S., (2004). Efficient discovery of DNA polymorphisms in natural populations by ecotilling. *The Plant Journal, 37*(5), 778–786.

Comastri, A., Janni, M., Simmonds, J., Uauy, C., Pignone, D., Nguyen, H. T., & Marmiroli, N., (2018). Heat in wheat: Exploit reverse genetic techniques to discover new alleles within the *Triticum durum* sHsp26 family. *Frontiers in Plant Science, 9*, 1337.

Crain, J. L., Mondal, S., Rutkoski, J., Singh, R. P., & Poland, J., (2018). Combining high-throughput phenotyping and genomic information to increase prediction and selection accuracy in wheat breeding. *Plant Genome*.

Deryng, D., Conway, D., Ramankutty, N., Price, J., & Warren, R., (2014). Global crop yield response to extreme heat stress under multiple climate change futures. *Environmental Research Letters, 9*(3), 034011.

Gornall, J., Betts, R., Burke, E., Clark, R., Camp, J., Willett, K., & Wiltshire, A., (2010). Implications of climate change for agricultural productivity in the early twenty-first century. *Philosophical Transactions of the Royal Society B: Biological Sciences, 365*(1554), 2973–2989.

Hickey, L. T., Germán, S. E., Pereyra, S. A., Diaz, J. E., Ziems, L. A., Fowler, R. A., & Dieters, M. J., (2017). Speed breeding for multiple disease resistance in barley. *Euphytica, 213*(3), 64.

Hoffmann, U., (2013). *Section B: Agriculture: A Key Driver and a Major Victim of Global Warming.* Lead article in 3–5.

Ihsan, M. Z., El-Nakhlawy, F. S., Ismail, S. M., & Fahad, S., (2016). Wheat phenological development and growth studies as affected by drought and late season high temperature stress under arid environment. *Frontiers in Plant Science, 7*, 795.

Li, C., Li, W., Zhou, Z., Chen, H., Xie, C., & Lin, Y., (2020a). A new rice breeding method: CRISPR/Cas9 system editing of the Xa13 promoter to cultivate transgene-free bacterial blight-resistant rice. *Plant Biotechnology Journal, 18*, 313–315. doi: 10.1111/pbi.13217.

Li, T., Liu, B., Spalding, M. H., Weeks, D. P., & Yang, B., (2012). High-efficiency TALEN-based gene editing produces disease-resistant rice. *Nature Biotechnology, 30*(5), 390–392.

Mason, A. S., & Batley, J., (2015). Creating new interspecific hybrid and polyploid crops. *Trends in Biotechnology, 33*(8), 436–441.

Maxted, N., Dulloo, E., Ford-Lloyd, B. V., Iriondo, J. M., & Jarvis, A., (2008). Gap analysis: A tool for complementary genetic conservation assessment. *Diversity and Distributions, 14*(6), 1018–1030.

Mba, C., (2013). Induced mutations unleash the potentials of plant genetic resources for food and agriculture. *Agronomy, 3*(1), 200–231.

Milius, S., (2017). Worries grow that climate change will quietly steal nutrients from major food crops. *Science News Year in Review Nutrition, Climate, Sustainability.*

Nuruddin, M. M., Madramootoo, C. A., & Dodds, G. T., (2003). Effects of water stress at different growth stages on greenhouse tomato yield and quality. *HortScience, 38*(7), 1389–1393.

Ogata, T., Ishizaki, T., Fujita, M., & Fujita, Y., (2020). CRISPR/Cas9-targeted mutagenesis of OsERA1 confers enhanced responses to abscisic acid and drought stress and increased primary root growth under nonstressed conditions in rice. *PLoS One, 15*, 0243376.

Ottman, M. J., Kimball, B. A., White, J. W., & Wall, G. W., (2012). Wheat growth response to increased temperature from varied planting dates and supplemental infrared heating. *Agronomy Journal, 104*(1), 7–16.

Pan, Z., Liu, M., Zhao, H., Tan, Z., Liang, K., Sun, Q., Gong, D., et al., (2020). ZmSRL5 is involved in drought tolerance by maintaining cuticular wax structure in maize. *J. Integr. Plant Biology, 62*, 1895–1909.

Pfeiffer, N. E., (1926). Microchemical and morphological studies of effect of light on plants. *Botanical Gazette, 81*(2), 173–195.

Ray, D. K., Ramankutty, N., Mueller, N. D., West, P. C., & Foley, J. A., (2012). Recent patterns of crop yield growth and stagnation. *Nature Communications, 3*(1), 1–7.

Riaz, A., Athiyannan, N., Periyannan, S., Afanasenko, O., Mitrofanova, O., Aitken, E. A., & Hickey, L. T. (2017). Mining Vavilov's treasure chest of wheat diversity for adult plant resistance to *Puccinia triticina. Plant Disease, 101*(2), 317–323.

Rosenzweig, C., Iglesius, A., Yang, X. B., Epstein, P. R., & Chivian, E., (2001). *Climate Change and Extreme Weather Events-Implications for Food Production, Plant Diseases, and Pests, 2*, 90–104.
Scarascia-Mugnozza, G., D'amato, F., Avanzi, S., Bagnara, D., Bell, M. L., Bozzini, A., Cervigni, T., Devreux, M., Donini, B., Giorgi, B., et al., (1990). Mutation breeding for durum wheat (*Triticum turgidum* ssp. *Durum* desf.) improvement in Italy. In: *Proceedings of the International Symposium on the Contribution of Plant Mutation Breeding to Crop Improvement* (pp. 1–28). Vienna, Austria.
Tadele, Z., Chikelu, M. B. A., & Till, B. J., (2010). TILLING for mutations in model plants and crops. In: *Molecular Techniques in Crop Improvement* (pp. 307–332). Springer, Dordrecht.
Thudi, M., Upadhyaya, H. D., Rathore, A., Gaur, P. M., Krishnamurthy, L., Roorkiwal, M., et al., (2014). Genetic dissection of drought and heat tolerance in chickpea through genome-wide and candidate gene-based association mapping approaches. *PLoS One, 9*(5), e96758.
Tsai, H., Howell, T., Nitcher, R., Missirian, V., Watson, B., Ngo, K. J., & Comai, L., (2011). Discovery of rare mutations in populations: TILLING by sequencing. *Plant Physiology, 156*(3), 1257–1268.
Waltz, E., (2018). With a free pass, CRISPR-edited plants reach market in record time. *Nature Biotechnology, 36*, 6–7. doi: 10.1038/nbt0118-6b.
Watson, A., Ghosh, S., Williams, M. J., Cuddy, W. S., Simmonds, J., Rey, M. D., & Hickey, L. T., (2018). Speed breeding is a powerful tool to accelerate crop research and breeding. *Nature Plants, 4*(1), 23–29.
Zhao, C., Piao, S., Wang, X., Huang, Y., Ciais, P., Elliott, J., & Peñuelas, J., (2016). Plausible rice yield losses under future climate warming. *Nature Plants, 3*(1), 1–5.

CHAPTER 7

Reverse Breeding: A Novel Tool for Crop Improvement

MANDAKINI KABI,[1] S. K. TRIPATHY,[2] MANASI DASH,[3] BHABENDRA BAISAKH,[3] TAPAS RANJAN DAS,[4] and MAHIPAL SINGH KESAWAT[1]

[1]*Faculty of Agriculture, Sri Sri University, Cuttack, Odisha, India*

[2]*Department of Agricultural Biotechnology, College of Agriculture, OUAT, Bhubaneswar, India*

[3]*Department of Plant Breeding and Genetics, College of Agriculture, OUAT, Bhubaneswar, Odisha, India*

[4]*Indian Agricultural Research Institute, Regional Station, Pusa, Samastipur, Bihar, India*

ABSTRACT

Hybrid vigor is the most important event in crop improvement, making F_1 typically superior to either of their homozygous parents. The development of the best hybrid requires the evaluation of a large number of crosses derived from inbred lines. Breeders often confront considerable difficulties during evaluation due to the unpredictable nature of heterosis and characterization of heterozygotes. It leads to the loss of favorable alleles combinations in the next generation due to the segregation of genes. Apomixis is a way to preserve heterozygosity but has not been fully exploited in food crops due to mechanisms, i.e., apo-meiosis, parthenogenesis, and endosperm formation, which need to be operationalized and synchronized. However, reverse breeding seems to be a powerful and versatile technique to fix the heterozygous genotypes and develop complementary homozygous parental lines

Smart Breeding: Molecular Interventions and Advancements for Crop Improvement. Kailash Chandra, Subhash Chand, Ravi Prakash Saini, and Rohit Sharma (Eds.)

to reconstitute even better plant types than even original hybrid. Further, it also helps to develop homozygous chromosome substitution lines (CSL) and the construction of heterozygous inbred families (HIFs). Reverse breeding comprises selection of desirable heterozygous genotype, suppression of crossing over (CO), production of doubled haploids followed by selection, and crossing of complimentary lines through marker-assisted selection (MAS). Suppression of CO can be done using RNAi technology for the genes responsible for meiotic recombination or applying chemical compounds that inhibit recombination. Despite its potentiality, this technique is limited to diploid crops with basic chromosome number 12 or less and a high response to double haploid (DH) production. Besides, it necessitates standardization of protocol for suppression of meiotic CO.

7.1 INTRODUCTION

Plant breeding has continuously evolved to introduce new varieties with better quality and yield after extensive field trials to feed the expanding population (Chand et al., 2020). Heterosis is the most crucial phenomenon in plant breeding activities making F_1 plants typically superior in yield contributing characteristics compared to their respective homozygous parents (Shull, 1908). Breeders evaluate a large number of crosses to get superior hybrids; however, the phenomenon is poorly understood due to the involvement of multiple underlying mechanisms (Springer, 2007; Stupar et al., 2008; Fernandez-Silva et al., 2009; Wei et al., 2009). Unpredictability nature of heterosis may drive difficulty to a breeder for selecting the best crosses for the development of hybrid (Dirks et al., 2009). Suppose a cross is identified as best due to the presence of a desirable combination of an unknown allele in the parent. In that case, this hitted heterozygous F_1 is a bonus for a breeder to improve that crop. However, the superiority of the F_1 hybrid is lost in subsequent generations due to segregation and recombination, and the farmers need to procure hybrid seeds every year. Besides, the fantastic favorable allele combinations cannot be achieved again in the resulting inbreds. Hence, breeders need to maintain the parental lines of the identified hybrids by preventing outcrossing and go for hybrid seed production each year to cater for the need of farmers. In this context, developing a new plant breeding tools for easy preservation of heterozygous may be considered most rewarding. Apomixis has repeatedly been proposed to preserve heterozygosity (Perotti et al., 2004) and exploit heterosis (Bicknell & Catanach, 2015). It maintains the genetic make-up of the mother plant by producing identical offspring (Gaafar et al., 2017). The success of apomixis depends upon three mechanisms to be operational and synchronized, i.e., apomeiosis, parthenogenesis, and endosperm formation (Koltnow & Grossniklaus, 2003). Transfer of apomictic genes to the crop is possible by genetic engineering. But, it is not yet widely used due to the complexity of most apomictic crops' traits and polyploidy nature. Vegetative propagation is also a common way to produce true to the type plants and maintain a genotypic constitution (Murashige, 1974; Priestly, 1929). However, it is only applicable for some horticultural crops (Jankowicz-Cieslak et al., 2012).

Further, vegetatively propagated plants are facultative, where the competence of sexual reproduction for hybrid production is available and follows up fixation of heterozygous status is possible. But this facility is limited to some crops, so retaining hybrid vigor is challenging (Van Harten, 1998). However, the loss of hybrid vigor due to segregation can be restricted by inhibition of meiotic recombination. Knocking down of cross over event at meiotic prophase-I is now possible by RNAi and dominant-negative mutation. The non-recombinant gametes can be regenerated and double haploid (DH) produced by doubling the chromosome number. This chapter focuses on reverse breeding as a new breeding tool that works reversely compared to traditional breeding by reconstructing the heterozygous genotype from complementary homozygous lines (Dirks et al., 2003). It is a fast-track breeding technique that provides more flexibility to combine desirable parents for the development of heterozygous genotypes.

7.2 CROSS OVER MANAGEMENT: BREEDERS DESIRE

Meiosis-1 in F_1 brings about the random distribution of genes between homologous chromosomes due to crossing over (CO) and also results in the random shuffling of parental chromosomes due to chromosome separation. Combining the right chromosomes with desired alleles resulting from meiotic recombination is always the plant breeder's desire (Wijnker, 2013). Two factors largely determined the extent of meiotic recombination to occur, i.e., the number of chromosomes and the number and position of crossovers on homologous chromosomes pair. Manipulation of the number of chromosomes is not possible except for the addition of artificial chromosomes, but the breeder can manipulate the number and position of cross over on chromosome (Carlson et al., 2007). Cross-over assurance ensure that at least one crossover per chromosome occurs at meiosis-I for proper segregation (Shinohara et al., 2008). Crossover preferentially occurs at recombination hotspots (Drouaud et al., 2006; Mezard et al., 2006). The total number and position of cross-over are generally limited by interference (Sturtevant et al., 1915). Several factors, such as genetic constitution, morphological, and developmental differences, sexes, and external factors, including environmental and chemical treatment, influence cross-over frequency (Lenormand et al., 2005). Cross-over is desirable when breeders aim to isolate useful recombinants in the segregating generation. However, it is frustrating when breeders attempt to integrate valuable traits through introgressive hybridization. So, it is imperative to identify genes that either increase or suppress CO.

7.3 IDENTIFICATION OF MEIOTIC GENES

Recent advancement of molecular biology has led to understanding many aspects of meiotic recombination in fungi, animals, and plants (Osman et al., 2011; Keeney et al., 2014; Hunter et al., 2015; Mercier et al., 2015; Gray et al., 2016; Lake et al., 2016; Yu et al., 2016). Across eukaryotes, the molecular events of meiotic recombination are highly conserved (Roeder, 1997; Zickler & Kleckner, 1999). *Arabidopsis* is an excellent crop for the study of meiosis, which enabled forward and reverses genetics approaches to identify and characterize many meiotic genes (Rodder, 1997; Klenkner, 1999). Recently, the study of meiotic recombination has been extended to other crop species (Mercier et al., 2015), and many meiotic genes have been identified in different crops (Table 7.1). Individual gene functions involved in meiotic recombination can be identified using various techniques. It is mainly assessed by functional analysis, which follows mutating the gene of interest with T-DNA and isolation of mutant phenotype by screening (Caryl et al., 2003). Direct gene disruption by homologous recombination (HR) may not be possible in crops as in yeast and mouse. However, antisense RNA or RNA-interference (RNAi) can be successfully applied to down regulate a gene of interest (Higgins et al., 2005; Siaud et al., 2004). Gain-of-function mutants can be produced by inserting a transcriptional enhancer known as activation tagging. Applicability of proteomics for identifying meiotic genes showing little or no conservation at the primary sequence level is an alternative approach. Expression analysis using microarray technology is a potent one for identification of genes that are upregulated during meiosis or have meiosis-specific splice variants.

Out of several genes involved in meiotic CO, disrupted meiotic cDNA 1 (*DMC1*) is an essential meiotic gene for crossover formation and opens up the way to control meiotic recombination (Higgins et al., 2004; Siaud et al., 2004). The absence of CO in microspore mother cells (PMCs) increases the chances of getting DH lines (Link & Melchinger, 1995). Later, targeted suppression of CO for generation of the homozygous line from heterozygote was first proposed by Dirks (2006) and then experimentally validated by Wijnker (2014). In this context, the discovery of high throughput genotyping and the development of dense markers has led to efficient monitoring of recombination events, at least up to fix the whole chromosome to reconstruct parental lines (Gupta, 2008).

TABLE 7.1 List of Meiotic Genes with Their Proposed Function Identified in Crops

Gene	Identified in Crop	Function	References
SPO 11-1	*Arabidopsis thaliana*, *Oryza sativa*	Double-strand break formation	Grelon et al. (2001); Hartung et al. (2007)
SPO 11-2	*Arabidopsis thaliana*	Double-strand break formation	Stacey et al. (2006); Hartung et al. (2007)
MTOPVIB	*Arabidopsis thaliana*, *Oryza sativa*	Double-strand break formation	Vrielynck et al. (2016); Xue et al. (2016); Fu et al. (2016); Tang et al. (2017)
PRD1	*Arabidopsis thaliana*	Double-strand break formation	De Muyt et al. (2007)
PRD2	*Arabidopsis thaliana*	Double-strand break formation	De Muyt et al. (2009)
PRD3	*Arabidopsis thaliana*, *Oryza sativa*	Double-strand break formation	De Muyt et al. (2009)
DFO	*Arabidopsis thaliana*	Double-strand break formation	Zhang et al. (2012)
PCH2	*Oryza sativa*, *Arabidopsis thaliana*	Double-strand break formation	Jolivet et al. (2005)
PHS1	*Arabidopsis thaliana*, *Zea mays*	Double-strand break formation	Lu et al. (2012); Lao et al. (2013)
MRE11	*Arabidopsis thaliana*, *Oryza sativa*	Double-strand break end resection	Puizina et al. (2004)
RAD50	*Arabidopsis thaliana*	Double-strand break end resection	Bleuyard et al. (2004)

TABLE 7.1 *(Continued)*

Gene	Identified in Crop	Function	References
NBS1	*Arabidopsis thaliana, Oryza sativa*	Double-strand break end resection	Sakaguchi et al. (2009)
RAD51	*Arabidopsis thaliana, Oryza sativa, Zea mays*	Recombination intermediate formation	Li et al. (2004)
DMC1	*Arabidopsis thaliana, Oryza sativa*	Recombination intermediate formation	Couteau et al. (1999)
POL2A	*Arabidopsis thaliana*	DNA synthesis in meiotic recombination	Huang et al. (2015)
RFC1	*Arabidopsis thaliana*	DNA synthesis in meiotic recombination	Wang et al. (2012)
MUS81	*Arabidopsis thaliana, Oryza sativa, Zea mays*	Minor promotion of Cos	Higgins et al. (2004)
MSH4	*Arabidopsis thaliana, Oryza sativa, Zea mays*	Stabilization of dHjs; directing some to Cos	Higgins et al. (2004)
MSH5	*Arabidopsis thaliana, Oryza sativa, Zea mays*	Stabilization of dHjs; directing some to Cos	Higgins et al. (2004)
HEI10	*Arabidopsis thaliana, Oryza sativa, Zea mays*	Cross over formation	Wang et al. (2012)
MLH1	*Arabidopsis thaliana, Zea mays*	Cross over formation	Jackson et al. (2006)
MSH2	*Arabidopsis thaliana*	Cross over suppression	Emmanuel et al. (2006)

7.4 REVERSE BREEDING: THE WAYS TO PROCEED

Reverse breeding is considered a novel plant breeding method with the concept to reconstitute elite hybrids from genotypes that are not exactly similar to original parents but can complement each other to reproduce almost the same original hybrid (Dirks et al., 2006, 2009). It was first experimented with in the model plant, *Arabidopsis thaliana* having a small genome size (Bennett et al., 2003). By suppressing meiotic recombination, it allows the production of parental lines from selected heterozygous plants. The procedure includes certain steps which are discussed in subsections.

7.4.1 SELECTION OF DESIRABLE HETEROZYGOUS GENOTYPE

The starting material for reverse breeding is the selected desirable heterozygous genotype. Superior F_1 hybrids are selected when the objective is to recreate hybrid. While for the characterization of unknown heterozygotes, the genotypes from highly variable populations are selected as the starting materials followed by evaluation in various environmental conditions to identify the best one.

7.4.2 SUPPRESSION OF CROSSING OVER (CO)

In meiotic prophase-I, synaptonemal complex formation due to extensive and stable interaction between homologous chromosomes is essential for crossover (Moss, 1956). CO event occurs between bivalent chromosomes, each with duplicated sister chromatids. For proper segregation of homologous chromosome pairs, at least one CO should appear during the pachytene phase. The absence of cross over leads to an achiasmatic condition in which chromosome remains univalent and chances of production of unbalanced gametes increases due to segregation of homologous chromosome to the same pole due to non-disjunction (Couteau et al., 1999; Hartung et al., 2007). If each univalent has an equal chance to move the opposite pole, then the chances of getting a spore containing the usual chromosome complement is ½ x (x = haploid chromosome number) which implies the probability of getting balanced gametes are reduces with an increase in chromosome number. The production of non-recombinant gametes from the desired heterozygote by suppressing CO is an essential step in reverse breeding which can be achieved by dominantly suppressing one of the several meiotic genes keeping the chromosome structure intact. The function of such a meiotic gene can be completely knocked out by either mutating it or through the use of RNAi technology. Dominant-negative mutation of the target genes used for this purpose as recessive mutations can reintroduce the same mutation into the offspring, which is not desirable. An example in male mice, dominant-negative alleles of *DMC1* gene resulted in a loss of meiotic recombination (Bannister et al., 2007). RNAi is another such approach based on two steps, each involving the ribonuclease (RNAse) like protein. First, the dsRNA molecule is processed into the short-interfering RNAs (siRNA) by the RNase III enzymes Dicer. Then, the double-strand of siRNAs is unwound, leading to the guide and passenger strand. Subsequently, the passenger strand is degraded, and the guide strand is loaded into the effector complex, i.e., RNA-induced silencing complex (RISC). The RISC loaded with the guide strand then scans the cytoplasm to search the mRNA of the target gene and then binds with the mRNA in a homology-dependent manner. After binding to the target mRNA, the mRNA is cleaved by the action of the Argonaut enzyme, which is a part of the RISC. In this way, the transcripts of the target genes are degraded, leading to the knock-down of the target gene.

The use of RNAi technology in many crops as a single cassette to target meiotic gene is an efficient one as the dominant nature of RNAi cause recovery of progeny devoid of RNAi cassette that would otherwise cause sterility. Several researchers reported the knock-down of the targeted gene most efficiently by RNAi (Higgins et al., 2004; Siaud et al., 2004). The first use of RNAi is the knock-down of the RecA homolog *DMC1* (a meiosis-specific recombinase), which is essential for crossover formation (Wijnker et al., 2012). *DMC1* is one of the crucial genes involved in CO process by interacting with *Rad51*, *Tid1*, and other genes. Inhibition of *DMC1* gene can suppress crossovers during meiosis leading to the production of achiasmatic gametes (Bishop et al., 1992). *Brassica carinata DMC1* gene was found to be effective in silencing *Arabidopsis thaliana DMC1* due to 91.1% similarity of former with later. RNAi technique can also effectively suppress other meiotic cross over gene(s), e.g., *ASY1/SDS* responsible for homologous synapsis; and *SPO 11-1* meant for double-strand break formation during CO.

Suppression of crossover through RNAi requires a stable transgene in the genome of the heterozygote to generate chromosome substitution lines (CSLs) (Wijnker et al., 2012). So, to produce achiasmatic hybrid, one of the parental lines must be stably transformed with the transgene. When heterozygotes are selected from outcrossing populations for reverse breeding, this practice is not feasible because no parental lines are available (Baltanas et al., 2020). This problem can be overcome by a transiently silencing technique like virus-induced gene silencing (VIGS). Such a knock-down method for the meiotic gene is often practiced in plant species difficult to stably transform (Dung et al., 2007; Lieu et al., 2016; Huang et al., 2017). It involves a fragment of the host gene that is targeted to be silenced is engineered into the viral genome, allowing host infection by virus and silencing the target gene as part of the plant's defense mechanism against virus infection (Senthil et al., 2011). VIGS has been routinely used to modify the expression of genes involved in several processes involving meiosis in several plant species (Senthil-Kumar & Mysore, 2011b; Bennypaul et al., 2012; Bhullar et al., 2014).

Graft transmission may also be used for knock-down of target meiotic genes. Shoots of the plants in which genes are to be silenced would be grafted on transgenic rootstock (Shaharuddin et al., 2006). Besides, exogenous applications of chemicals that inhibit recombination during meiosis broaden the scope of reverse breeding. One such chemical, "Mirin," is reported to inhibit Mre11-Rad50-Nbs1 complex, which has a role in double-strand break end processing in yeast (Dupre et al., 2008). Other chemicals such as DIDS (4,4′-Diisothiocyano-2,2′-tilbenedisulfonic acid), PSF (phosphorus fluoride sulfide), and Halena-quinone are also considered potent enough to prevent CO by inhibiting RAD51 gene, which has a role in strand invasion. The use of cross over suppressor chemicals is advantageous as the outcomes of reverse breeding products being free of transgenes and do not need standard GMO regulatory clearance. The presence of dominant knock-down construct in a hemizygous state, causes half of the gametes will be transgene-free and DH lines produced from it would be transgene-free. A full array of complementary DH lines with no transgene can be generated by using many transgenic lines, each with knock-down constructs on different chromosomes (Wijnker & de Jong, 2008). Irrespective of the ways to completely prevent cross over, a single residual cross over may still occur in any chromosome pair, increasing two-fold chances of obtaining a balanced gamete. In this context, the single two strands cross over generates two non-recombinant (parental) chromatids, resulting in its potential to be used in reverse breeding (Dirk et al., 2006).

7.4.3 PRODUCTION OF DOUBLED HAPLOID LINES

Haploids from plants engineered with achiasmatic meiosis can be obtained either through gynogenesis or androgenesis. However, the production of haploids through androgenesis is safe to avoid a greater chance of chimerism and mixoploidy. Protocols for getting haploid plants vary from crop to crop (Jain et al., 1996). Further, haploid production becomes easy if haploid inducer lines are available. Centromere mediated genome elimination techniques require engineered centromere to produce haploid lines (Ravi & Chan, 2010). Haploid inducer lines have been created in *Arabidopsis thaliana*, *Brassica juncea*, and *Zea mays* using engineered centromeric histone 3 (CENH3) variants. It involves crossing CENH3 mutant with one parental accession, which need not necessarily be achiasmatic. The whole genome of the mutant would undergo self-elimination at a high frequency after crossing, and hence, the gametes produced in F_1 would contain all the chromosomes from wild-type parents only. These F_1 plants would generate DHs by selfing (Wijnen & Keurentjes, 2014). However, such an amazing technique for haploid induction has not been fully exploited in the crop improvement program (Figure 7.1).

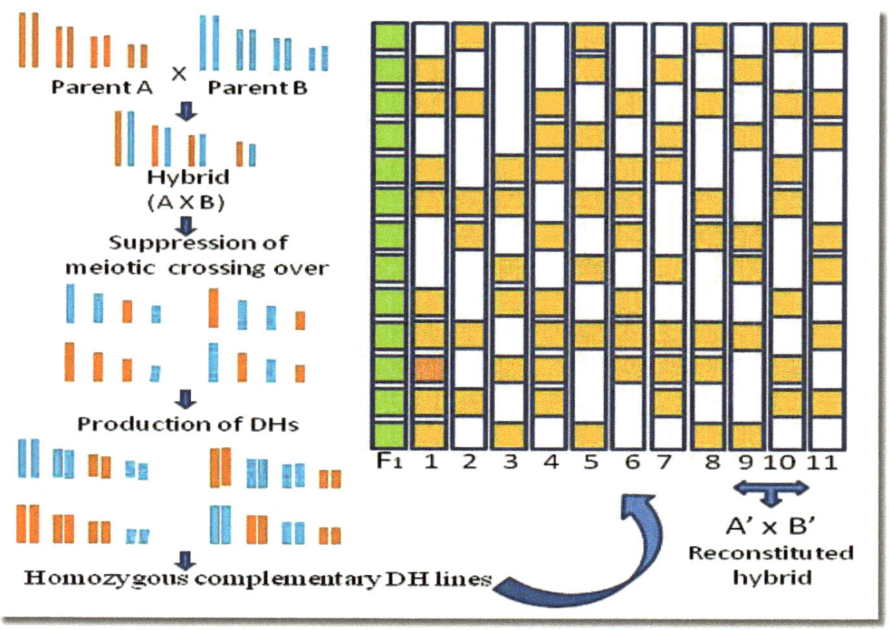

FIGURE 7.1 Reverse breeding for reconstitution of hybrids using complementary doubled haploid (DH) lines.

The efficiency of DH production from haploid spore is species and genotype-dependent (Forster et al., 2007). DHs can be produced from recombinant gamete, and non-recombinant gametes resulted from achiasmatic meiosis. The major limitation of DH production technology is limited only to crop species where well-established protocols are available and routinely used in plant breeding (Maluszynski et al., 2003; Forster et al., 2007). Applications of reverse breeding technique is restricted in crops; soybean, cotton, lettuce, and tomato, where DH plants are rarely formed.

7.4.4 SELECTION OF COMPLEMENTARY LINES THROUGH MARKER-ASSISTED SELECTION

The use of polymorphic markers is indispensable to display genetic variation among DH plants. Marker-assisted reverse breeding (MARB) efficiently produces homozygous complementary lines within two to three years (De Vienne, 2003) than reverse breeding alone. It can be applied in the crop where enough co-dominant molecular markers such as SSR and SNP are available. It helps to identify DH lines at early stages and select complementary parents for the development of hybrid. One polymorphic marker per chromosome is sufficient to genotype every DH produced from achiasmatic meiosis since each chromosome's behavior serves as a single linkage block. However, two molecular markers are required if any residual CO occurs. As chip technology becomes cheap, it would be expected that polymorphism screening following marker-assisted selection could be done with one all-purpose high-density chip.

The DH technique usually produces homozygous parental lines within one to one and a half years. With RMRB (RNAi mediated reverse breeding), an additional six months to one year is required to produce homozygous lines because of the extra transformation step. While, Marker Assisted Reverse Breeding will take a similar time as RMRB due to population development and chip screening (Xin et al., 2015). All above reverse breeding methods have higher efficiency than conventional breeding, usually requiring 8–10 years to produce homozygous parental lines.

7.4.5 CROSSING OF COMPLEMENTARY LINES: ATTEMPT TO RECONSTRUCT THE ORIGINAL HETEROZYGOTES

DH lines are the critical components for the reconstruction of hybrids. The chances of finding DH from a heterozygous plant in reverse breeding is 2x, where 'x' is the basic chromosome number. Further, the probability that two DH forms a pair of complementary parents is $(1/2)^x$. Therefore, the haploid chromosome number is the most important determining factor for the generation of DH lines to find complementary parental pairs. Chromosome number also determines the probability of regular disjunction in a species, which is (n):P (balanced segregation) = 1/2n, where n is the chromosome number of a crop. Therefore, the chances of getting viable gamete formation in crops decrease with an increase in chromosome number. For reconstruction of starting hybrids, the number of DH lines required at a different level of probability has been demonstrated in various crops (Table 7.2). Indeed, the number of DH lines required to reconstruct heterozygotes (similar to the original hybrid) is low in *Arabidopsis*, spinach/corn, and rye as just 18,25 and 35 DH lines, respectively at 99% probability are expected to contain two common genotypes with an exactly complementary match of genomic regions.

TABLE 7.2 The Number of Non-Recombinant DH Lines Required for Reconstructing the Original Stating Hybrid at Different Probability Levels (Dirk et al., 2006)

Haploid Chromosome Number	Number of DH Lines Required at a Different Level of Probability				Crop
–	**0.90**	**0.95**	**0.99**	**1.00**	
5	13	14	18	47	*Arabidopsis*
6	18	20	25	67	Spinach, corn
7	25	28	35	94	Cucumber, barley, rye
8	35	40	49	133	Onion
9	49	56	69	188	Carrot, sugar beet, brassica, lettuce
10	69	79	98	266	Maize, sorghum, asparagus, cacao
11	98	111	138	377	Banana, watermelon, celery, fennel, common bean

7.5 APPLICATION OF REVERSE BREEDING

This new concept has enormous potential and is a reliable method to maintain superior hybrid in crop plants independent of access to the parental lines. Few significant applications of reverse breeding are worth mentioning.

7.5.1 *RECONSTRUCTION OF HETEROZYGOUS LINE*

Plant breeding has gifted many crop varieties of the heterozygous genotypic constitution; however, the parental information eroded by time due to improper handling of breeding materials. Further, in some crops, several germplasm lines are insufficient to produce potential hybrids, and the available heterozygous populations are uncharacterized. In this context, even if some excellent heterozygotes (high vigor and yield potential) are sorted out from above heterozygous populations through rigorous selection, it is entirely impossible to reconstitute these for seed multiplication and general cultivation. Thus, there is a dearth need for parental information of uncharacterized heterozygotes to reconstitute the elite heterozygous line(s). So reverse breeding can suitably be used for this purpose.

7.5.2 *DEVELOPMENT OF CHROMOSOME SUBSTITUTION LINE (CSL)*

Most of the essential characters in plants are governed by polygenes that are mainly distributed on different chromosomes. Substitution lines can be used to detect QTL (Nadeau et al., 2000; Singer et al., 2004; Spiezio et al., 2012; Ziolkowski et al., 2017). In conventional breeding methods, CSLs are generated through the crossing of F_1 hybrid with one of the parents, followed by several rounds of backcrossing and selection of progeny carrying specific non-recombinant chromosomes. It takes a longer time to develop CSLs, particularly in crops with a longer life cycle (Nadeau et al., 2000; Koumproglou et al., 2002). In these aspects, reverse breeding generates CSLs more efficiently by partial or complete suppression of CO in the hybrids (Dirks et al., 2009; Wijnker et al., 2012). Haploid plants with non-recombinant chromosomes produced from such hybrids can be directly used to develop DHs (Dirks et al., 2009; Wijnker et al., 2012, 2018), and subsequently, CSLs are recovered in the progeny of DHs. Breeders can bring targeted characteristics on a single chromosome by selfing plants that carry a substituted chromosome. However, finding a CSL, especially in a crop with a high chromosome number is very difficult. In such a case, repeated backcrossing a DH line carrying the desired substitution (in addition to undesired substitution) with one of the original parents might be helpful.

7.5.3 BACKCROSSING IN CMS BACKGROUND

The presence of male sterility is a great challenge to the breeder for reverse breeding. In many vegetable crops such as cabbage and carrots, Cytoplasmic male sterility is used to produce hybrid seeds (Chase, 2007). To circumvent this problem, use of gynogenesis rather than androgenesis is the better option to recover DH plants. Due to occurrence of male sterility in several crops; Brassica, maize, sugar beet, cucumber, melon, onion, sunflower, and barley, gynogenic DHs production is a method of choice (Keller & Korzun, 1996). It is perfectly compatible with reverse breeding in that the chromosome from the maintainer line (B-line) can be recovered directly in the cytoplasm of the sterile line (A-line) in one step. Crossing of a male sterile line (A) with the maintainer line (B) that carries one copy of the restorer gene is followed, where the efficiency of gynogenesis is too low (Dirks et al., 2006). The AB combination will be fertile, and reverse breeding can be performed. For this restorer gene and a gene for CO suppressor is designed in the same vector.

7.6 MAJOR BOTTLENECKS OF REVERSE BREEDING

The novel technique of reverse breeding to reconstitute the heterozygous status of the original hybrid demonstrates proof-of-concept. Still, it often faces two major hurdles-requirement of a stable transgene for complete knock-down of CO and recovery of shallow frequency (3% in *Arabidopsis thaliana* and much lower in other crops) of balanced gametes (due to CO suppression) that may compromise plant fertility. Besides, reverse breeding is practically not feasible for plant species with more chromosome pairs (>12) (Dirks, 2003; Dun, 2009). Though the RNAi technique to knock down meiotic cross over gene(s) in F_1 has been successful, the follow-up high throughput DH plant regeneration system of F_1 gametes is a limiting factor in many plant species. The most frequently used technology is CENH3 modified haploid inducer (Ravi & Chan, 2010) to produce non-recombinant DH lines. But such genetic stock (haploid inducer) is not readily available for reverse breeding except in maize and a few other crops.

7.7 NEAR REVERSE BREEDING: CONQUER THE BOTTLENECKS OF REVERSE BREEDING

Two perfectly complementing genotypes (i.e., two CSLs) from the progeny of reverse breeding off-spring can be used as parental pairs to recreate the original heterozygous genotype (hereafter referred to as a full hybrid) (Figure 7.2). The occurrence of CO in any one of the parents leads to the production of near full hybrids, which are not exactly similar to the full hybrid. The performance of the near full hybrid may not be affected as compared to the full hybrid due to a decrease in heterozygosity unless if the specific regions that become homozygous affect hybrid performance (Huang et al., 2012). Therefore, low-recombinant offspring can be expected to use as parental lines in hybrid breeding despite not knowing beforehand. Experimentally by phenotyping of several near-full hybrids, it is possible to determine which homozygous regions are affecting hybrid performance. Choosing reduction of cross over than completely suppression is an alternate possibility to maintain fertility in reverse breeding progeny (Dirks et al., 2009). Reduction of cross over formation still leads to spores recovery, which is helpful for the generation of CSLs through backcross (BC) in reverse breeding.

Generation of near full hybrids through incomplete suppression of CO in parental lines open up the way of its feasibility in species with higher chromosome number. So, near reverse breeding is a modification of reverse breeding that utilizes SDR gametes to reconstruct heterozygote plants. The selection of complementing DH plants that are generated from SDR gametes is made through SNP markers. SDR gametes (2n) are unreduced gametes formed due to skipping anaphase-II of meiotic CO and retain homozygosity if no CO occurred at prophase-I (Ashman et al., 2013). It has been reported that SDR generally leads to a 60% loss in heterozygosity (Carputo, 2003). In contrast, the occurrence of two or more CO can lead to retention of heterozygosity and thus limit its use for the generation of DH lines (Dirks, 2006;

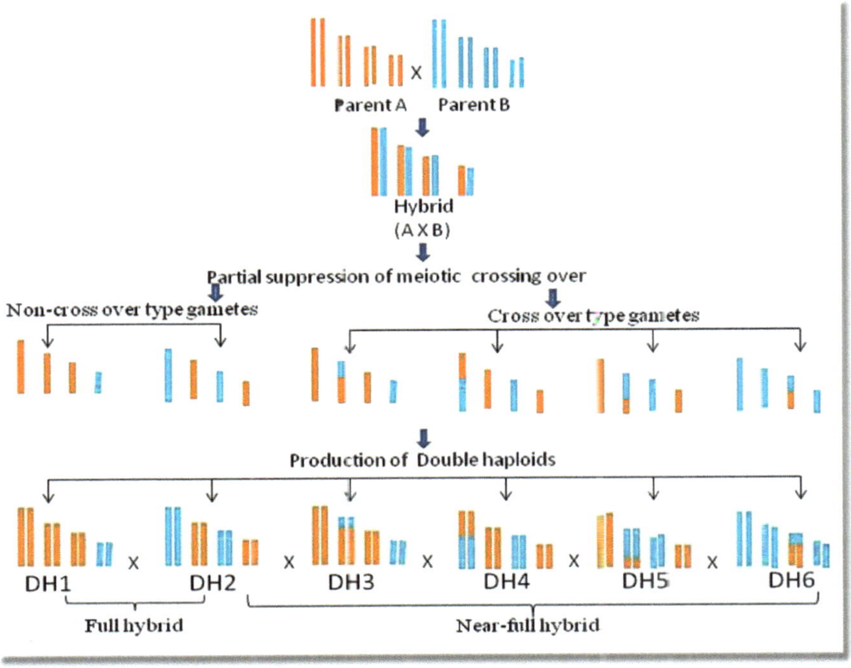

FIGURE 7.2 Reconstitution of near full hybrids through near reverse breeding.

Dun, 2008). In order to search for the most complementing lines, DH lines are tested by marker-assisted selection. It has its own advantages over GMO technology and reverse breeding as it suffers less from chromosome number effects, speeds up backcrossing, and can produce small introgressions, called heterogeneous inbred families for QTL and plant trait studies (Dun, 2008). Near-reverse-breeding requires four steps: (i) induction of natural SDR in plant either due to environmental or genetic ways; (ii) isolation and separation of unreduced spore; (iii) haploid plants are either regenerated through *in-vitro* or through CENH3 haploid inducer followed by the production of SDR plants; and (iv) use of sets of co-dominant PCR markers for genotyping SDR plants. SDR gametes can be produced by suppressing CO through ZMM protein pathway through the action of the heterodimer formed by MUTs HOMOLOG 4 and 5 (MSH4 and MSH5) (Higgins et al., 2004; Mercier et al., 2015). The ZMM protein is required during prophase-I for stabilization of recombination intermediate. The MSH4 and MSH5 mutants show 1,2, and 1.6 chiasmata per cell on an average during meiosis (Higgins et al., 2004; Wijnker et al., 2019). The SDR can also be produced with the application of nitrous oxide (N_2O) that is reported in *Arabidopsis*.

7.8 CONCLUSION

Reverse breeding allows reconstruction of the parental lines from the selected heterozygote by complete knock-down of meiotic CO using dominantly acting RNAi transgene targeting the genes involved in meiotic CO. The suppression of CO enriches the recovery of gametes with non-recombinant chromosomes. The rational use of this technique depends on the efficiency of CO suppression and the generation of DH lines. This method has two main disadvantages; it sacrifices the plant fertility due to suppression of CO, and the success rate is limited to the crops with haploid chromosome numbers less than 12. The near reverse breeding approach can suitably overcome the major limitation of reverse breeding by allowing only partial suppression of CO instead of complete suppression.

Therefore, the research needs to focus on the high efficiency of SDR gamete production and SDR plants. The proof of the concept for reverse breeding has been widely demonstrated in the model plant, *Arabidopsis*. The same may be worth trying in many other agronomically important crops, e.g., rice, maize, sorghum, pea, tomato, pepper, eggplant, cucumber, onion, broccoli, cauliflower, sugar beet, watermelon, and many more plant species with basic chromosome number 12 or less. Further, genes that are crucial for meiotic CO are widely conserved among plant species. So, its practical application can be expanded to other crops.

KEYWORDS

- **crossover suppressor**
- **disrupted meiotic cDNA 1**
- **heterosis**
- **interfering RNAs**
- **marker-assisted selection**
- **meiotic recombination**
- **reverse breeding**
- **RNA-induced silencing complex**
- **RNA-interference**

REFERENCES

Ashman, T. L., Kwok, A., & Husband, B. C., (2013). Revisiting the dioecy-polyploidy association: Alternate pathways and research opportunities. *Cytogenetic and Genome Research, 140*, 241–255.

Baltanas, V. C., Wijnen, C. L., Yang, C., Lukhovitskaya, N. C., Bastiaan De, S. C., Linus, H. L., Joost, J. B., et al., (2020). Meiotic crossover reduction by virus-induced gene silencing enables the efficient generation of chromosome substitution lines and reverse breeding in *Arabidopsis thaliana. The Plant Journal, 104*, 1437–1452.

Bannister, L. A., Pezza, R. J., Donaldson, J. R., Rooij, D. Gd., Schimenti, K. J., Camierini-Otero, R. D., & Schimenti, J. C., (2007). A dominant, recombination-defective allele of DMC1 causing male-specific sterility. *PLoS Biology, 5*, 1016–1025.

Bennett, M. D., & Leitchl, I. J., (2003). Comparisons with *Caenorhabditis* (100 Mb) and *Drosophila* (175 Mb) using flow cytometry show genome size in *Arabidopsis* to be 157 Mb and thus 25 % larger than the *Arabidopsis* genome initiative estimate of 125 Mb. *Annals of Botany, 91*(5), 547–557.

Bennypaul, H. S., Mutti, J. S., Rustgi, S., Kumar, N., Okubara, P. A., & Gill, K. S., (2012). Virus-induced gene silencing (VIGS) of genes expressed in root, leaf, and meiotic tissues of wheat. *Functional and Integrative Genomics, 12*, 143–156.

Bhullar, R., Nagarajan, R., Bennypaul, H., Sidhu, G. K., Sidhu, G., Rustgi, S., Wettstein, D., & Gill, K. S., (2014). Silencing of a metaphase I-specific gene results in a phenotype similar to that of the pairing homeologous 1 (Ph1) gene mutations. *Proceedings of National Academy of Sciences. USA, 111*(39), 14187–14192.

Bicknell, R., & Catanach, A., (2015). Apomixis: The asexual formation of seed. *Somatic Genome Manipulation*, 147–167. Springer.

Bishop, D. K., Park, D., Xu, L., & Kleckner, N., (1992). DMC1: A meiosis-specific yeast homolog of *E. coli* recA required for recombination, synaptonemal complex formation, and cell cycle progression. *Cell, 69*, 439–456.

Bleuyard, J. Y., & White, C. I., (2004). The *Arabidopsis* homologue of Xrcc3 plays an essential role in meiosis. *EMBO Journal, 23*, 439–449.

Carlson, S. R., (2007). Meiotic transmission of an *in vitro*-assembled autonomous maize minichromosome. *PLoS Genetics,* 3, e179.

Carputo, D., & Frusciante, L., (2003). The role of 2n gametes and endosperm balance number in the origin and evolution of polyploids in the tuber-bearing solanums. *Genetics, 163*(1), 287–294.

Caryl, A. P., Jones, G. H., & Franklin, F. C. H., (2003). Dissecting plant meiosis using *Arabidopsis thaliana* mutants. *Journal of Experimental Botany 54,* 25–38.

Chand, S., Chandra, K., & InduandKhatik, C. L., (2020). Varietal release, notification and denotification system in India. In: *Plant Breeding-Current and Future Views*. Intech Open.

Chase, C. D., (2007). Cytoplasmic male sterility: A window to the world of plant mitochondrial nuclear interactions. *Trends Genetics 23*, 81–90.

Couteau, F., Belzile, F., Horlow, C., Grandjean, O., Vezon, D., & Doutriaux, M. P., (1999). Random chromosome segregation without meiotic arrest in both male and female meiocytes of a dmc1 mutant of *Arabidopsis. Plant Cell, 11*, 1623–1634.

De Muyt, A., Pereira, L., Vezon, D., Chelysheva, L., & Gendrot, G., (2009). A high throughput genetic screen identifies new early meiotic recombination functions in *Arabidopsis thaliana. PLOS Genetics, 5*, e1000654.

De Muyt, A., Vezon, D., Gendrot, G., Gallois, J. L., Stevens, R., & Grelon, M., (2007). AtPRD1 is required for meiotic double strand break formation in *Arabidopsis thaliana*. *EMBO Journal, 26*(18), 4126–4137.

De Vienne, D., (2003). *Molecular Markers in Plant Genetics and Biotechnology*. Enfield, NH, USA: Science Publishers.

Dirks, R. H. G., Dun, C. M. P. V., Reinink, K., & Wit, J. P. C. D., (2006). *Reverse Breeding*, United States Patent and Trademark Office.

Dirks, R., (2003). *Reverse Breeding* (p. 47). The Netherlands.

Dirks, R., Van, D. K., & De Snoo, C. B., (2009). Reverse breeding: A novel breeding approach based on engineered meiosis. *Plant Biotechnology Journal, 7*, 837–845.

Dong, Y., Burch-Smith, T. M., Liu, Y., Mamillapalli, P., & Dinesh-Kumar, S. P., (2007). A ligation-independent cloning tobacco rattle virus vector for high throughput virus-induced gene silencing identifies roles for NbMADS4, 1 and 2 in floral development. *Plant Physiology, 145*, 1161–1170.

Drouaud, J., & Camilleri, C., (2006). Variation in crossing-over rates across chromosome 4 of *Arabidopsis thaliana* reveals the presence of meiotic recombination "hot spots". *Genome Research, 16*(1), 106–114.

Dun, C. M. P. V., (2008). *Near-Reverse Breeding*. United States Patent Application 20080098496. USA. 20080098496: 26.

Dupre, A., Boyer-Chatenet, L., Sattler, R. M., Modi, A. P., Lee, J. H., Nicolette, M. L., Kopelovich, L., et al., (2008). A forward chemical genetic screen reveals an inhibitor of the mre11-rad50-nbs1 complex. *Nature Chemical Biology, 4*, 119–125.

Emmanuel, E., Yehuda, E., Melamed-Bessudo, Avivi-Ragolsky, C. N., & Levy, A. A., (2005). The role of AtMSH2 in homologous recombination in *Arabidopsis thaliana*. *EMBO Report*, 100–105.

Fernandez-Silva, I., Moreno, E., Eduardo, I., Arus, P., Alvarez, J. M., & Monforte, A. J., (2009). On the genetic control of heterosis for fruit shape in melon (*Cucumismelo* L.). *Journal of Heredity, 100*, 229–235.

Forster, B. P., Heberle-Bors, E., Kasha, K. J., & Touraev, A., (2007). The resurgence of haploids in higher plants. *Trends Plant Science, 12*, 368–375.

Fu, M., Wang, C., Xue, F., Higgins, J., & Chen, M., (2016). The DNA topoisomerase VI-B subunit OsMTOPVIB is essential for meiotic recombination initiation in rice. *Molecular Plant, 9*(11), 1539–1541.

Gaafar, R. M., Shanshoury, A. R. E., Hisseiwy, A. A. E., AbdAlhak, M. A., Omar, A. F., Abd, E. W. M. M., & Nofal, R. S., (2017). Induction of apomixis and fixation of heterosis in Egyptian rice hybrid-1 line using colchicine mutagenesis. *Annals of Agricultural Sciences, 62*(1), 51–60.

Gray, S., & Cohen, P. E., (2016). Control of meiotic crossovers: From double-strand break formation to designation. *Annual Review Genetics, 50*(1), 175–210.

Grelon, M., Vezon, D., Gendrot, G., & Pelletier, G., (2001). AtSPO11-1 is necessary for efficient meiotic recombination in plants. *EMBO Journal, 20*, 589–600.

Gupta, P. K., (2008). Array-based high-throughput DNA markers for crop improvement. *Heredity, 101*, 5–18.

Hartung, F., Wurz-Wildersinn, R., Fuchs, J., Schubert, I., Suer, S., & Puchta, H., (2007). The catalytically active tyrosine residues of both SPO11-1 and SPO11-2 are required for meiotic double strand break induction in *Arabidopsis*. *Plant Cell, 19*, 3090–3099.

Higgins, J. D., Armstrong, S. J., Franklin, F. C. H., & Jones, G. H., (2004). The *Arabidopsis* MutS homolog AtMSH4 functions at an early step in recombination: Evidence for two classes of recombination in *Arabidopsis*. *Genes Development 18*, 2557–2570.

Higgins, J. D., Sanchez-Moran, E., Armstrong, S. J., Jones, G. H., & Franklin, F. C. H., (2005). The *Arabidopsis* synaptonemal complex protein ZYP1 is required for chromosome synapsis and normal fidelity of crossing over. *Genes Development, 19*, 2488–2500.

Huang, C., Qian, Y., Li, Z., & Zhou, X., (2012). Virus-induced gene silencing and its application in plant functional genomics. *Science China Life Science 21*, 99–108.

Huanga, J., Cheng, Z., Wang, C., Hong, Y., Su, H., Wang, J., Copenhaver, G. P., et al., (2015). Formation of interference-sensitive meiotic cross-overs requires sufficient DNA leading-strand elongation. *PNAS, 112*(40), 12534–12539.

Hunter, N., (2015). Meiotic recombination: The essence of heredity. *Cold Spring Harb Perspect. Biology, 7*, a016618.

Jackson, N., Sanchez-Moran, E., Buckling, E., Armstrong, S. J., Jones, G. H., & Franklin, F C. H., (2006). Reduced meiotic crossovers and delayed prophase I progression inAtMLH3-deficient *Arabidopsis*. *EMBO Journal,* 1315–1323.

Jain, S., Sopory, S. K., & Veilleux, R. E., (1996). *In vitro* haploid production in higher plants. *Current Plant Science and Biotechnology in Agriculture.* Dordrecht, Boston, London: Kluwer Academic Publishers.

Jankowicz-Cieslak, J., Huynh, O. A., Brozynska, M., Nakitandwe, J., & Bradley, J., (2012). Induction, rapid fixation and retention of mutations in vegetatively propagated banana. *Plant Biotechnology Journal, 10*(9), 1056–1066.

Jolivet, S., Vezon, D., Froger, N., & Mercier, R., (2006). Non conservation of the meiotic function of the Ski8/Rec103 homolog in *Arabidopsis*. *Genes Cells, 11*(6), 15–22.

Keeney, S., Lange, J., & Mohibullah, N., (2014). Self-organization of meiotic recombination initiation: General principles and molecular pathways. *Annual Review Genetics, 48*, 187–214.

Keller, E. R. J., & Korzun, L., (1996). Ovary and ovule culture for haploid production. *In Vitro Haploid Production in Higher Plants* (pp. 217–235). Dordrecht: Kluwer Academic Publisher.

Koltunow, A. M., & Grossniklaus, U., (2003). Apomixis: A developmental perspective. *Annual Review of Plant Biology, 54*, 547–574.

Koumproglou, R., Wilkes, T. M., Townson, P., Wang, X. Y., Beynon, J., Pooni, H. S., Newbury, H. J., & Kearsey, M. J., (2012). STAIRS: A new genetic resource for functional genomic studies of *Arabidopsis*. *Plant Journal, 31*(3), 355–364.

Lake, C. M., & Hawley, R. S., (2016). Becoming a crossover-competent DSB. *Seminar in Cell Development Biology, 54*, 117–125.

Lao, J. P., Cloud, V., Huang, C. C., Grubb, J., & Thacker, D., (2013). Meiotic crossover control by concerted action of Rad51-Dmc1 in homolog template bias and robust homeostatic regulation. *PLOS Genetics, 9*, e1003978.

Lenormand, T., & Dutheil, J., (2005). Recombination difference between sexes: A role for haploid selection. *PLoS Biology, 3*, 63.

Li, W., Chen, C., Markmann-Mulisch, U., Timofejeva, L., & Schmelzer, E., (2004). The *Arabidopsis* AtRAD51 gene is dispensable for vegetative development but required for meiosis. *PNAS*, 10596–10601.

Link, W., & Melchinger, A. E., (1995). An approach to the genetic improvement of clonal cultivars via backcrossing. *Crop Science, 35*, 931.

Liu, N., Xie, K., Jia, Q., Zhao, J., & Chen, T., (2016). Foxtail mosaic virus induced gene silencing in monocot plants. *Plant Physiology, 171*, 1801–1807.

Lu, P., Han, X., Qi, J., Yang, J., & Wijeratne, A. J., (2012). Analysis of *Arabidopsis* genome-wide variations before and after meiosis and meiotic recombination by resequencing Landsberg erecta and all four products of a single meiosis. *Genome Research*, 508–518.

Maluszynski, M., Kasha, K. J., Forster, B. P., & Szarejko, I., (2003). *Doubled Haploid Production in Crop Plants.* A manual. Kluwer Academic Publishers, Dordrecht/Boston/London.

Mercier, R., Mezard, C., Jenczewski, E., Macaisne, N., & Grelon, M., (2015). The molecular biology of meiosis in plants. *Annual Review Plant Biology, 66*, 297–327.

Mezard, C., (2007). The road to crossovers: Plants have their say. *Trends Genetics, 23*, 91–99.

Moses, M. J., (1956). Chromosomal structures in crayfish spermatocytes. *Journal of Biophysics, Biochemistry, Cytology, 2*, 215–218.

Murashig, T., (1974). Plant propagation through tissue-cultures. *Annual Review Plant Physiology, 25*, 135–166.

Nadeau, J. H., Singer, J. B., Matin, A., & Lander, E. S., (2000). Analyzing complex genetic traits with chromosome substitution strains. *Nature Genetics, 24*, 221–225.

Osman, K., Higgins, J. D., Sanchez-Moran, E., Armstrong, S. J., & Franklin, F. C., (2011). Pathways to meiotic recombination in *Arabidopsis thaliana*. *New Phytology*, 523–544.

Perotti, E., Grimanelli, D., John, P., Hoisington, D., & Leblanc, O., (2004). Why is transferring apomixis to crops still a dream? In new directions for a diverse planet. *Proceedings for the 4th International Crop Science Congress.*

Priestley, J. H., (1929). *Vegetative Propagation from the Standpoint of Plant Anatomy* (pp. 1–89). In Technical Bulletin No. 151 (Agriculture U.S.D.O. ed.), Washington, DC: United States Department of Agriculture.

Puizina, J., Siroky, J., Mokros, P., Schweizer, D., & Riha, K., (2004). MRE11 deficiency in *Arabidopsis* is associated with chromosomal instability in somatic cells and SPO11-dependent genome fragmentation during meiosis. *Plant Cell, 16*(8), 1968–1978.

Ravi, M., & Chan, S. W. L., (2010). Haploid plants produced by centromere-mediated genome elimination. *Nature, 464*, 615–618.

Roeder, G. S., (1997). Meiotic chromosomes: It takes two to tango. *Gen. Dev., 11*, 2600–2621.

Sakaguchi, K., Ishibashi, T., Uchiyama, Y., & Iwabata, K., (2009). The multi-replication protein A (RPA) system—A new perspective. *FEBS Journal, 276*(4), 943–963.

Sanchez-Moran, E., Armstrong, S. J., Santos, J. L., Franklin, F. C. H., & Jones, G. H., (2002). Variation in chiasma frequency among eight accessions of *Arabidopsis thaliana*. *Genetics*, 1415–1422.

Senthil-Kumar, M., & Mysore, K. S., (2011a). New dimensions for VIGS in plant functional genomics. *Trends Plant Science*, 656–665.

Senthil-Kumar, M., & Mysore, K. S., (2011b). Virus-induced gene silencing can persist for more than 2 years and also be transmitted to progeny seedlings in *Nicotiana benthamiana* and tomato. *Plant Biotechnology Journal*, 797–806.

Shaharuddin, N. A., Yuanhuai, H., Hongying, L., & Grierson, D., (2006). The mechanism of graft transmission of sense and antisense gene silencing in tomato plants. *FEBS Letter, 500*, 6579–6586.

Shinohara, M., (2008). Crossover assurance and crossover interference are distinctly regulated by the ZMM proteins during yeast meiosis. *Nature Genetics, 40*, 299–309.

Shull, G. H., (1908). *A Pure Line Method in Corn Breeding* (Vol. 5, pp. 51–59.). Report American Breeders Association.

Siaud, N., Dray, E., Gy, I., Gérard, E., Takvorian, N., & Doutriaux, M. P., (2004). Brca2 is involved in meiosis in *Arabidopsis thaliana* as suggested by its interaction with Dmc1. *EMBO Journal, 233*, 1392–1401.

Singer, J. B., Hill, A. E., & Burrage, L. C., (2004). Genetic dissection of complex traits with chromosome substitution strains of mice. *Science, 16*(304), 445–448.

Spiezio, S. H., Takada, T., Shiroishi, T., & Nadeau, J. H., (2012). Genetic divergence and the genetic architecture of complex traits in chromosome substitution strains of mice. *BMC Genetics, 13*, 13–38.

Springer, N. M., & Stupar, R. M., (2007). Allelic variation and heterosis in maize: How do two halves make more than a whole. *Genome Research, 17*, 264–275.

Stacey, N. J., Kuromori, T., Azumi, Y., Roberts, G., & Breuer, C., (2006). *Arabidopsis* SPO11-2 functions with SPO11-1 in meiotic recombination. *Plant Journal, 48*(2), 206–216.

Stupar, R., Gardiner, J., Oldre, A., Haun, W., Chandler, V., & Springer, N., (2008). Gene expression analyses in maize inbreds and hybrids with varying levels of heterosis. *BMC Plant Biology, 8*.

Sturtevant, A. H., (1915). The behavior of the chromosomes as studied through linkage. *Z. Indukt. Abstamm. Vererbungsl., 13*, 234–287.

Tang, W., Wu, J. Q., Guo, Y., Hansen, D. V., & Perry, J. A., (2008). Cdc2 and Mos regulate Emi2 stability to promote the meiosis I-meiosis II transition. *Molecular Biology, Cell, 19*(8), *3536*–3543.

Van, H. A. M., (1998). *Mutation Breeding: Theory and Practical Applications*. Cambridge: Cambridge University Press.

Vrielynck, N., Chambon, A., Vezon, D., Pereira, L., & Chelysheva, L., (2016). A DNA topoisomerase VI–like complex initiates meiotic recombination. *Science, 351*(6276), 939–943.

Wang, K., Wang, M., Tang, D., Shen, Y., & Miao, C., (2012b). The role of rice HEI10 in the formation of meiotic crossovers. *PLOS Genetics, 8*, e1002809.

Wang, Y., Cheng, Z., Huang, J., Shi, Q., & Hong, Y., (2012a) The DNA replication factor RFC1 is required for interference-sensitive meiotic crossovers in *Arabidopsis thaliana*. *PLOS Genetics., 8*, e1003039.

Wei, G., Tao, Y., Liu, G., Chen, C., Luo, R., Xia, H., Gan, Q., et al., (2009). A transcriptomic analysis of super hybrid rice LYP9 and its parents. *PNAS*, 7695–7701.

Wijnen, C. L., Botet, R., & Van De, B., (2018). *A Complete Chromosome Substitution Mapping Panel Reveals Genome-Wide Epistasis in Arabidopsis, 19*(11), e47153. BioRxiv.

Wijnker, E., & De Jong, H., (2008). Managing meiotic recombination in plant breeding. *Trends Plant Science., 13*(12), 640–646.

Wijnker, E., Deurhof, L., Van De, B. J., De Snoo, C. B., Blankestijn, H., Becker, F., & Keurentjes, J. J., (2014). Hybrid recreation by reverse breeding in *Arabidopsis thaliana*. *Nature Protocols, 9*(4), 761–772.

Wijnker, E., Harashima, H., Muller, K., Parra-Nu € nez, P., De Snoo, C. B., Van ~ De Belt, J., Dissmeyer, N., Bayer, M., et al., (2019). The Cdk1/Cdk2 homolog CDKA;1 controls the recombination landscape in *Arabidopsis*. *Proceedings of Nature Academic Science, 116*(25), 12534–12539.

Wijnker, E., Van Dun, K., De Snoo, C. B., Lelivelt, C. L. C., Keurentjes, J. J. B., Naharudin, N. S., Ravi, M., et al., (2012). Reverse breeding in *Arabidopsis thaliana* generates homozygous parental lines from a heterozygous plant. *Nature Genetics, 44*(4), 467–470.
Winjen, C. L., & Keurenties, J. J., (2014). Genetic resources for quantitative trait analysis: Novelty and efficiency in design from an *Arabidopsis* perspective. *Current Opinion in Plant Biology, 18*, 103–109.
Xin, G. Y., Bao-Hua, W., Yan, F., & Ping, L., (2015). Development and application of marker-assisted reverse breeding using hybrid maize germplasm. *Journal of Integrative Agriculture*, 2538–2546.
Xue, Z., Li, Y., Zhang, L., Shi, W., & Zhang, C., (2016). OsMTOPVIB promotes meiotic DNA double-strand break formation in rice. *Molecular Plant, 9*(11), *1535*–1538.
Yu, Z., Kim, Y., & Dernburg, A. F., (2016). Meiotic recombination and the crossover assurance checkpoint in *Caenorhabditis elegans*. *Semin. Cell Development Biology, 54*, 106–116.
Zhang, C., Song, Y., Cheng, Z. H., Wang, Y. X., & Zhu, J., (2012). The *Arabidopsis thaliana* DSB formation (AtDFO) gene is required for meiotic double-strand break formation. *Plant Journal, 72*(2), 271–281.
Zickler, D., & Kleckner, N., (1999). Meiotic chromosomes: Integrating structure and function. *Annual Review Genetics, 33*, 603–754.
Zickler, D., & Kleckner, N., (2015). Recombination, pairing, and synapsis of homologs during meiosis. *Cold Spring Harb. Perspective Biology, 33*(7), a0166.
Ziolkowski, P. A., Underwood, C. J., & Lambing, C., (2017). Natural variation and dosage of the HEI10 meiotic E3 ligase control *Arabidopsis* crossover recombination. *Genes Development, 31*, 306–317.

CHAPTER 8

Pre-Breeding to Molecular Breeding for Biotic Stress Management in Crop Plants

GURVINDER SINGH,[1] REENA RANI,[2] SUKHDEEP SINGH SIVIA,[3] SUNAYANA,[4] and ARCHANA SANYAL[2]

[1]*Department of Plant Pathology, Dr. Y.S. Parmar University of Horticulture and Forestry, Nauni, Solan, Himachal Pradesh, India*

[2]*Scientist, ICAR–Central Arid Zone Research Institute, Jodhpur, Rajasthan, India*

[3]*Department of Genetics and Plant Breeding, CCSHAU, Hisar, Haryana, India*

[4]*Plant Breeder, RRS, Abohar, Punjab Agricultural University, Ludhiana, Punjab, India*

ABSTRACT

Climate change is adversely affecting agriculture and food production worldwide by increasing frequency and severity of both biotic and abiotic stresses, posing serious production constraints. Although genetic resources in the form of wild species, wild relatives and landraces are cornerstones for competitive plant breeding but in present agricultural conditions, to meet the food demand and nutritional security of ever-rising population, traditional crop breeding methods alone are inadequate and incompetent. Gene pyramiding using marker-assisted selection, marker-assisted backcrossing (MABC) and marker-assisted recurrent selection (MARS) have enhanced the development of durable stress-resistant/tolerant lines for agricultural sustainability. The current improvements in genomics, high throughput

Smart Breeding: Molecular Interventions and Advancements for Crop Improvement. Kailash Chandra, Subhash Chand, Ravi Prakash Saini, and Rohit Sharma (Eds.)

phenomics, sequencing, and breeding methodologies in conjunction with cutting edge genome-editing tools open up new ways for accelerated climate resiliency in crop improvement. QTL mapping, genome-wide association study, next-generation sequencing (NGS) and high-throughput (HT) genotyping technologies have facilitated the availability of genome sequence assemblies, generation of high-density genetic maps, a series of marker genotyping platforms and recognition of markers associated with a number of desired traits in the crops. Therefore, holistic smart breeding approaches being successfully harbored in present-day agriculture for intensifying crop tolerance to biotic stresses can be an encouraging way out to address the climate change for sustainable crop improvement.

8.1 INTRODUCTION

The contemporary agriculture will face massive challenges with an ever-increasing global population which is expected to surpass 9 billion by 2050 (Lenaerts et al., 2019). Climate change and agriculture have inextricable links; both take place on a global scale. Sustainable agricultural production is vulnerable to various environmental factors, including abiotic (drought, heat, cold, and salinity) and biotic stresses (pests, diseases, and weeds) that had adversely affected agricultural production since last two decades and severe consequences are still awaited. It is predicted that annual losses due to biotic stresses are about 25% globally (Pandey et al., 2017; Waddington et al., 2010). The complex and essential relationship between biotic stresses and climate change had emerged a dire need to increase and sustain agricultural crop production without yield and economic penalty globally (Hasan et al., 2015). In order to meet the food demand of the escalating population, the Food and Agriculture Organization (FAO) anticipated that food production must be increased by 70% by 2050, globally (Mwobobia et al., 2020).

Chemical pesticides have been used to manage the pests in crops since long back to reduce yield losses. Although, these are very effective, selective, and efficient to control pests, but their use have recently been discouraged due to high product cost, regular applications and being mostly hazardous to human health and ecological balance (Vincelli, 2016). Another major problem to use chemical pesticides is the development of pest resistance and the emergence of super-weeds against them. In comparison to the chemical control, development of the disease and insect resistant crop varieties is one of the cheap and ecofriendly way to deal with plant diseases and insects (Hanson et al., 2016). Plant breeding has always rendered a fundamental role to feed the burgeoning human population through the development of genetically superior genotypes/varieties that suits for different agro-climatic conditions (Falconer & Mackay, 1996; Chand et al., 2020). But in the recent time of climate change, there is an imperative requirement to breed the crop plants with distinct genetic backgrounds for better crop adaptation.

During the phase of domestication, the process of selection resulted in modern crops with narrow genetic backgrounds, limited environmental adaptation and low breeding capacity (Smykal et al., 2018). In order to nourish the rapidly increasing population, there is a dire need to develop varieties carrying resistance gene/s from crop wild relatives (CWR) and landraces thereby possessing broader genetic base and adaptation to climate change. Although the conventional plant breeding is labor intensive, environment sensitive and takes several years for screening genotypes for the development of commercial varieties, it is currently the most widely used approach in crop improvement (Chand et al., 2020). In the process of crop improvement through traditional breeding methods, some unwanted genes, continue to exist along with the desired ones even after several backcross (BC) generations. Moreover, conventional crop breeding methods are inadequate for speeding up the crop improvement at the required rate (Barrett, 2010; Atique et al., 2018). In contrast to the conventional breeding, molecular breeding technologies have resulted into accurate, refined, and rapid breeding cycles using of molecular markers (Kage et al., 2016; Hasan et al., 2015; Kumawat et al., 2020).

After the genesis of molecular biology and biotechnology techniques for proliferating the breeding programs, an enormous amount of genetic data associated with genes and QTLs (quantitative trait loci) were generated (Wang, 2007). The phenotyping and genotyping methodologies offered remarkable opportunities to develop crop varieties suitable for changing climatic conditions (Gobu et al., 2020). The surge in the incidence of different crop stresses has demanded the stacking of several resistance genes into high yielding cultivars to confer a wider spectrum of resistance to crops under stressful growing conditions. This enhanced capability will help them to survive in multiple biotic and abiotic stresses simultaneously. Marker-assisted selection (MAS) approach includes the selection of traits indirectly with the help of markers linked with the preferred gene/s for tagging of some desired trait which is otherwise difficult to track for resistance against biotic stresses (Kumawat et al., 2020). The different molecular techniques have hastened the durable resistant lines development with high accuracy in the shortest time interval.

8.2 PRE-BREEDING: EXPLORING UNTAPPED HUB OF GENETIC DIVERSITY

The modern agricultural practices, emerging new pest and pathogen populations, changing climate, and accelerated need for food and nutrition necessitated the identification and utilization of new gene/s and allele/s sources in the diverse germplasm (wild species and landraces) for the development of high-yielding cultivars having resistance to biotic and abiotic stresses. Under these circumstances, pre-breeding provides an exclusive platform for enhancing the germplasm utilization into the breeding pipeline (Lenaerts et al., 2019). Pre-breeding is an integral part of germplasm diversification strategies where the promising germplasm accessions possessing desirable characters and/or genes are crossed with already existing varieties to create new populations with a high frequency of useful genes. Lastly, these pre-breeding populations are assessed for identifying the desirable introgression lines carrying traits and/or genes donated by wild species for their future usage in breeding programs. It can be inferred that pre-breeding is a connector for connecting the desirable traits (e.g., disease resistance) of CWR to the modern cultivar development (Sharma et al., 2013).

8.2.1 MAJOR ROLES OF PRE-BREEDING

1. Broadening the genetic base and reducing the vulnerability of crops to different types of stresses.
2. Identifying and moving the desirable traits from exotic materials and wild species into plant material available to breeders.
3. Identification and transfer of the elite genes from unrelated species into the cultivated or elite plant materials.

8.2.2 UTILIZATION OF PLANT GENETIC RESOURCES (PGRS)

The traditional approaches using plant genetic resources (PGRs) in crop improvement (Cooper et al., 2001) through pre-breeding are discussed in subsections.

8.2.2.1 INTERSPECIFIC AND INTERGENERIC HYBRIDIZATION

In crop breeding programs, wild species or relatives are being used mostly for introgressing the disease and insect resistance into the cultivated crop varieties using methods of interspecific and intergeneric hybridization. Genes for biotic stress resistance have been investigated in wild relatives of many crops such as chickpea, barley, and maize (Von Wettberg et al., 2018; Brozynska et al., 2016; Fustier et al., 2017; Liu et al., 2020). To deal with the lethal wheat rust diseases in the perspective of changing climate, some resistant genes have been successfully recognized and transferred to hexaploid wheat from different *Triticum spp.* and *Aegilops spp.* (Table 8.1). In this context, Arora et al. (2019) had reported and transferred stem rust resistance genes from three or more species.

TABLE 8.1 Stem Rust Resistance Genes in Wheat and Their Source

Sl. No.	Genes	Source
1.	*Sr 21, Sr 22*	*Triticum monococcum*
2.	*Sr 2, Sr 9e, Sr 13*	*T. dicoccum*
3.	*Sr 14, Sr 17, Sr 33*	*Ae. squarrosa*

The wild chickpea species have contributed to crop improvement through provision of resistance against cyst nematode (Malhotra et al., 2001), wilt, foot rot, root rot (Singh et al., 2005) and phytophthora root rot diseases (Knights et al., 2008). Similarly, resistance for Botrytis gray mold have been developed from crosses involving *Cicer reticulatum*, *C. echinospermum*, *C judaicum*, and *C. pinnatifidum* (Sharma et al., 2016; Singh et al., 2005; Upadhyaya, 2008; Chaturvedi & Nadarajan, 2010; Kaur et al., 2013). In pigeon pea improvement, resistance against phytophthora blight is being introgressed from wild species into popular high yielding varieties (Pande et al., 2011).

8.2.2.2 DE NOVO DOMESTICATION OF WILD RELATIVES

The cultivation of most staple crops is being done in the geographic locations that are different from their domesticated areas, where climatic conditions vary. Yet, their wild relatives and landraces have better adaptation in local regions. *De novo* domestication technique was initiated to develop climate resilient crops by avoiding the harmful effect of foreign allele/s introgressed from wild species into cultivars (Fernie & Yan, 2019). In some crops like quinoa, some strategies have been projected for domestication of wild germplasm (DeHaan et al., 2016). Besides the cultivation of non-domesticated crop plants directly, the domestication of wild germplasm is assisted by advance methodology of CRISPR/Cas9 by editing the domesticated genes, as in case of wild tomatoes (*Solanum pimpinellifolium*) and ground cherry (*Physalis pruinosa*) (Li et al., 2018; Lemmon et al., 2018). From such successful instances of domesticated gene introgressions, it becomes very clear that the employment of wild relatives is reasonably significant in traditional plant breeding in changing climatic conditions for ensuring food security.

8.2.2.3 INTROGRESSIOMICS APPROACH

In crop improvement, the actual potential of the CWR (including wild species and wild relatives) mainly remains underutilized due to linkage drag and other breeding barriers with the cultivated crop varieties. In introgressiomics approach, with the generation of introgression lines from CWR into the different genetic backgrounds of crops, plant populations have been developed at mass scale (Prohens et al., 2017). Molecular markers characterized and developed introgressiomics populations can be easily used in ongoing breeding programs for stress resistance breeding. MAS can be used for generation of chromosome substitution lines (CSLs) and introgression lines for developing genetically characterized elite material. Systematic pre-breeding efforts have resulted in the formation of elite gene pools possessing a higher frequency of valuable genes and alleles in crops like chickpea, pigeon pea, and groundnut.

8.2.3 EXAMPLES OF PROMISING PRE-BREEDING MATERIALS

Pre-breeding has been used effectively for crop improvement in different crops such as maize (Maxted & Kell, 2009), tomato (Tanksley et al., 1996), late blight in potato (Jansky et al., 2013), tolerance to root rot pathogen in bean, for anthracnose and wilt resistance in lentil (Fiala et al., 2009). Use of CWR for improvement of different qualitative and quantitative traits have been done in several crops like soybean, cotton, tobacco, sugarcane, lentil, eggplant, carrot, sweet potato, barley, and sorghum (Sebolt et al., 2000; Iqbal et al., 2001; Singh et al., 2005, 2020; Seetharam, 2007).

1. **Pigeon Pea:** A high-yielding line (ICPL 15028) having resistance to manifold diseases (wilt, sterility mosaic and phytophthora blight) have been selected at the International Crops Research Institute for the Semi-Arid Tropics (ICRISAT). Pod borer resistance has also

been achieved from the utilization of *C. acutifolius* in pigeon pea breeding program (Mallikarjuna et al., 2007).

2. **Groundnut:** High levels of resistance to different diseases and insect pests have been developed in cultivars using wild *Arachis* species (Sharma et al., 2013).
3. **Cowpea:** A number of wild accessions of *Vigna* species such as *V. vexillata*, *V. davyi*, *V. oblongifolia,* and *V. luteola* possessed good levels of resistance to insect pests (Fatokun, 2002).
4. **Wheat:** Pre-breeding activities in wheat (Valkoun, 2001) aimed at transferring resistances from wild relatives into cultivated forms against major diseases and insects, using backcrossing and embryo rescue techniques. Germplasm enrichment of wheat by transfer of genes from rye through translocations has enabled gene transfer for disease/insect resistance (Table 8.2).

TABLE 8.2 Transfer of Genes Through Translocations for Disease and Insect Resistance in Wheat

Sl. No.	Translocations	Resistance for
1.	1B. R – Tr	Powdery mildew; stem, leaf, and stripe rust
2.	2BS. 2RL – Tr	Hessian fly
3.	Irradiation-induced translocations	Hessian fly

5. **Rice:** Pre-breeding activities in rice (Brar & Khush, 2002) aimed at transferring resistances to major diseases and insects as germplasm enhancement activity into cultivated rice through homoeologous pairing as shown in Table 8.3.

TABLE 8.3 Genes Transferred in Rice for Disease and Insect Resistance

Sl. No.	Species	Resistance for
1.	*Oryza australiensis*	Bacterial leaf blight, brown plant hopper
2.	*Oryza brachyantha*	Bacterial leaf blight
3.	*Oryza latifolia*	Bacterial leaf blight, brown plant hopper, white backed plant hopper
4.	*Oryza minuta*	Blast, bacterial leaf blight, brown plant hopper
5.	*Oryza officinalis*	Bacterial leaf blight, brown plant hopper, white backed plant hopper, green leaf hopper
6.	*Oryza nivara*	Grassy stunt

Homologous and homoeologous pairing of chromosome induced polyploidy, chromosome substitutions and interspecific and intergeneric gene transfers have significantly helped for gene introgression in the following crops (Table 8.4).

TABLE 8.4 Gene Introgression in Crops Through Chromosome Substitution Method

Sl. No.	Crops	Cross	Resistance for
1.	Wheat	*Triticum aestivum* × Haynaldia	Powdery mildew
2.	Berseem	*Trifolium repens* × *T. nigrescens*	Cyst nematode
3.	Chickpea	*Cicer arietinum* × *C. reticulatum*	Cyst nematode
4.	Soybean	*Glycine max* × *G. tomentella*	Cyst nematode

8.2.4 CHALLENGES IN ADOPTING PRE-BREEDING

Pre-breeding has tremendous potential to create genetic variability that could be used in future breeding programs for various purposes to develop climate smart and disease tolerance cultivars. However, the following factors become hurdles in taking up pre-breeding as the foremost step of crop improvement:

- Lack of ability to characterize and evaluate;
- Knowledge of the genetic diversity;
- Relationship between different species;
- Fund resources;
- Linkage drag;
- Cross incompatibility in inter-specific crosses;
- Barriers instability and chromosome pairing in hybrids;
- Problems of inviability and sterility conditions in hybrids;
- Small sample size of inter-specific hybrid population;
- Lack of availability of donors for biotic stress resistance;
- Legal restrictions like IPR difficulties in exchanging and accessing the cultivated germplasm material (Loknathan et al., 2003).

8.2.5 FUTURE PROSPECTS OF PRE-BREEDING

Owing to the increased probability of extinction of species with narrow adaptation, there is an urgent need for collecting, characterizing, and documenting the wild species, including CWR for adapting agriculture to biotic and abiotic stresses. Pre-breeding is a prolonged and resource-demanding research venture that involves unadapted as well as incompatible germplasm to develop useful breeding materials. Therefore, strategies to conquer such problems, e.g., development of large BC populations and precise introgression of desirable genes using molecular markers with reduced, or no linkage drag should be employed. Further, new molecular breeding tools are necessary to deal with complex traits more precisely and effectively, including the transfer of desired gene(s) from the tertiary gene pool and/or beyond.

8.3 MOLECULAR BREEDING APPROACHES

The recent advancements in biotechnology have opened up vast possibilities for base broadening in pre-breeding as well as for introgression of specific traits. Moreover, the new genomic tools assure a speedy progress and have the ability to solve some earlier unmanageable breeding issues. In the recent years, there has been a great progress in crop improvement programs with the introduction of modernized molecular techniques. Molecular markers are simple parts of DNA that are tightly linked with a trait of interest, i.e., the resistant gene(s) (also referred to as QTLs) present on the same chromosome (Kumawat et al., 2020). Marker-based screening is quite more proficient than conventional crop breeding methods, thereby, leading to improved accuracy coupled with reduced breeding duration to develop crop varieties (William et al., 2007). Molecular marker-based technology is independent of environmental factors for plant selection. The QTLs have been known to be linked with resistance to several plant diseases, therefore, can be used for the selection and crop improvement through modern breeding tools (Ribaut et al., 2010). The deployment of molecular markers in plant breeding as selection tools in the past 30 years has led to significant betterment in efficiency and the subsequent release of new varieties (Collard & Mackill, 2008). Marker-assisted selection (MAS), marker-assisted backcrossing (MABC) and marker-assisted recurrent selection (MARS) approaches are being used to breed for stress resilience using the molecular markers that are associated with traits of interest (Varshney et al., 2012b). There have been quite improvement in legumes regarding the development of numerous molecular breeding products (Table 8.5) regardless of the late arrival of genomic resources (Varshney et al., 2013b; Pandey et al., 2016; Varshney, 2016).

TABLE 8.5 Biotic Stress Resistance in Legumes Through Molecular Approaches

Crops	Resistance to	References
Chickpea	Fusarium wilt and Ascochyta blight	Varshney et al. (2013b)
Groundnut	Nematode	Chu et al. (2011)
	Leaf rust	Varshney et al. (2014); Janila et al. (2016)
Soybean	Rust disease	Khanh et al. (2013)
	Soybean mosaic virus	Shi et al. (2009); Parhe et al. (2017)
Cowpea	Striga	Lucas et al. (2015); Boukar et al. (2016)
Lentil	Ascochyta blight and anthracnose	Taran et al. (2003)
Pea	Powdery mildew	Ghafoor & McPhee (2012)
	Aphanomyces root rot	Lavaud et al. (2015)

8.3.1 MARKER-ASSISTED SELECTION (MAS)

Molecular markers have emerged as powerful and efficient tools to harmonize the traditional plant breeding methods for improving crops (Kumawat et al., 2020). This approach has been particularly used in plant breeding to screen the plants for several traits individually or simultaneously at the initial stage (seedling stage) and minimizes linkage drag (Tanksley et al., 1989). MAS employ closely linked DNA-based markers with the target gene to facilitate phenotypic evaluation by selecting the target genes (Chukwu et al., 2019). For effective applications of DNA-based markers in crop improvement programs, some of the major factors include their reliability, reproducibility, high polymorphism, co-dominance inheritance, cost effectiveness and closeness to the trait of interest (genetic distance of less than 5 cM) (Shirasawa & Hirakawa, 2013; Nadeem et al., 2018; Tong et al., 2020).

The steps involved in MAS are as follows:

i. **Development of Mapping Population:** It involves parental selection and hybridization between them.

ii. **QTL Mapping:** It includes the construction of a linkage map followed by phenotypic evaluation for trait(s) and QTL analysis.

iii. **QTL Validation:** This step involves the confirmation of position and effects of QTLs, their verification in independent populations and testing in different genetic backgrounds. Finally fine mapping is done.

iv. **Marker Validation:** It includes the testing of markers in important breeding material and finally the identification of polymorphic markers.

v. Correlation with morphological traits.

For marker-assisted varietal development of major crops, MAS has now been regularly enforced (Gregorio et al., 2013; Gupta et al., 2010). MAS has been efficiently applied for stress resilience breeding in crops, such as rice, wheat, and cotton (Das et al., 2017). MAS in breeding for resistance to different diseases and insect pests in common bean have been summarized in Table 8.6.

TABLE 8.6 MAS in Breeding for Resistance in Common Bean

Diseases and Insect Pests	References
Fungal Diseases	
Anthracnose	Zuiderveen et al. (2016)
Angular leaf spot	Keller et al. (2015)
Rust	Hurtado-Gonzales et al. (2017)
Leafspot	Hanson et al. (1993)
Pythium root rot	Campa et al. (2010)
Fusarium wilt	Kamfwa et al. (2013)
Web blight	Takegami et al. (2004)
Bacterial Diseases	
Common blight	Miklas et al. (2017)
Halo blight	Tock et al. (2017)
Bacterial brown spot	Muedi et al. (2015)
Viruses	
Bean common mosaic virus and mosaic necrosis virus	Strausbaugh et al. (1999)
Bean golden yellow mosaic virus	Blair et al. (2007)
Insects	
Leafhopper	Brisco (2012)
Bean pod weevil	Blair et al. (2006b)
Bean fly or bean stem maggot	Miklas et al. (2006)
Seed weevils	Blair et al. (2010b)

8.3.2 MARKER-ASSISTED BACKCROSSING (MABC)

The MABC involves the use of molecular markers for introgressing particular trait(s) from a donor parent into the genetic background of a recurrent parent (usually a popular variety) (Hospital, 2005). Therefore, the product (variety/cultivar) retains the complete genome of the recurrent parent with gene of interest from the donor parent (Gupta et al., 2010). In MABC, the use of

DNA markers significantly increases the selection efficiency, while reducing the time required for varietal development by many years. This approach is also useful to create near-isogenic lines (NILs) or chromosome segment substitution lines (CSSLs), that are often used for analyzing genes/QTLs and alien gene introgressions (Lorieux, 2005; Varshney et al., 2010b). One of the key applications of MABC is marker-assisted gene pyramiding, in which a few different genes for the same or different traits are combined together in one line/variety using molecular markers.

The steps involved in MABC are as follows:

i. Crossing between recurrent and donor parents (to generate F_1);
ii. Backcrossing of F_1 with the recurrent parent while performing foreground and background selections;
iii. Selfing of backcross generation to get the gene of interest in homozygous form.

MABC breeding program has been used very efficiently for enhancing stress tolerance in rice where, Pusa1460 variety was used for donating introgressing two bacterial blight resistance genes, *Xa13* and *Xa21* into the parental lines of the superfine grain aromatic rice hybrid Pusa RH 10 (Basavaraj et al., 2010). Similarly, a simultaneous but stepwise transfer method was adopted for transferring the blast resistance genes, *Pi-Kh* and *Piz5* from the donor Tetep and C101A51, respectively. In cowpea, Striga resistance was introgressed into popular but otherwise susceptible varieties using MABC (Beebe et al., 2016). In groundnut, the genes for nematode resistance (Chu et al., 2011) and rust resistance (Varshney et al., 2014) were combined by deploying the MABC approach. In bean breeding, MABC has been found effective for the improvement of traits controlled by major genes and gene stacking (Kelly, 2004; Carneiro et al., 2010). However, most of the traits desired by breeders are controlled by multiple genes, e.g., resistance to multiple races of pathogens. In such cases, MABC approach is found to be a bit challenging in retaining the desirable gene combinations (Peleman & Van der Voort, 2003).

8.3.3 MARKER-ASSISTED RECURRENT SELECTION (MARS)

In this approach, the genotypic selection and intercrossing are enabled in the same cropping season that leads to enhanced effectiveness of recurrent selection, expedition of selection and help in integrating numerous desirable genes (Cai et al., 2019). MARS approach includes selection of parents from similar and dissimilar populations (Ali et al., 2020) and further there is involvement of two to three recombination cycles depending upon the presence of marker alleles

(Eathington et al., 2007). During the initial phase of MARS, QTLs are identified in the population derived after crossing the better parents. Consequently, the lines which carry superior alleles for maximum QTLs are crossed for pyramiding of the alleles in single genetic background. The best F_{1s} are categorized by genotyping during each cycle for further use during next recombination cycle (Bankole et al., 2017). In the next steps, MARS of F_3-derived plants is carried out through single-seed descent strategy and final phenotypic screening of recombined lines is done so as to select the best lines for multi-location testing to release them as varieties. Several crop scientists have declared that the genetic advance obtained through MARS was almost double in compared to the phenotypic selection (PS) in the case of maize (Ribaut & Ragot, 2006). MARS is being used by several multinational companies in crops like maize and soybean, while it has been used in crops likes wheat, sorghum (Abdallah et al., 2009) and rice (Grenier et al., 2012) by a few public sector institutes. In cowpea, Striga resistance and *Macrophomina* resistance were improved significantly through MARS approach (Boukar et al., 2016).

8.3.4 MARKER-ASSISTED GENE PYRAMIDING

The majority of crop breeding programs for stress breeding are based on the principle of inserting single resistant or tolerant gene into susceptible plants and resistance tends to last only for very less time interval. Hence, it is required to develop novel and efficient strategies to boost stress resistance in crops on a sustainable basis (Sedeek et al., 2019; Bailey-Serres et al., 2019). In some instances of biotic stress, a particular trait is controlled by more than one gene; therefore, it becomes essential to manipulate multiple genes for building up resistance against such biological agents (Gupta & Pandey, 2014). The recent breeding methodologies are concentrating on pyramiding of various diverse resistance genes or QTLs into a single genotype to assure long-lasting and durable resistance (Kottapalli et al., 2010; Devi et al., 2017; Nelson et al., 2018; Steiner et al., 2019; Lv et al., 2020).

Gene pyramiding is a very important technique to transfer a number of desired genes from diverse genotypes into a single genotype using closely associated markers in the shortest possible time as compared to conventional breeding. So, it could be deduced unquestionably that marker-assisted gene pyramiding in crops is rapid, proficient, and cost-effective technique to combine desired genes for building up multiple stress tolerances (Collard & Mackill, 2008; Hasan et al., 2015; Kumar et al., 2018; Angeles-Shim et al., 2020). Marker-assisted gene pyramiding includes three different techniques for pyramiding of genes.

In the first technique, recurrent and donor parents are crossed together for getting the F_1 hybrid, which is then backcrossed with recurrent parent up to three generations for getting the best parent. For pyramiding two or more genes of interest, the BCF_1 is further crossed with another donor parent. In the second technique, the recurrent and donor parents are crossed to obtain the F_1 hybrids separately. These F_{1s} are then crossed with each other to get improved F_1, which is further backcrossed with the recurrent parent to get a better version of the recurrent parent. The third technique is a combination of the first two techniques where recurrent parent is crossed with different donor parents simultaneously and then backcrossed up to the BC_3 generation. Finally, gene pyramiding is achieved after intercrossing the backcrossed populations with each other. This technique is thought to be the most suitable as it takes less time. MABC-based gene pyramiding can lead to durable and sustainable crop improvement (Hu et al., 2016; Rana et al., 2019; Rezene et al., 2019). This technique has been successfully utilized against biotic stresses in many major crops (Table 8.7).

TABLE 8.7 Gene Pyramiding for Biotic Stress Improvement in Different Crops

Crop	Traits	Pyramided Genes	References
Potato	Late blight resistance	*Rpi-phu1, Rpi-rzc*	Sliwka et al. (2012)
Cotton	Bacterial blight resistance	*Chi11, tlp, Xa21*	Maruthasalam et al. (2007)
	Bollworm resistance	*Cry1Ac, Cry2Ab*	Jackson et al. (2004)
	Weed and pathogen resistance	*ptxD/Phi*	Pandey et al. (2017)
	Insect and pest resistance	*Cry1Ac, Cry2Ac*	Puspito et al. (2015)
Wheat	Leaf and stem rust resistance	*SrCad, Sr33, Lr34, Fhb*	Zhang et al. (2019)
	Cereal cyst nematode resistance	*CreX, CreY*, CRISPR-Cas9	Ali et al. (2019)
	Aphid resistance	*Gn2, Gn4*	Liu & Santi, (2000)
Rice	Gall midge resistance	*Gm1, Gm2, Gm4*	Das & Rao, (2015)
	Blast resistance	*Pi(2)t, Pi25, Pi(t)a, Xa4, Xa5, Xa13*, and *Xa21*	Joseph et al., 2004); Singh et al. (2001)
	Brown plant hopper resistance	*Bph1, Bph2*	Liu et al. (2016)
	Blight resistance	*Xa5, Xa13*, and *Xa21*	Narayanan et al. (2002)
	Bacterial sheath blight, stem borer	*Xa12, Rc7, Cry1AB1, Cry14c*	Datta et al. (2002)
Soybean	Mosaic virus resistance	*Rsv1, Rsv3, Rsv4*	Zhu et al. (2012)
Tomato	Leaf curl/spotted virus	*Ty-1, Ty-3, Sw-5*	Verlaan et al. (2013)
Barley	Mosaic virus resistance	*rym4, rym5, rym9, rym11*	Werner et al. (2005)
Corn	Corn borer resistance	*Cry1Ac, Cry1le*	Jiang et al. (2016)
Chickpea	Lepidopteran resistance	*Cry1Ac, Cry1Ab*	Mehrotra et al. (2011)
Pepper	Root-knot nematode resistance	*Me1, Me2*	Djian-Caporalino et al. (2014)

8.3.5 OMICS TECHNIQUES

Omics includes several techniques which are mostly used for classification and measurement of the functions and relationships of different molecules in an organism. These techniques include transcriptomics (sequencing) (Gupta & Varshney, 2004), proteomics (structure and function of the proteins) (Bhushan et al., 2007; Pandey et al., 2008) and metabolomics (detection and quantification of metabolites in cellular regulatory processes of plants) (Dawid & Hille, 2018; Razzaq et al., 2019; Piasecka et al., 2019). In metabolomics, the metabolome performs an important role in the exploration of interactions between environment and gene, phenotyping, and biomarker identification in any species. Furthermore, metabolomics could be one of the best innovative approaches owing to its metabolic networks linked with biotic stress tolerance in plants (Sauer et al., 2017). However, metabolomics needs a cavernous understanding of secondary metabolic pathways in the plants (Heinig et al., 2013). Utilization of metabolic QTL-assisted breeding might be helpful to get a clearer picture of the function of specific desired genes for their use in breeding of crops (Tugizimana et al , 2018; Agarrwal & Nair, 2020).

8.3.6 ADVANCED BACKCROSS (BC) QTL ANALYSIS-BASED BREEDING (AB-BREEDING)

The molecular breeding approaches like MABC and MARS are helpful only when the superior alleles for the desired trait exist in the germplasm collection itself. On the other hand, there are cases when genetic variability for a specific trait may not be accessible in the primary gene pool, such as resistance to pod borer in the case of chickpea and pigeon pea. Under such conditions, the introduction of elite alleles from wild relatives is best done by advanced-BC QTL-based breeding approach. In this method, a selected crop wild species is backcrossed to a cultivar/variety and QTL mapping is postponed up to BC_2 or BC_3 generations to circumvent the incompatibility and pollen fertility problems in the initial BC populations. Selection is then imposed in the BC_2F_2 or BC_2F_3 population for identification of plants carrying desirable traits. This segregating population is used to generate the genotypic and phenotypic data, which is then subjected to QTL analysis to identify QTL and to ensure the role of any of these QTLs in trait improvement in the progenies. Hence, there is involvement of the simultaneous finding and transfer of desired alleles/QTL from an unadapted germplasm into selected breeding lines, e.g., superior alleles for traits like disease resistance (Tanksley & Nelson, 1996). Initially, AB-breeding was practiced in tomatoes, where elite tomato lines for various fruit characters were generated from crosses between wild tomato species (Fulton et al., 2000). This approach is now being used in semi-arid tropical legume crops for improving various traits.

8.3.7 GENOMICS-ASSISTED BREEDING (GAB)

Genomics-assisted breeding (GAB) deals with integrating and using genomic tools for developing superior lines having improved stress tolerance. The main objective of GAB is to establish and utilize the relationship between genotype and phenotype for crop improvement. GAB includes different approaches such as genomics, transcriptomics, and proteomics for identifying the molecular markers associated with the desired traits that facilitate prediction of phenotype from the genotype to aid crop breeding. GAB has been quite robust technique for numerous traits in cereals (Varshney et al., 2006; Septiningsih et al., 2013) and legumes (Varshney, 2016; Pratap et al., 2017). With the introduction of next-generation sequencing (NGS) (Varshney et al., 2009c) and high-throughput (HTP) genotyping technologies (Varshney, 2011), it became possible to use the marker profile/allele data for forecasting the phenotype of progenies in breeding programs.

8.3.8 NEXT GENERATION GAB APPROACHES

Due to the advent of cost-effective genotyping and availability of diagnostic molecular markers, some technologies such as, early generation selection, genomic selection, and genome editing (GE) are expected to be used in future GAB approaches. Genomic selection (GS), a budding molecular approach, is a form of MAS that enables breeders to enlarge the genetic gain very quickly (Heffner et al., 2009; Kumar et al., 2021). It includes the identification of better genotypes directly *via* predicted breeding value (BV), with the help of markers having genome-wide distribution. GS has been comprehensively used breeding approach for complex polygenic traits in the recent past for climate resilience in agriculture. Highly accurate prediction models are involved in GS that can lead to increased crop yields by decreasing the breeding cycles. Next generation artificial intelligence (AI) holds promise for GS as it can categorize and enlighten the complex biological interactions with large scale genomics and phenomics data (Harfouche et al., 2019). GE being an alternative approach for modification of targeted genomic regions, may be more appropriate for improving the deficiency of one or two traits in the popular varieties. In several legume species, due to genetic barriers like ploidy differences between wild relatives and crop plants, the exchange of genetic material gets restricted by natural means. In such cases, GE assists in targeted editing of the well-established cultivars without fearing about linkage drag. The genome-editing technologies have witnessed significant developments over the passing years (Wang et al., 2017). Out of these technologies, CRISPR/Cas9 has got wider recognition among researchers due to its massive potential role in crop improvement in the times to come (Kim, 2016). GE technology has been used in the development of improved plant varieties of many crops including Arabidopsis, tobacco, maize, rice, wheat, canola, soybean, camelina, tomato, and cucumber (Kamburova et al., 2017).

8.4 CONCLUSION

Burgeoning global population and the human accelerated climate change are two major challenges to meet future food and nutrition demand. Moreover, there are several biotic and abiotic stresses that hamper the crop cultivation and productivity all over the world. Under such circumstances, there is an immense requirement to protect crop plants from these stresses to feed the world's growing population. Plant breeding has constantly played a crucial role in mitigating hunger in the past utilizing the genetic resources available in the form of wild species, wild relatives and landraces. But the utilization of these resources through pre-breeding is time-taking and resource demanding. In recent times, the molecular marker-based technologies have emerged as powerful, rapid, and efficient approaches to complement traditional plant breeding for improving crops against different types of biotic stresses for ensuring nutritional and food security. MAS can play a significant role in pyramid traits of interest in elite germplasm lines or cultivars to combat prevailing stresses, particularly biotic, in a short time period. The latest advancements in genomics coupled with cutting-edge GE tools provide new ways for stress tolerance breeding programs. Thus, smart breeding technologies can be promising for dealing with changing climatic conditions and for the development of better-adapted varieties of major crops like potato, rice, wheat, common bean, tomato, chickpea, pigeon pea, and barley.

KEYWORDS

- **biotic stress**
- **breeding**
- **chromosome substitution lines**
- **crop wild relatives**
- **Food and Agriculture Organization**
- **marker-assisted recurrent selection**
- **markers**
- **molecular and pre-breeding**

REFERENCES

Abdallah, A. A., Ali, A. M., Geiger, H. H., & Parzies, H. K., (2009). Marker-assisted recurrent selection for increased out crossing in *Caudatum* race sorghum. In: *International Conference on Applied Biotechnology (ICAB)* (pp. 28–30).

Agarrwal, R., & Nair, S., (2011). Metabolomics-assisted crop improvement. In: *Advancement in Crop Improvement Techniques* (pp. 263–274). Woodhead Publ.

Ali, M. A., Shahzadi, M., Zahoor, A., Dababat, A. A., Toktay, H., Bakhsh, A., Nawaz, M. A., & Li, H., (2019). Resistance to cereal cyst nematodes in wheat and barley: An emphasis on classical and modern approaches. *International Journal of Molecular Sciences, 20*(2), 432.

Ali, M., Zhang, L., DeLacy, I., Arief, V., Dieters, M., Pfeiffer, W. H., Wang, J., & Li, H., (2020). Modeling and simulation of recurrent phenotypic and genomic selections in plant breeding under the presence of epistasis. *The Crop Journal, 8*(5), 866–877.

Angeles-Shim, R. B., Reyes, V. P., Del Valle, M. M., Lapis, R. S., Shim, J., Sunohara, H., Jena, K. K., et al., (2020). Marker-assisted introgression of quantitative resistance gene Pi21 confers broad spectrum resistance to rice blast. *Rice Science, 27*(2), 113–123.

Arora, S., Steuernagel, B., Gaurav, K., Chandramohan, S., Long, Y., Matny, O., Johnson, R., et al., (2019). Resistance gene cloning from a wild crop relative by sequence capture and association genetics. *Nature Biotechnology, 37*(2), 139–143.

Atique, R., Farooq, M., Rashid, A., Nadeem, F., Stuerz, S., Asch, F., Bell, R. W., & Siddique, K. H. M., (2018). Boron nutrition of rice in different production systems: A review. *Agronomy for Sustainable Development, 38,* 25.

Bailey-Serres, J., Parker, J. E., Ainsworth, E. A., Oldroyd, G. E., & Schroeder, J. I., (2019). Genetic strategies for improving crop yields. *Nature, 575*(7781), 109–118.

Bankole, F., Menkir, A., Olaoye, G., Crossa, J., Hearne, S., Unachukwu, N., & Gedil, M., (2017). Genetic gains in yield and yield related traits under drought stress and favorable environments in a maize population improved using marker-assisted recurrent selection. *Frontiers in Plant Science, 8*, 808.

Barrett, C. B., (2010). Measuring food insecurity. *Science, 327*(5967), 825–828.

Basavaraj, S. H., Singh, V. K., Atul Singh, Ashutosh Singh, Anita Singh, Anand, D., Yadav, S., et al., (2010). Marker-assisted improvement of bacterial blight resistance in parental lines of Pusa RH10, a superfine grain aromatic rice hybrid. *Molecular Breeding, 26*(2), 293–305.

Beebe, S., Chirwa, R., Rubyogo, J. C., Katungi, E., Mukankusi, C., Bodo, R., & Bekele, A., (2016). Enhancing common bean productivity and production in Sub-Saharan Africa. *Seven Seasons of Learning and Engaging Smallholder Farmers in the Drought-Prone Areas of Sub-Saharan Africa and South Asia through Tropical Legumes, 2007–2014* (p. 236).

Bhushan, D., Pandey, A., Choudhary, M. K., Datta, A., Chakraborty, S., & Chakraborty, N., (2007). Comparative proteomics analysis of differentially expressed proteins in chickpea extracellular matrix during dehydration stress. *Molecular and Cellular Proteomics, 6*(11), 1868–1884.

Blair, M. W., Muñoz, C., Buendía, H. F., Flower, J., Bueno, J. M., & Cardona, C., (2010b). Genetic mapping of microsatellite markers around the Arcelin bruchid resistance locus in common bean. *Theoretical and Applied Genetics, 121*(2), 393–402.

Blair, M. W., Muñoz, C., Garza, R., & Cardona, C., (2006b). Molecular mapping of genes for resistance to the bean pod weevil (*Apion godmani* Wagner) in common bean. *Theoretical and Applied Genetics, 112*(5), 913–923.

Blair, M. W., Rodriguez, L. M., Pedraza, F., Morales, F., & Beebe, S., (2007). Genetic mapping of the bean golden yellow mosaic Gemini virus resistance gene bgm-1 and linkage with poty virus resistance in common bean (*Phaseolus vulgaris* L.). *Theoretical and Applied Genetics, 114*(2), 261–271.

Boukar, O., Fatokun, C. A., Huynh, B. L., Roberts, P. A., & Close, T. J., (2016). Genomic tools in cowpea breeding programs: Status and perspectives. *Frontiers in Plant Science, 7*, 757.

Brar, D. S., & Khush, G. S., (2002). Transferring Genes from Wild Species into Rice. In: Kang, M. S. (eds.), *Quantitative Genetics, Genomics and Plant Breeding* (pp. 197–217). 10.1079/9780851996011.0197

Brisco, E. I., (2012). *Empoasca Leafhopper Species Resistance in Common Bean, Phaseolus Vulgaris: Field Screening and QTL Identification.* Michigan State University, Plant Breeding, Genetics and Biotechnology Program/Crop and Soil Sciences.

Brozynska, M., Furtado, A., & Henry, R. J., (2016). Genomics of crop wild relatives: Expanding the gene pool for crop improvement. *Plant Biotechnology Journal, 14*(4), 1070–1085.

Cai, J., Wang, S., Su, Z., Li, T., Zhang, X., & Bai, G., (2019). Meta-analysis of QTL for *Fusarium* head blight resistance in Chinese wheat landraces. *The Crop Journal, 7*(6), 784–798.

Campa, A., Pérez-Vega, E., Pascual, A., & Ferreira, J. J., (2010). Genetic analysis and molecular mapping of quantitative trait loci in common bean against *Pythium ultimum*. *Phytopathology, 100*(12), 1315–1320.

Carneiro, F. F., Santos, J. B. D., & Leite, M. E., (2010). Marker-assisted backcrossing using microsatellites and validation of SCAR Phs marker for resistance to white mold in common bean. *Electronic Journal of Biotechnology, 13*(6), 9–10.

Chand, S., Chandra, K., & Khatik, C. L., (2020). Varietal release, notification and denotification system in India. In: *Plant Breeding-Current and Future Views* (pp. 1–12). IntechOpen.

Chaturvedi, S. K., & Nadarajan, N., (2010). Genetic enhancement for grain yield in chickpea–accomplishments and resetting research agenda. *Electronic Journal of Plant Breeding, 1*(4), 611–615.

Chu, Y., Wu, C. L., Holbrook, C. C., Tillman, B. L., Person, G., & Ozias-Akins, P., (2011). Marker-assisted selection to pyramid nematode resistance and the high oleic trait in peanut. *The Plant Genome, 4*(2), 110–117.
Chukwu, S. C., Rafii, M. Y., Ramlee, S. I., Ismail, S. I., Hasan, M. M., Oladosu, Y. A., Magaji, U. G., et al., (2019). Bacterial leaf blight resistance in rice: A review of conventional breeding to molecular approach. *Molecular Biology Reports, 46*(1), 1519–1532.
Collard, B. C., & Mackill, D. J., (2008). Marker-assisted selection: An approach for precision plant breeding in the twenty-first century. *Philosophical Transactions of the Royal Society B: Biological Sciences, 363*(1491), 557–572.
Cooper, H. D., Spillane, C., Hodgkin, T., & Cooper, H., (2001). Broadening the genetic base of crops: An overview. In: *Broadening the Genetic Base of Crop Production* (p. 1–23).
Das, G., & Rao, G. J. N., (2015). Molecular marker assisted gene stacking for biotic and abiotic stress resistance genes in an elite rice cultivar. *Frontiers in Plant Science, 6*, 698.
Das, G., Patra, J. K., & Baek, K. H., (2017). Insight into MAS: A molecular tool for development of stress resistant and quality of rice through gene stacking. *Frontiers in Plant Science, 8*, 985.
Datta, K., Baisakh, N., Thet, K. M., Tu, J., & Datta, S., (2002). Pyramiding transgenes for multiple resistance in rice against bacterial blight, yellow stem borer and sheath blight. *Theoretical and Applied Genetics, 106*(1), 1–8.
Dawid, C., & Hille, K., (2018). Functional metabolomics – a useful tool to characterize stress-induced metabolome alterations opening new avenues towards tailoring food crop quality. *Agronomy, 8*(8), 138.
DeHaan, L. R., Van, T. D. L., Anderson, J. A., Asselin, S. R., Barnes, R., Baute, G. J., Cattani, D. J., et al., (2016). A pipeline strategy for grain crop domestication. *Crop Science, 56*(3), 917–930.
Devi, E. L., Devi, C. P., Kumar, S., Sharma, S. K., Beemrote, A., Chongtham, S. K., Singh, C. H., et al., (2017). Marker assisted selection (MAS) towards generating stress tolerant crop plants. *Plant Gene., 11*, 205–218.
Djian-Caporalino, C., Palloix, A., Fazari, A., Marteu, N., Barbary, A., Abad, P., Sage-Palloix, A. M., et al., (2014). Pyramiding, alternating or mixing: Comparative performances of deployment strategies of nematode resistance genes to promote plant resistance efficiency and durability. *BMC Plant Biology, 14*(1), 1–13.
Eathington, S. R., Crosbie, T. M., Edwards, M. D., Reiter, R. S., & Bull, J. K., (2007). Molecular markers in a commercial breeding program. *Crop Science, 47*, S-154.
Falconer, D. S., & Mackay, T. F. C., (1996). *Introduction to Quantitative Genetics*. Essex. UK: Longman Group.
Fatokun, C. A., (2002). Breeding cowpea for resistance to insect pests: Attempted crosses between cowpea and *Vigna vexillata. Challenges and Opportunities for Enhancing Sustainable Cowpea Production* (p. 52–61.
Fernie, A. R., & Yan, J., (2019). De novo domestication: An alternative route toward new crops for the future. *Molecular Plant, 12*(5), 615–631.
Fiala, J. V., Tullu, A., Banniza, S., Séguin-Swartz, G., & Vandenberg, A., (2009). Interspecies transfer of resistance to anthracnose in lentil (*Lens culinaris* Medic.). *Crop Science, 49*(3), 825–830.
Fulton, T. M., Grandillo, S., Beck-Bunn, T., Fridman, E., Frampton, A., Lopez, J., Petiard, V., et al., (2000). Advanced backcross QTL analysis of a *Lycopersicon esculentum*× *Lycopersicon parviflorum* cross. *Theoretical and Applied Genetics, 100*(7), 1025–1042.

Fustier, M. A., Brandenburg, J. T., Boitard, S., Lapeyronnie, J., Eguiarte, L. E., Vigouroux, Y., Manicacci, D., & Tenaillon, M. I., (2017). Signatures of local adaptation in lowland and highland teosintes from whole-genome sequencing of pooled samples. *Molecular Ecology, 26*(10), 2738–2756.

Ghafoor, A., & McPhee, K., (2012). Marker assisted selection (MAS) for developing powdery mildew resistant pea cultivars. *Euphytica, 186*(3), 593–607.

Gobu, R., Shiv, A., Kumar, C. A., Basavaraj, P. S., Harish, D., Adhikari, S., Vinita, R., Umesh, H., & Sujatha, M., (2020). Accelerated crop breeding towards development of climate resilient varieties. In: Srinivasarao, C., Srinivas, T., Rao, R. V. S., Rao, N. S., Vinayagam, S. S., & Krishnan, P., (eds.), Climate change and Indian agriculture: Challenges and adaptation strategies (pp. 49–69). ICAR–National Academy of Agricultural Research Management, Hyderabad, Telangana, India.

Gregorio, G. B., Islam, M. R., Vergara, G. V., & Thirumeni, S., (2013). Recent advances in rice science to design salinity and other abiotic stress tolerant rice varieties. *SABRAO Journal of Breeding and Genetics, 45*(1), 31–40.

Grenier, C., Châtel, M., Ospina, Y., Cao, T. V., Guimaraes, E. P., Martinez, C., Tohmé, J., et al., (2012). Population improvement through recurrent selection in rice. Prospect for maker assisted recurrent selection and genome-wide selection. *Plant and Animal Genome XX*. San Diego, USA.

Gupta, P. K., & Varshney, R. K., (2004). Cereal genomics: An overview. *Cereal Genomics*, 1–18.

Gupta, P. K., Kumar, J., Mir, R. R., & Kumar, A., (2010). Marker-assisted selection as a component of conventional plant breeding. *Plant Breeding Reviews, 33*, 145.

Gupta, S. K., & Pandey, M. K., (2014). Breeding for disease and insect-pest resistance. In: *Integrated Pest Management* (pp. 401–417). Academic Press.

Hanson, P. M., Pastor-Corrales, M. A., & Kornegay, J. L., (1993). Heritability and sources of *Ascochyta* blight resistance in common bean. *Plant Disease* (Vol. 77, p. 711). St Paul.

Hanson, P., Lu, S. F., Wang, J. F., Chen, W., Kenyon, L., Tan, C. W., Tee, K. L., et al., (2016). Conventional and molecular marker-assisted selection and pyramiding of genes for multiple disease resistance in tomato. *Scientia Horticulturae, 201*, 346–354.

Harfouche, A. L., Jacobson, D. A., Kainer, D., Romero, J. C., Harfouche, A. H., Mugnozza, G. S., Moshelion, M., et al., (2019). Accelerating climate resilient plant breeding by applying next-generation artificial intelligence. *Trends in Biotechnology, 37*(11), 1217–1235.

Hasan, M. M., Rafii, M. Y., Ismail, M. R., Mahmood, M., Rahim, H. A., Alam, M. A., Ashkani, S., et al., A., (2015). Marker-assisted backcrossing: A useful method for rice improvement. *Biotechnology and Biotechnological Equipment, 29*(2), 237–254.

Heffner, E. L., Sorrells, M. E., & Jannink, J. L., (2009). Genomic selection for crop improvement. *Crop Science, 49*, 1–12.

Heinig, U., Gutensohn, M., Dudareva, N., & Aharoni, A., (2013). The challenges of cellular compartmentalization in plant metabolic engineering. *Current Opinion in Biotechnology, 24*(2), 239–246.

Hospital, F., (2005). Selection in backcross programs. *Philosophical Transactions of the Royal Society B: Biological Sciences, 360*(1459), 1503–1511.

Hu, J., Xiao, C., & He, Y., (2016). Recent progress on the genetics and molecular breeding of brown plant hopper resistance in rice. *Rice, 9*(1), 1–12.

Hurtado-Gonzales, O. P., Valentini, G., Gilio, T. A., Martins, A. M., Song, Q., & Pastor-Corrales, M. A., (2017). Fine mapping of Ur-3, a historically important rust resistance locus in common bean. *G3: Genes, Genomes, Genetics, 7*(2), 557–569.

Iqbal, M. J., Reddy, O. U. K., El-Zik, K. M., & Pepper, A. E., (2001). A genetic bottleneck in the 'evolution under domestication' of upland cotton *Gossypium hirsutum* L. examined using DNA fingerprinting. *Theoretical and Applied Genetics, 103*(4), 547–554.

Jackson, R. E., Bradley, Jr. J. R., & Van, D. J. W., (2004). Performance of feral and Cry1Ac-selected *Helicoverpa zea* (Lepidoptera: Noctuidae) strains on transgenic cottons expressing one or two *Bacillus thuringiensis* ssp. kurstaki proteins under greenhouse conditions. *Journal of Entomological Science, 39*(1), 46–55.

Janila, P., Pandey, M. K., Shasidhar, Y., Variath, M. T., Sriswathi, M., Khera, P., Manohar, S. S., et al., (2016). Molecular breeding for introgression of fatty acid desaturase mutant alleles (ahFAD2A and ahFAD2B) enhances oil quality in high and low oil containing peanut genotypes. *Plant Science, 242*, 203–213.

Jansky, S. H., Dempewolf, H., Camadro, E. L., Simon, R., Zimnoch-Guzowska, E., Bisognin, D. A., & Bonierbale, M., (2013). A case for crop wild relative preservation and use in potato. *Crop Science., 53*(3), 746–754.

Jiang, F., Zhang, T., Bai, S., Wang, Z., & He, K., (2016). Evaluation of Bt corn with pyramided genes on efficacy and insect resistance management for the Asian corn borer in China. *PLoS One, 11*(12), e0168442.

Joseph, M., Gopalakrishnan, S., Sharma, R. K., Singh, V. P., Singh, A. K., Singh, N. K., & Mohapatra, T., (2004). Combining bacterial blight resistance and basmati quality characteristics by phenotypic and molecular marker-assisted selection in rice. *Molecular Breeding, 13*(4), 377–387.

Kage, U., Kumar, A., Dhokane, D., Karre, S., & Kushalappa, A. C., (2016). Functional molecular markers for crop improvement. *Critical Reviews in Biotechnology, 36*(5), 917–930.

Kamburova, V. S., Nikitina, E. V., Shermatov, S. E., Buriev, Z. T., Kumpatla, S. P., Emani, C., & Abdurakhmonov, I. Y., (2017). Genome editing in plants: An overview of tools and applications. *International Journal of Agronomy*. https://doi.org/10.1155/2017/7315351.

Kamfwa, K., Mwala, M., Okori, P., Gibson, P., & Mukankusi, C., (2013). Identification of QTL for *Fusarium* root rot resistance in common bean. *Journal of Crop Improvement, 27*(4), 406–418.

Kaur, L., Sirari, A., Kumar, D., Sandhu, J. S., Singh, S., Kapoor, K., Singh, I., et al., (2013). Combining *Ascochyta* blight and botrytis grey mould resistance in chickpea through interspecific hybridization. *Phytopathologia Mediterranea, 52*(1), 157–165.

Keller, B., Manzanares, C., Jara, C., Lobaton, J. D., Studer, B., & Raatz, B., (2015). Fine-mapping of a major QTL controlling angular leaf spot resistance in common bean (*Phaseolus vulgaris* L.). *Theoretical and Applied Genetics, 128*(5), 813–826.

Kelly, J. D., (2004). Advances in common bean improvement: Some case histories with broader applications. *Acta Horticulturae, 637*, 99–122.

Khanh, T. D., Linh, T. H., & Xuan, T. D., (2013). Rapid and high-precision marker assisted backcrossing to introgress the SUB1 QTL into the Vietnamese elite rice variety. *Journal of Plant Breeding and Crop Science, 5*(2), 26–33.

Kim, J. S., (2016). Genome editing comes of age. *Nature Protocols, 11*(9), 1573–1578.

Knights, E. J., Southwell, R. J., Schwinghamer, M. W., & Harden, S., (2008). Resistance to *Phytophthora medicaginis* Hansen and Maxwell in wild Cicer species and its use in breeding root rot resistant chickpea (*Cicer arietinum* L.). *Australian Journal of Agricultural Research, 59*(4), 383–387.

Kottapalli, K. R., Narasu, M. L., & Jena, K. K., (2010). Effective strategy for pyramiding three bacterial blight resistance genes into fine grain rice cultivar, Samba Mahsuri, using sequence tagged site markers. *Biotechnology Letters, 32*(7), 989–996.

Kumar, A., Sandhu, N., Dixit, S., Yadav, S., Swamy, B. P. M., & Shamsudin, N. A. A., (2018). Marker-assisted selection strategy to pyramid two or more QTLs for quantitative trait-grain yield under drought. *Rice, 11*(1), 1–16.

Kumar, N., Rana, M., Kumar, B., Chand, S., Shiv, A., Wani, S. H., & Kumar, S., (2021). Genomic selection for wheat improvement. *Physiological, Molecular and Genetic Perspectives of Wheat Improvement*, 175–207.

Kumawat, G., Kumawat, C. K., Chandra, K., Pandey, S., Chand, S., Mishra, U. N., Lenka, D., & Sharma, R., (2020). Insights into marker assisted selection and its applications in plant breeding. In: *Plant Breeding-Current and Future Views* (pp. 1–21). Intech Open.

Lavaud, C., Lesne, A., Piriou, C., Le Roy, G., Boutet, G., Moussart, A., Poncet, C., et al., (2015). Validation of QTL for resistance to *Aphanomyces euteiches* in different pea genetic backgrounds using near-isogenic lines. *Theoretical and Applied Genetics, 128*(11), 2273–2288.

Lemmon, Z. H., Reem, N. T., Dalrymple, J., Soyk, S., Swartwood, K. E., Rodriguez-Leal, D., Van, E. J., & Lippman, Z. B., (2018). Rapid improvement of domestication traits in an orphan crop by genome editing. *Nature Plants, 4*(10), 766–770.

Lenaerts, B., Collard, B. C., & Demont, M., (2019). Improving global food security through accelerated plant breeding. *Plant Science, 287*, 110207.

Li, T., Yang, X., Yu, Y., Si, X., Zhai, X., Zhang, H., Dong, W., et al., (2018). Domestication of wild tomato is accelerated by genome editing. *Nature Biotechnology, 36*(12), 1160–1163.

Liu, M., Li, Y., Ma, Y., Zhao, Q., Stiller, J., Feng, Q., Tian, Q., et al., (2020). The draft genome of a wild barley genotype reveals its enrichment in genes related to biotic and abiotic stresses compared to cultivated barley. *Plant Biotechnology Journal, 18*(2), 443–456.

Liu, Y., & Santi, D. V., (2000). m5C RNA and m5C DNA methyl transferases use different cysteine residues as catalysts. *Proceedings of the National Academy of Sciences, 97*(15), 8263–8265.

Liu, Y., Chen, L., Liu, Y., Dai, H., He, J., Kang, H., Pan, G., et al., (2016). Marker assisted pyramiding of two brown planthopper resistance genes, Bph3 and Bph27 (t), into elite rice cultivars. *Rice, 9*(1), 1–7.

Lokanathan, T. R., Singh, P., Agarwal, D. K., Mohan, P., Singh, S. B., Gotmare, V., & Singh, V. V., (2003). *Genetic Enhancement in Cotton*. Technical Bulletin from CICR Report No. 26.

Lorieux, M., (2005). *CSSL Finder: A Free Program for Managing Introgression Lines*. http://mapdisto.free.fr/.

Lucas, M. R., Huynh, B. L., Roberts, P. A., & Close, T. J., (2015). Introgression of a rare haplotype from Southeastern Africa to breed California blackeyes with larger seeds. *Frontiers in Plant Science, 6,* 126.

Lv, H., Fang, Z., Yang, L., Zhang, Y., & Wang, Y., (2020). An update on the arsenal: Mining resistance genes for disease management of brassica crops in the genomic era. *Horticulture Research, 7*(1), 1–18.

Malhotra, R. S., Singh, K. B., Di Vito, M., Greco, N., & Saxena, M. C., (2002). Registration of ILC 10765 and ILC 10766 chickpea germplasm lines resistant to cyst nematode. (registrations Of germplasm). *Crop Science, 42*(5), 1756–1757.

Mallikarjuna, N., Sharma, H. C., & Upadhyaya, H. D., (2007). Exploitation of wild relatives of pigeonpea and chickpea for resistance to *Helicoverpa armigera. Journal of SAT Agricultural Research, 3*(1), 4.

Maruthasalam, S., Kalpana, K., Kumar, K. K., Loganathan, M., Poovannan, K., Raja, J. A. J., Kokiladevi, E., et al., (2007). Pyramiding transgenic resistance in elite indica rice cultivars against the sheath blight and bacterial blight. *Plant Cell Reports, 26*(6), 791–804.

Maxted, N., & Kell, S. P., (2009). *Establishment of a Global Network for the in Situ Conservation of Crop Wild Relatives: Status and Needs* (p. 112). FAO Commission on Genetic Resources for Food and Agriculture, Rome, Italy.

Mehrotra, M., Singh, A. K., Sanyal, I., Altosaar, I., & Amla, D. V., (2011). Pyramiding of modified cry1Ab and cry1Ac genes of *Bacillus thuringiensis* in transgenic chickpea (*Cicer arietinum* L.) for improved resistance to pod borer insect (*Helicoverpa armigera*). *Euphytica, 182*(1), 87–102.

Miklas, P. N., Fourie, D., Chaves, B., & Chirembe, C., (2017). Common bacterial blight resistance QTL BC420 and SU91 effect on seed yield, seed weight, and canning quality in dry bean. *Crop Science, 57*(2), 802–811.

Miklas, P. N., Kelly, J. D., Beebe, S. E., & Blair, M. W., (2006). Common bean breeding for resistance against biotic and abiotic stresses: From classical to MAS breeding. *Euphytica, 147*(1), 105–131.

Muedi, H. T., Fourie, D., & McLaren, N. W., (2015). Distribution and severity of bacterial brown spot on dry beans in South Africa: An update. *South African Journal of Science, 111*(3, 4), 1–6.

Mwobobia, E. G., Sichangi, A. W., & Thiong'o, K. B., (2020). Characterization of wheat production using earth-based observations: A case study of Meru County, Kenya. *Modeling Earth Systems and Environment, 6*(1), 13–25.

Nadeem, M. A., Nawaz, M. A., Shahid, M. Q., Doğan, Y., Comertpay, G., Yıldız, M., Hatipoğlu, R., et al., (2018). DNA molecular markers in plant breeding: Current status and recent advancements in genomic selection and genome editing. *Biotechnology and Biotechnological Equipment, 32*(2), 261–285.

Narayanan, N. N., Baisakh, N., Vera, C. C. M., Gnanamanickam, S. S., Datta, K. B., & Datta, S. K., (2002). Molecular breeding for the development of blast and bacterial blight resistance in rice cv. IR50. *Crop Science, 42*(6), 2072–2079.

Nelson, R., Wiesner-Hanks, T., Wisser, R., & Balint-Kurti, P., (2018). Navigating complexity to breed disease-resistant crops. *Nature Reviews Genetics, 19*(1), 21–33.

Pande, S., Sharma, M., Mangla, U. N., Ghosh, R., & Sundaresan, G., (2011). Phytophthora blight of pigeon pea [*Cajanus cajan* (L.) Millsp.]: An updating review of biology, pathogenicity and disease management. *Crop Protection, 30*(8), 951–957.

Pandey, A., Chakraborty, S., Datta, A., & Chakraborty, N., (2008). Proteomics approach to identify dehydration responsive nuclear proteins from chickpea (*Cicer arietinum* L.). *Molecular and Cellular Proteomics, 7*(1), 88–107.

Pandey, M. K., Roorkiwal, M., Singh, V. K., Ramalingam, A., Kudapa, H., Thudi, M., Chitikineni, A., et al., (2016). Emerging genomic tools for legume breeding: Current status and future prospects. *Frontiers in Plant Science., 7*, 455.

Pandey, P., Irulappan, V., Bagavathiannan, M. V., & Senthil-Kumar, M., (2017). Impact of combined abiotic and biotic stresses on plant growth and avenues for crop improvement by exploiting physio-morphological traits. *Frontiers in Plant Science, 8*, 537.

Parhe, S. D., Chimote, V. P., Deshmukh, M. P., Chandra, K., & Akash, M., (2017). Marker-assisted pyramiding of four QTL/genes for Asian rust (*Phakopsora pachyrhizi*) resistance in soybean. *Journal of Crop Improvement, 31*(5), 689–711.

Peleman, J. D., & Van, D. V. J. R., (2003). Breeding by design. *Trends in Plant Science, 8*(7), 330–334.

Piasecka, A., Kachlicki, P., & Stobiecki, M., (2019). Analytical methods for detection of plant metabolomes changes in response to biotic and abiotic stresses. *International Journal of Molecular Sciences, 20*(2), 379.

Pratap, A., Chaturvedi, S. K., Tomar, R., Rajan, N., Malviya, N. Thudi, M., Saabale, P. R., et al., (2017). Marker-assisted introgression of resistance to fusarium wilt race 2 in Pusa 256, an elite cultivar of desi chickpea. *Molecular Genetics and Genomics 292*(6), 1237–1245.

Prohens, J., Gramazio, P., Plazas, M., Dempewolf, H., Kilian, B., Diez, M. J., Fita, A., et al., (2017). Introgressiomics: A new approach for using crop wild relatives in breeding for adaptation to climate change. *Euphytica., 213*(7), 1–19.

Puspito, A. N., Rao, A. Q., Hafeez, M. N., Iqbal, M. S., Bajwa, K. S., Ali, Q., Rashid, B., et al., (2015). Transformation and evaluation of Cry1Ac + Cry2A and GT gene in *Gossypium hirsutum* L. *Frontiers in Plant Science, 6*, 943.

Rana, M., Sood, A., Hussain, W., Kaldate, R., Sharma, T. R., Gill, R. K., Kumar, S., & Singh, S., (2019). Gene pyramiding and multiple character breeding. In: *Lentils* (pp. 83–124) Academic Press.

Razzaq, A., Sadia, B., Raza, A., Khalid, H. M., & Saleem, F., (2019). Metabolomics: A way forward for crop improvement. *Metabolites, 9*(12), 303.

Rezene, Y., Tesfaye, K., Mukankusi, C., Ratz, B., & Gepts, P., (2019). Marker-assisted pyramiding resistance genes against angular leaf spot and common bacterial blight disease into preferred common bean cultivar "Redwolaita". *Advances in Crop Science and Technology, 7*(1), 416. doi: 10.4172/2329-8863.1000416.

Ribaut, J. M., & Ragot, M., (2006). Marker-assisted selection to improve drought adaptation in maize: The backcross approach, perspectives, limitations, and alternatives. *Journal of Experimental Botany, 58*(2), 351–360.

Ribaut, J. M., De Vicente, M. C., & Delannay, X., (2010). Molecular breeding in developing countries: Challenges and perspectives. *Current Opinion in Plant Biology, 13*(2), 213–218.

Sauer, U. G., Deferme, L., Gribaldo, L., Hackermüller, J., Tralau, T., Van, R. B., Yauk, C., et al., (2017). The challenge of the application of omics technologies in chemicals risk assessment: Background and outlook. *Regulatory Toxicology and Pharmacology, 91*, S14–S26.

Sebolt, A. M., Shoemaker, R. C., & Diers, B. W., (2000). Analysis of a quantitative trait locus allele from wild soybean that increases seed protein concentration in soybean. *Crop Science, 40*(5), 1438–1444.

Sedeek, K. E., Mahas, A., & Mahfouz, M., (2019). Plant genome engineering for targeted improvement of crop traits. *Frontiers in Plant Science, 10*, 114.

Seetharam, A., (2007). Pre-breeding: An important step in the effective utilization of conserved germplasm. In: *National Workshop on Utilization of Wild Mulberry Genetic Resources* 2nd and 3rd Nov (pp. 9–16).

Septiningsih, E. M., Collard, B. C., Heuer, S., Bailey-Serres, J., Ismail, A. M., & Mackill, D. J., (2013). Applying genomics tools for breeding submergence tolerance in rice. *Translational Genomics for Crop Breeding, 2*, 9–30.

Sharma, S., Upadhyaya, H. D., Roorkiwal, M., Varshney, R. K., & Mason, A. S., (2016). Interspecific hybridization for chickpea (*Cicer arietinum* L.) improvement. *Polyploidy and Hybridization for Crop Improvement* (pp. 445–469). CRC Press. Boca Raton FL.

Sharma, S., Upadhyaya, H. D., Varshney, R. K., & Gowda, C. L. L., (2013). Pre-breeding for diversification of primary gene pool and genetic enhancement of grain legumes. *Frontiers in Plant Science, 4*, 309.

Shi, A., Chen, P., Li, D., Zheng, C., Zhang, B., & Hou, A., (2009). Pyramiding multiple genes for resistance to soybean mosaic virus in soybean using molecular markers. *Molecular Breeding, 23*(1), 113–124.

Shirasawa, K., & Hirakawa, H., (2013). DNA marker applications to molecular genetics and genomics in tomato. *Breeding Science, 63*(1), 21–30.

Singh, G., Singh, I., Taggar, G. K., Rani, U., Sharma, P., Gupta, M., & Singh, S., (2020). Introgression of productivity enhancing traits, resistance to pod borer and *Phytopthora* stem blight from *Cajanus scarabaeoides* to cultivated pigeonpea. *Physiology and Molecular Biology of Plants, 26*, 1399–1410.

Singh, S., Gumber, R. K., Joshi, N., & Singh, K., (2005). Introgression from wild *Cicer reticulatum* to cultivated chickpea for productivity and disease resistance. *Plant Breeding, 124*(5), 477–480.

Singh, S., Sidhu, J. S., Huang, N., Vikal, Y., Li, Z., Brar, D. S., Dhaliwal, H. S., & Khush, G. S., (2001). Pyramiding three bacterial blight resistance genes (Xa5, Xa13 and Xa21) using marker-assisted selection into indica rice cultivar PR106. *Theoretical and Applied Genetics, 102*(6), 1011–1015.

Sliwka, J., Jakuczun, H., Chmielarz, M., Hara-Skrzypiec, A., Tomczyńska, I., Kilian, A., & Zimnoch-Guzowska, E., (2012). Late blight resistance gene from solanum ruiz-ceballosii is located on potato chromosome X and linked to violet flower color. *BMC Genetics, 13*(1), 1–12.

Smykal, P., Nelson, M. N., Berger, J. D., & Von, W. E. J., (2018). The impact of genetic changes during crop domestication. *Agronomy, 8*(7), 119.

Steiner, B., Michel, S., Maccaferri, M., Lemmens, M., Tuberosa, R., & Buerstmayr, H., (2019). Exploring and exploiting the genetic variation of *Fusarium* head blight resistance for genomic-assisted breeding in the elite durum wheat gene pool. *Theoretical and Applied Genetics, 132*(4), 969–988.

Strausbaugh, C. A., Myers, J. R., Forster, R. L., & McClean, P. E., (1999). bc-1 and bc-u—two loci controlling bean common mosaic virus resistance in common bean are linked. *Journal of the American Society for Horticultural Science, 124*(6), 644–648.

Takegami, J. C., Beaver, J. S., Godoy-Lutz, G., Echávez-Badel, R., & Steadman, J. R., (2004). Inheritance of web blight resistance in common bean. *The Journal of Agriculture of the University of Puerto Rico., 88*(1, 2), 45–54.

Tanksley, S. D., & Nelson, J. C., (1996). Advanced backcross QTL analysis: A method for the simultaneous discovery and transfer of valuable QTLs from unadapted germplasm into elite breeding lines. *Theoretical and Applied Genetics, 92*(2), 191–203.

Tanksley, S. D., Grandillo, S., Fulton, T. M., Zamir, D., Eshed, Y., Petiard, V., Lopez, J., & Beck-Bunn, T., (1996). Advanced backcross QTL analysis in a cross between an elite processing line of tomato and its wild relative *L. pimpinellifolium*. *Theoretical and Applied Genetics, 92*(2), 213–224.

Tanksley, S. D., Young, N. D., Paterson, A. H., & Bonierbale, M. W., (1989). RFLP mapping in plant breeding: New tools for an old science. *Biotechnology, 7*(3), 257–264.

Taran, B., Buchwaldt, L., Tullu, A., Banniza, S., Warkentin, T. D., & Vandenberg, A., (2003). Using molecular markers to pyramid genes for resistance to *Ascochyta* blight and anthracnose in lentil (*Lens culinaris* Medik). *Euphytica, 134*(2), 223–230.

Tock, A. J., Fourie, D., Walley, P. G., Holub, E. B., Soler, A., Cichy, K. A., Pastor-Corrales, M. A., et al., (2017). Genome-wide linkage and association mapping of halo blight resistance in common bean to race 6 of the globally important bacterial pathogen. *Frontiers in Plant Science, 8*, 1170.

Tong, H., Küken, A., & Nikoloski, Z., (2020). Integrating molecular markers into metabolic models improves genomic selection for *Arabidopsis* growth. *Nature Communications, 11*(1), 1–9.

Tugizimana, F., Mhlongo, M. I., Piater, L. A., & Dubery, I. A., (2018). Metabolomics in plant priming research: The way forward. *International Journal of Molecular Sciences, 19*(6), 1759.

Upadhyaya, H. D., (2008). Crop germplasm and wild relatives: A source of novel variation for crop improvement. *Korean Journal of Crop Science, 53*, 12–17.

Valkoun, J. J., (2001). Wheat pre-breeding using wild progenitors. *Euphytica, 119*(1), 17–23.

Varshney, R. K., (2011). Application of next generation sequencing and genotyping technologies to develop large-scale genomic resources in SAT legume crops. In: Muralidharan, K., & Siddiq, E. A., (eds.), *Genomics and Crop Improvement: Relevance and Reservations* (pp. 1–10). Hyderabad, India: Acharya NG Ranga Agricultural University.

Varshney, R. K., (2016). Exciting journey of 10 years from genomes to fields and markets: Some success stories of genomics-assisted breeding in chickpea, pigeonpea and groundnut. *Plant Science, 242*, 98–107.

Varshney, R. K., Glaszmann, J. C., Leung, H., & Ribaut, J. M., (2010b). More genomic resources for less-studied crops. *Trends in Biotechnology, 28*(9), 452–460.

Varshney, R. K., Hoisington, D. A., & Tyagi, A. K., (2006). Advances in cereal genomics and applications in crop breeding. *Trends in Biotechnology, 24*(11), 490–499.

Varshney, R. K., Kudapa, H., Roorkiwal, M., Thudi, M., Pandey, M. K., Saxena, R. K., Chamarthi, S. K., et al., (2012). Advances in genomics research and molecular breeding applications in SAT legume crops by using next generation sequencing and high-throughput genotyping technologies. *Journal of Biosciences, 37*, 811–820.

Varshney, R. K., Nayak, S. N., May, G. D., & Jackson, S. A., (2009c). Next-generation sequencing technologies and their implications for crop genetics and breeding. *Trends in Biotechnology, 27*(9), 522–530.

Varshney, R. K., Pandey, M. K., Janila, P., Nigam, S. N., Sudini, H., Gowda, M. V. C., Sriswathi, M., et al., (2014). Marker-assisted introgression of a QTL region to improve rust resistance in three elite and popular varieties of peanut (*Arachis hypogaea* L.). *Theoretical and Applied Genetics, 127*(8), 1771–1781.

Varshney, R. K., Song, C., Saxena, R. K., Azam, S., Yu, S., Sharpe, A. G., Cannon, S., et al., (2013b). Draft genome sequence of chickpea (*Cicer arietinum*) provides a resource for trait improvement. *Nature Biotechnology, 31*(3), 240–246.

Verlaan, M. G., Hutton, S. F., Ibrahem, R. M., Kormelink, R., Visser, R. G., Scott, J. W., Edwards, J. D., & Bai, Y., (2013). The tomato yellow leaf curl virus resistance genes Ty-1 and Ty-3 are allelic and code for DFDGD-class RNA–dependent RNA polymerases. *PLoS Genetics, 9*(3), e1003399.

Vincelli, P., (2016). Genetic engineering and sustainable crop disease management: Opportunities for case-by-case decision-making. *Sustainability, 8*(5), 495.

Von, W. E. J., Chang, P. L., Başdemir, F., Carrasquila-Garcia, N., Korbu, L. B., Moenga, S. M., Bedada, G., et al., (2018). Ecology and genomics of an important crop wild relative as a prelude to agricultural innovation. *Nature Communications, 9*(1), 1–13.

Waddington, S. R., Li, X., Dixon, J., Hyman, G., & De Vicente, M. C., (2010). Getting the focus right: Production constraints for six major food crops in Asian and African farming systems. *Food Security, 2*(1), 27–48.

Wang, J. K., (2007). Simulation modeling in plant breeding: Principles and applications. *Agricultural Sciences in China, 6*(8), 908–921.

Wang, Y., Jiang, W., Liu, H., Zeng, Y., Du, B., Zhu, L., He, G., & Chen, R., (2017). Marker assisted pyramiding of Bph6 and Bph9 into elite restorer line 93–11 and development of functional marker for Bph9. *Rice, 10*(1), 1–13.

Werner, K., Friedt, W., & Ordon, F., (2005). Strategies for pyramiding resistance genes against the barley yellow mosaic virus complex (BaMMV, BaYMV, BaYMV-2). *Molecular Breeding, 16*(1), 45–55.

William, H. M., Morris, M., Warburton, M., & Hoisington, D. A., (2007). *Technical, Economic and Policy Considerations on Marker-Assisted Selection in Crops: Lessons from the Experience at an International Agricultural Research Centre* (p. 381). Marker-Assisted Selection.

Zhang, B., Chi, D., Hiebert, C., Fetch, T., McCallum, B., Xue, A., Cao, W., Depauw, R., & Fedak, G., (2019). Pyramiding stem rust resistance genes to race TTKSK (Ug99) in wheat. *Canadian Journal of Plant Pathology, 41*(3), 443–449.

CHAPTER 9

Perspectives of MutMap and Its Variants Towards Next-Generation Plant Breeding

S. L. SOUMYA[1] and S. MUKESH SANKAR[2]

[1]*Protection of Plant Varieties and Farmers' Rights Authority (PPVFRA), Ministry of Agriculture and Farmers Welfare, Government of India, New Delhi, India*

[2]*Division of Genetics, ICAR–Indian Agricultural Research Institute, New Delhi, India*

ABSTRACT

During the recent years, crop genomics has exploded in popularity with the dawn of next-generation sequencing (NGS) backed by high performance computing techniques and pipelines which can handle the big data in genomics. Whole genome sequencing (WGS) based on various NGS techniques are becoming widely exploiting technique and draft genomes of several crop species are now easily available. This technology provides a major positive impact on crop improvement in the most threatening era of food crisis associated with increasing world population. Draft genome sequence accessibility leads to the precise mapping of the genes associated with particular phenotypic traits in plant breeding. Conventional technique of gene mapping is very costly, slow, and low throughput nature demanding large mapping populations, cloning onto large sized vectors such as BAC and YAC followed by library construction which can yield only limited number of molecular markers spanning the whole linkage groups (LGs). Nowadays, quick mapping of genes based on resequencing techniques become effective within a short span of time and low cost. Currently, trait mapping approaches such as SHOREmap, NGM, QTL-Seq, MutMap, MutMap+, MutMap-Gap,

Smart Breeding: Molecular Interventions and Advancements for Crop Improvement. Kailash Chandra, Subhash Chand, Ravi Prakash Saini, and Rohit Sharma (Eds.)

Seq-BSA, Indel-Seq, BSR-Seq, and SLAF-Seq in which variant analysis demands pooled samples over individual samples of whole population. Among these techniques, MutMap is widely used technique for the identification of causal genes because it requires minimum time and effort. MutMap techniques use only the pooled DNA samples from a segregating population to resequence the entire genome to detect the causal SNP accountable for the mutant phenotype. For which a backcross (BC) population of the mutant line with the corresponding wild type is generated comparative SNP analysis in both DNA pools contrasting for the target phenotype were carried out to discard the non-causal SNPs, thereby pinpointing SNPs associated with the mutant phenotype. MutMap techniques and its variants were now exploited in a large way in a large number of crop plants, such as rice, chickpea, wheat, sorghum, maize, soybean, cucumber, etc., for the candidate gene identification and its mapping.

9.1 INTRODUCTION

Next-generation sequencing (NGS) technologies have advanced greatly from the days of Sanger sequencing to deliver enhanced data output, efficiency, and applications. The pioneering work of Frederick Sanger and colleagues enabled the development of new techniques that have opened up new avenues for low-cost, rapid DNA sequencing (Zhang et al., 2011). The NGS has been categorized into various generations based on the read length that has been used in a wide range of molecular biology applications to overcome the constraints of traditional DNA sequencing technologies. Second generation or Short-read technology and 3rd generation or long-read sequencing technologies (3GS) are the two main categories. Illumina and Ion Torrent are the most common short-read sequencing approaches, whereas Pacific Biosciences and Oxford Nanopore technology are the most popular long-read sequencing approaches. Second generation sequencing was considered to provide excellent accuracy to its cost but was restricted by read length, whereas long-read technologies allowed for much greater read lengths but at the cost of accuracy. With the coexistence of longer read lengths and excellent accuracy, emerging third-generation technologies provide promise for the next phase of sequencing evolution (Hu et al., 2021). DNA-sequencing and assembly to discover an unknown genome without any preparation or search for variances among genome samples are just some of the applications of NGS technologies (Wang et al., 2009; Morin et al., 2008) to investigate gene expression (Furey, 2012) and in particular, to identify regions of DNA binding proteins, such as transcription factors and others. Identifying mutations is the most important application of NGS. Short NGS reads (50–250 bp) are commonly mapped to a reference genome first, and then changes are discovered from the mapped data. NGS also detects mutations that are undetected by traditional sequencing methods (Wadapurkar & Vyas, 2018).

The success of any crop improvement program depends on the extent of availability of natural variations. Spontaneous mutations along with genetic recombination of new genes through meiosis and subsequent act of environmental selection all leads to new variety (Novak & Brunner, 1992). But these processes occur very slowly in nature and one can hasten this through artificial means. Mutation breeding involves the creation and exploitation of the genetic variability resulting from chemical, physical or insertional mutagenesis in the crop improvement leading to new crop varieties (Kharkwal & Shu, 2009; Forster & Shu, 2012). Along with this, recombinant breeding and transgenic breeding, forms the foundation for the modern plant breeding (Shu et al., 2012). The popularity of a certain variety of a particular crop over the other is due to its presence of a novel gene or a combination of genes. Understanding the number of genes, their structure, location, genetic inheritance, functions, interaction with other genes and environment all plays a crucial role in the genetic improvement of crop for the target trait. Traditionally, these data were derived through morphological observation of traits and their statistical analysis. However, these efforts gained only limited success due to complexity in certain traits including yield and tolerance to abiotic stresses. With the advent of molecular marker technology during 1987 followed by the considerable refinements in statistical techniques, paved the way for discovery of the quantitative trait loci (QTLs) which enables speedy and more precise selections and their introgression in the desired genetic background (Jaganathan et al., 2020).

QTLs mapping refers to the process of identifying the genomic regions responsible for the expression of quantitative trait. The major factor responsible for successful genetic dissection of QTLs through QTL mapping approaches were: (i) number of molecular markers employed; (ii) type of mapping population; (iii) size of mapping population; and (iv) precision in phenotyping. Utilization of limited number molecular markers will lead to coarse mapping which often extend to few centimorgan (which may be several Mbp in physical map) which may contain several unwanted genes along with the target gene. Thus, while trait introgression *via* broad QTLs segments, it can often lead to linkage drag and may lead to many false positives for the marker presence but not the trait. Hence, QTL mapping advocate the utilization of large number of molecular markers preferably genome-wide SNPs to pinpoint the causal genetic variation in genome. In this context, new generation sequencing techniques such as NGS and 3GS can comes as rescue. These technologies considerably reduced the cost of genotyping/sequencing per sample over a period of time and also increased throughput dramatically. These techniques can yield huge amounts of SNPs via sequence comparison or alignment with reference genome thereby improving QTL mapping resolution from 10 to 30 cM to 0.5–1 cM.

The traditional QTL mapping approach demands genotyping a huge amount of genotypes in a segregating population which is often less economic, time demanding and labor-intensive (Salvi & Tuberosa, 2005). In this context, Bulked Segregant Analysis (BSA) suggested by Michelmore et al., 1991 can offer more easily and effectual alternative to shortlist those DNA markers associated to QTL or genes affecting a trait of interest, using just one pair of collected DNA samples from two groups of individuals with discrete or contradicting phenotypes. NGS allows for the discovery of SNPs, indels, and structural variants across the entire genome (Ossowski et al., 2008) possible to compare entire genomes even in crops with minimal genome sequence information. While comparing mutant and wild-type genomes, induced mutations (induced through artificial means) can be detected (Nordström et al., 2013) the huge number of mutations in mutant genomes makes it difficult to identify causal mutations, or a small collection of candidate mutations, via forward genetic approaches. Genetic linkage analysis can be employed in sequencing-based analysis of pools of recombinant mutant genomes to discriminate between causative and non-causal mutations (Schneeberger et al., 2009) in one step, which is generally termed as mapping by sequencing (Schneeberger & Weigel, 2011) to minimize the list of possible causal mutations.

A series of techniques was developed employing NGS platform to decipher the genomic location of mutant genes based on the principle BSA. These included the following: SHOREmap (SHOrt REad map); NGM (next generation mapping); QTL-Seq; MutMap; MutMap+; MutMap-Gap; Seq-BSA; Indel-Seq; BSR-Seq; and SLAF-Seq techniques which would be able to successfully pinpoint the candidate gene based mutant responsible for the expression of particular phenotype (Strope et al., 2007; Schneeberger et al., 2009; Austin et al., 2011; Abe et al., 2012; Liu et al., 2012; Fekih et al., 2013; Sun et al., 2013; Takagi et al., 2013a, b). Prior to discussing all the above methods one by one, let us to see the principle of bulk segregants analysis and how these methods are associated with BSA.

9.2 BULK SEGREGANT ANALYSIS (BSA)

Michelmore et al. (1991) was first to devise bulk segregant analysis (BSA) to clearly spot genetic markers putatively related to a genomic region associated with disease resistance in rice. The mapping approach based on the principle of near-isogenic lines (NILs). NILs have identical genotypic composition but differ from each other for a few loci associated with the target gene to be mapped. For conducting BSA, parents with contrasting characters, such as a virus resistant and a susceptible line, are crossed to generate appropriate mapping population, such as an F_2, backcross (BC), DH, RIL, etc. Individual plants with extreme phenotypes were found in the mapping population, and an equal quantity of DNA was collected from the 10 most resistant and 10 most vulnerable plants to produce two bulks, resistant and susceptible bulks. Trait linked molecular marker can be delineated through screening all the available markers available for particular species using the DNAs from the two parents and the two bulk. "Any marker that shows polymorphism between parents as well as the resistant and susceptible DNA bulks is most likely linked to the target gene/trait, in this case, disease resistance. Further, these traits linked polymorphic markers run on entire population to understand the degree of linkage based on recombination fraction" (Michelmore et al., 1991). As a result, by preparing two DNA batches from samples obtained from mapping population individuals produced from a cross between parents differing in trait of interest, we were finally able to develop a pair of isogenic lines differing in trait of interest.

In Figure 9.1, marker A represent the unlinked marker condition, whereas marker B and C represent the polymorphic marker between contrasting parents which is linked to the trait of interest in which marker B is linked with a dominant resistance gene and marker C is linked with recessive resistance gene.

9.3 SHOREMAP (SHORT READ MAP)

SHOREmap is a gene mapping approach formulated by Schneeberger et al. (2009) to locate the site of mutation responsible for sluggish growth and light green leaves in *Arabidopsis* using NGS platform. SHOREmap analyzes

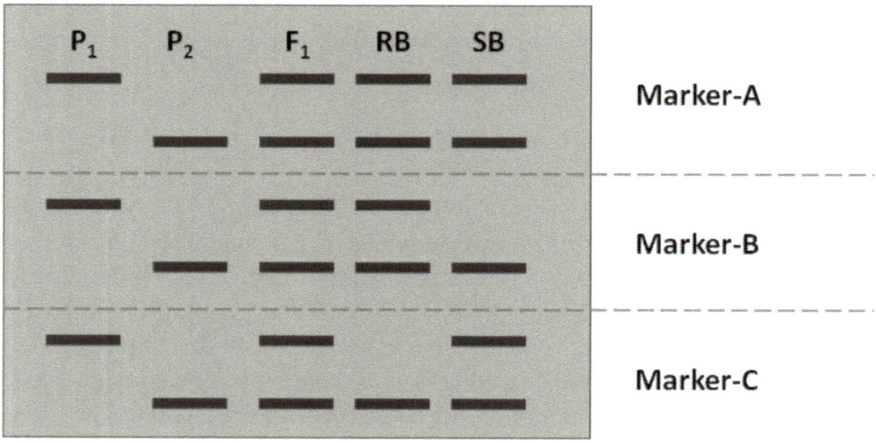

FIGURE 9.1 Schematic representation of the working principle of bulk segregant analysis, where; P_1 represent Parent 1, showing resistance to concern disease whereas; P_2 represents Parent 2, showing susceptible reaction. RB represents the bulked DNA of segregant showing resistance reaction and SB represents susceptible bulk generated from the DNA of susceptible segregant.

combines' genome-wide genotyping and reference-based mapping of short reads to identify the candidate gene from a large pool of recombinants (Lister et al., 2009). Initially, random mutations on the genome were created though ethyl methane sulfonate (EMS) mutagenesis. Further, homozygous mutant lines producing desired target phenotype were generated through selfing the M_1 generation which was then employed for gene mapping. These homozygous lines were further cross to a divergent accession (a distant relative line) to weaken the distribution of false single nucleotide polymorphism (SNP) all over the genome to obtain F_2 population (Leshchiner et al., 2012). As similar to BSA, a pooled DNA prepared from around 500 mutant genotypes with target trait were sequenced to obtain causal SNP between divergent accession and mutant bulk. Progenies representing wild type phenotype are mostly likely to carry SNP distribution similar to wild type (parent) for loci controlling wild type phenotype. Outcrossing the mutant to a wild type strain with a varied genetic background creates lots of natural polymorphisms into the genomes of recombinant, allowing for precise allele frequency calculations (Schneeberger et al., 2009). The result of a typical resequencing study of the recombinant pool is used for SHORE mapping. This approach was utilized to locate and limit the locus identified by the underlying changes in the chromosome to identify causative SNPs. SHOREMap could also map QTLs, recessive lethal or dominant mutations and large deletions. The disadvantage regarding SHOREMap approach is that it requires a sequenced reference genome; to overcome this shortcoming, it was proposed to use the genome of related organisms as "bridge" between the mutant and wild-type genomes. Refined algorithms are being developed by various labs for the analysis of reference-free homozygous mutation discovery through SHOREMap approach.

9.4 NGM (NEXT GENERATION MAPPING)

Next generation mapping (NGM) is extremely similar to SHOREmap, with the exception that it improves its ability to distinguish causal mutations by utilizing as little as 10 mutant plants for mutant bulk. The NGM utilizes the principles of linkage disequilibrium (LD) quantifies the parent-mutant relative contribution and maps lines with chastity statistics for each SNP in the pooled F_2 population. This information is then used to objectively identify the alleged nucleotide alteration by mutation based on the parent-mutant line's exclusive segregation. The main advantage of NGM is that it does not need any prior mapping data and requires only a small F_2 population. Austin et al. (2011) used NGM mapping to unravel the genes related to cell wall biosynthesis and their maintenance in *A. thaliana*. The NGM could be able to identify the non-recombinant, mutation-harboring region of the chromosome regardless of the type of mutation (SNP or insertion/deletion) that causes the phenotype.

NGM has an easy-to-use interface for rapidly locating a small number of candidate SNPs that are accountable for a causal trait. It uses data obtained from NGS short-read mapping and SNP calling software packages to create an interactive set of processes that result in an annotated list of SNPs that might be relevant for the trait observed (Austin et al., 2011).

9.5 MUTMAP

MutMap technique was devised by Abe et al. (2012) for mapping recessive genes showing monogenic inheritance, such as pale leaf color in rice. MutMap is based on bulked-segregant analysis and forms a quick method for fishing the mutant gene. A mutant (showing altered phenotype) chosen from the M_2 or later generation is crossed to a wild type of parent. The resultant F_1 plants are self-pollinated to generate F_2 mapping population. A similar amount of DNA from about 20–30 plants with the mutant and wild phenotypes is pooled and subjected to NGS-based WGS with substantial genomic coverage (10X). At the same time, a parental reference sequence is built by resequencing the wild type parent and aligning the reads to the species' publicly available reference genome (Figure 9.2). The alignment of F_2 bulked sequence reads to this parental reference sequence is then used to infer the genomic location of the causal SNPs responsible for the mutant phenotype. Unlinked or loosely linked SNP to phenotype will segregate in 1:1 mutant/wild type ratio whereas causal SNP will show 1:0 mutant/wild type ratios. Furthermore, an SNP index, which is the ratio of the number of reads corresponding to a mutant SNP to the total number of reads corresponding to the SNP, aids in the prediction of loci linked to the mutant phenotype. SNPs with an SNP index of 1 or close to 1 are linked to mutant phenotype, while those with an SNP index of 0.5 are unlinked loci. MutMap can detect SNPs, insertions, and deletions, as well as other nucleotide variations caused by mutagenesis.

MutMap is more feasible for crops with extended generation time since it does not require successive backcrosses. MutMap is especially beneficial for discovering genes that control quantitative minor effect traits, which is a difficult task in crop development. MutMap technique was successfully demonstrated in crop such as rice by Abe et al. (2012) to fish out causative SNP in the chlorophyllideea-oxygenase (*OsCAO1*) gene that lead to a pale-green phenotype. Similarly, technique was used for the characterization of salt tolerant rice mutant (*hst1*) from rice cultivar-Hitomebore (Takagi et al., 2015) and in sorghum for identification of a male sterile gene (*ms9*) (Chen et al., 2019).

9.6 MUTMAP+

The MutMap approach involves identifying concern mutants at M_3–M_5 generations, crossing them to the parental line, and then the segregating F_2 progenies are subjected to phenotyping. In case if mutants generated are not crossable, due to early development lethality or sterility, are not suited for MutMap. To solve this issue, Fekih et al. (2013) developed MutMap+, a versatile MutMap extension technique applied to rice mutants to quickly recognize the causal mutations by whole genome resequencing and comparison of SNP-index plots of two bulked DNA got from a segregating M_3 progeny. These investigations did not need crossing of candidate mutants with wild-type plants, allowing for novel applications not feasible with the original MutMap approach (Figure 9.3).

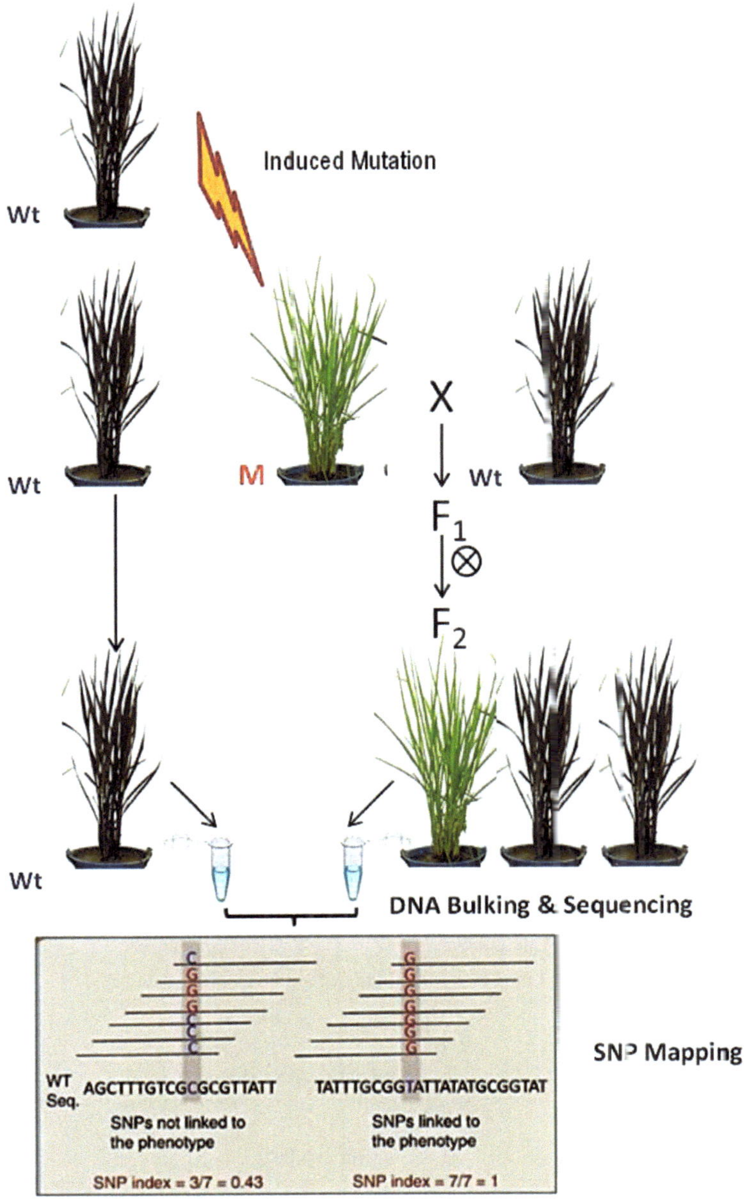

FIGURE 9.2 Schematic representation of MutMap approach given by Abe et al. (2012). Wild genotype (Wt) upon induced mutation produced mutant (M) for the concern trait. Later, both wild and mutant lines are crossed and the F_1 generation is selfed to F_2 population. The type of mutation associated with the change in phenotype can be detected from sequencing of F_2 population in which DNA from wild and mutant F_2 was sequenced, followed by SNP mapping.

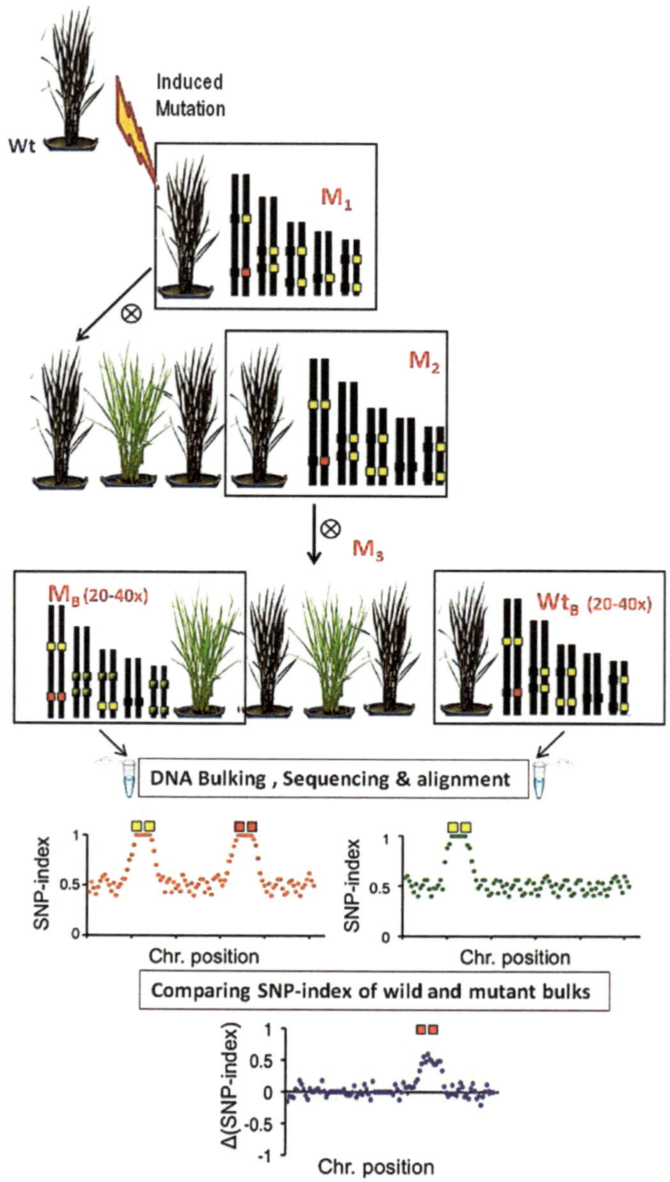

FIGURE 9.3 Schematic representation of MutMap+ approach given by Fekih et al. (2013). Wild genotype (Wt) upon induced mutation produced mutant (M_1) for the concern trait. Further, the heterozygous M_1 and M_2 generation are selfed (self) to obtain homozygous M_3 population. DNA from wild type M_3 bulks and mutant M_3 plant bulks are subjected to DNA sequencing followed by SNP mapping in comparison with reference genome to identify type of mutation associated with change in phenotype.

MutMap+ is ideal for isolating mutations in genes that prevent artificial crossover because it eliminates the need for backcrossing to the wild type of parent. MutMap+ thereby broadens the scope of MutMap's applicability in rice and other crop plants genetic improvement. MutMap− eliminates the need for both crossing and linkage analysis, allowing it to be implemented in less time. This technique can be used in a systematic way to quickly identify the genes that affect early plant development and fertility. This method could be used to find genes associated with a wide range of agronomically relevant traits.

There are several crop species such as millets (foxtail millet – *Setaria italica*, fonio-*Digitaria exilis* and tef – *Eragrostis tef*) for which effective artificial crossing procedures have not yet to be developed because of smaller flower size and difficulty in emasculation. MutMap+ is simple to use and can be used to a wide range of species crucial to agriculture in developing countries. Therefore, MutMap+ is positioned to be applicable for the improvement of orphan crops, given the predicted improvements in *de novo* sequencing technology and high sequencing costs. Improvement of these crops could boost up the food security by enhancing the crop quality (Fekih et al., 2013).

9.7 MUTMAP-GAP

The power of detecting mutant on gene by MutMap protocol depends on the quality of reference genome sequence. While induction of mutation, it often cause structural aberrations of chromosome in mutants as a result while aligning the mutant genome with reference sequence can arise missing genomic regions or gaps. To tackle the issue, Takagi et al. (2013) devised MutMap-Gap. The candidate genome region in the form of SNP clusters, matching the mutant phenotype with an SNP index of 1 are identified through MutMap method and then the gaps are identified between the reference genome and the mutant genome of the candidate region as follows. The sequence reads that did not map to the reference genome are first collected using sequence reads derived from a wild-type plant with the mutant's parental background. Second, scaffolds are created in a *de novo* assembly by combining sequence reads that have mapped near the candidate region with the unmapped reads. Finally, the SNP index is calculated using sequence reads derived from the bulk DNA of mutant F_2 progeny. Finally, the SNP index of one scaffold was being searched to locate the causal mutation (Yano et al., 2017). In rice, *japonica* cultivar, Nipponbare, is typically used as the reference genome (Kawahara et al., 2013). Takagi et al. (2013)

located mutations in the rice blast (*Magnaporthe oryzae*) resistance (R-) gene *Pii* from the rice cultivar Hitomebore using mutant lines that have lost *Pii* function.

9.8 MUTCHROMSEQ

This technique has the potential to be utilized in polyploid crops such as wheat, where a number of chromosomal cytogenetic stocks are available. In such species, chromosome number is high, and sequencing of each chromosome is additional cost and time taking. In this, the chromosome from the mutant individuals in the population generated using mutant stocks harboring the mutation is isolated using flow cytometry and then subjected to alignment over reference sequence and site is identified (Sánchez-Martin et al., 2016).

9.9 MUTRENSEQ

A novel technique which is a 3-step method (mutagenesis, exome capturing and cloning of gene(s)), has reduced the time required to identify the gene from the wild germplasm and making it amenable to incorporate in cultivated lines in a single breeding cycle. Steuernagel et al. (2016) have introduced and shown the potential of this technique by identifying stem rust resistance gene *Sr33* and *Sr22* in wheat from mutagenized source. In this technique, information on signature of disease resistance gene is utilized as most of the resistance genes are having nucleotide binding domain leucine rich repeats (NLRs). Availability of reference genome sequencing and low-cost sequencing platform is essential to this technology.

9.10 NIKS (NEEDLE IN THE K-STACK)

This is a reference free method on the basis of comparison of k-mers in the whole genome sequence of mutants. In this technique, there is no requirement of segregating population and genetics maps. Nordstrom et al. (2013) developed this technique to map mutations in rice.

9.11 QTL-SEQ

QTL mapping is normally done by genotyping a large number of individuals in segregating populations produced from bi-parental crossings, which is time-consuming, labor-intensive, and occasionally expensive (Salvi & Tuberosa, 2005). The BSA offers a simple and efficient alternate technique to recognize molecular markers linked to target genes or QTLs affecting a trait of interest by genotyping only a pair of bulked DNA samples from two sets of individuals with extreme phenotypes (Michelmore et al., 1991). QTL-seq combines BSA (Mansur et al., 1993; Darvasi & Soller, 1994) and whole-genome resequencing (WGRS) to quickly identify genomic areas that differ between the two parents utilized in a genetic cross, as well as contribute to the greater and lower values of the traits of interest among the progenies.

To use QTL-seq to map QTLs, create a mapping population by crossing two cultivars with different phenotypes for the traits of interest. Depending on the traits to be analyzed, different types of mapping populations can be employed for QTL-seq. Individuals in recombinant inbred lines (RILs) and doubled haploids (DH) have a high degree of homozygosity and can be regarded as proxy clones that allow replicated phenotypic measurements, making them appropriate for discovering minor effect QTLs. The advantage of using an F_2 population is the quickness with which it may be generated. For each genotype, however, no replicated measurements are possible. As a result, the method is ineffective for finding QTLs with small effects. Then subjected to phenotype segregation scoring and frequency distribution of measured values will be close to the normal distribution if there are numerous QTLs involved in the trait variation. Then the progenies showing the highest (Tall-bulk) and the lowest extreme (Dwarf-bulk) values are bulked. Whole genome resequencing was performed on each of the bulked DNAs. For all SNPs found between the two parents, the SNP-index is calculated, and SNP-index plots of tall-bulk and dwarf-bulk are compared. The sites of QTLs are indicated by genomic areas with differing patterns of SNP-index plots between the two bulks (Figure 9.4) (Takagi et al., 2013).

Although both MutMap+ and QTL-seq entail comparing SNP-index plots of two DNA bulks, MutMap+ is concerned with a single mutation in the genome that results in a defined phenotypic change. QTL-seq, on the other hand, focuses on many genetic alterations that have accumulated over time between two cultivars. As a result, the two approaches have different goals: MutMap+ is intended to find a single causal nucleotide alteration that causes the trait, whereas QTL-seq is used to map the approximate sites of many QTL.

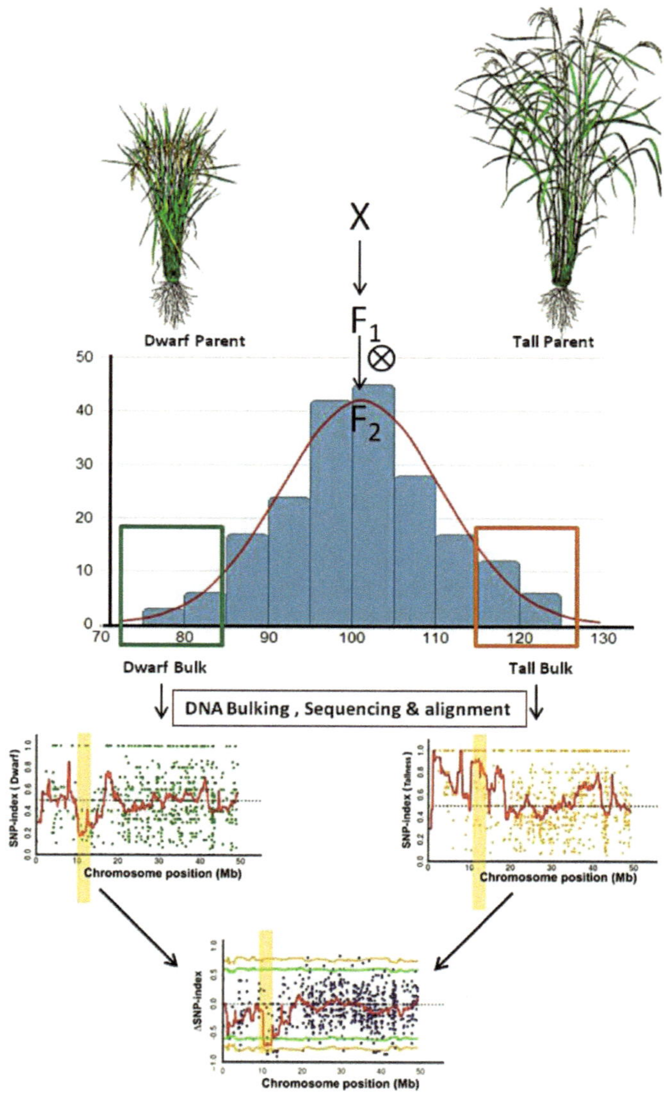

FIGURE 9.4 Schematic representation of QTL-seq approach given by Takagi et al. (2013). Parents showing contrasting characters were crossed and advanced to generate a mapping population in which the target traits show segregation. DNA from both extreme phenotypes (each with at least 100 genotypes) was bulked as lowest bulk and highest bulk. To calculate the single nucleotide polymorphism (SNP)-index, these DNA bulks are subjected to whole genome resequencing and aligned to the dwarf type reference sequence. Each dot represents a single nucleotide polymorphism, and the X-axis represents chromosomal position. Sliding window analysis was used to draw average values of SNP-index or (ΔSNP-index). The QTL can be found in the SNP-index plot as peaks or valleys.

9.12 SEQ-BSA

With the rapid advancement of NGS technology, combining the BSA and NGS techniques (Seq-BSA) is becoming a common approach in the mapping of linked QTLs and candidate gene identification (Wenger et al., 2010; Takagi et al., 2013). The principle of BSA-seq follows that of BSA where genomic DNA pools are used to identify and associate SNPs linked to the target genome phenotype. Limiting the chromosomal interval containing the target genes/QTLs to the shortest frame makes Seq-BSA technology much more powerful and resourceful strategy than traditional BSA. Zheng et al. (2016) mapped and identified a new wide spectrum resistance gene {*Pi65(t)*}, using a combination of the conventional QTL mapping with the Seq-BSA map. This method has been successfully demonstrated crop plants such as paddy (Abe et al., 2012; Wambugu et al., 2017), wheat (Trick et al., 2012), sorghum (Han et al., 2015), soybean (Song et al., 2017), pigeon pea (Singh et al., 2016), sunflower (Livaja et al., 2013) and in *Arabidopsis* (Schneeberger et al., 2009) lead to identification key QTLs and genes for important traits.

9.13 AGRENSEQ

Resistance genes in plants are characterized by the presence of peculiar signature sequences such as nucleotide binding domain leucine rich repeat (NLRs) which is most prevalent. Wild relatives of crop plants are reservoirs of such genes as they have been exposed to different variants of the pathogen as a result of trench warfare. Green revolution resulted in narrowing down of genetic base and the varieties or advanced lines of today are devoid of such NLR genes and hence, there is need to utilize wild resources. However, utilization of such wild genetic resources is restricted due to many genetic factors like crossing barriers, ploidy differences and linkage to undesirable genes. To overcome these limitations, a method of rapid identification and cloning of such genes has been developed utilizing germplasm panel, known as AgRenSeq (Arora et al., 2019). AgRenSeq combines enrichment of R genes using k-mer based association genetics without the use of sequence information. Arora et al. (2019) identified and cloned three genes *viz.*, *Sr33*, *Sr46*, and *SrTA1662* using AgRenSeq from 195 diploid accessions of wheat species *Aegilops tauschii* ssp. *strangulata* and Ae. *tauschii* ssp. *tauschii*. They also established that the panel size required for mapping does not need to be large as they could be able to map all the three genes using only 140 accessions. Also, it can identify multiple genes having epistatic effects and the time taken to map these genes is very less. AgRenSeq technique is able to tap disease resistant genes from the pan-genome variation representing a wide range of genetic resources. The technique has shown its potential in wheat and can be further extended to crops where there is no limitation germplasm availability. This technique gives importance to the utilization of wild germplasms in regular breeding program consistently.

9.14 INDEL-SEQ

Singh et al. (2016) proposed the Indel-seq technique, a novel approach which is a combination of WGRS and bulked segregant analysis (BSA) that depends on Indel (insertions and deletions) frequencies in extreme bulks for efficient usage of Indels in mapping target traits. Any segregating population (F_2/RILs/back-cross) for the desired traits could be used to commence the Indel-seq technique. The importance of Indel-seq in identifying important genomic areas linked with target attributes was underlined in a comparison of Indel-seq with other genetic mapping techniques. As a result, the Indel-seq technique could be used to identify candidate genomic areas for any target trait in any crop species quickly and precisely. Using the Indel-seq method, candidate genomic regions/genes for Fusarium wilt (FW) and sterility mosaic disease (SMD) resistance in pigeon pea were discovered. The Indels were also validated on a group of FW- and SMD-resistant and FW and SMD-susceptible genotypes. In conclusion, Indel-seq appears to be a viable technique for rapid and precise coarse and fine mapping of quantitative traits.

9.15 BSR-SEQ

The BSA is a powerful tool for quickly and accurately mapping the genes that cause mutant phenotypes. BSR-Seq combines the power of BSA with the simplicity of RNA-Seq and appropriate statistical procedures. Liu et al. (2012) created BSR-Seq (bulked segregant RNA-Seq), a BSA modification that uses RNA-Seq readings to efficiently map e-QTL regions and genes in populations where no polymorphic markers have previously been found. In this whole genome transcriptomes from specific tissues or stages were compared between pools showing extremely contrasting characters among a mapping population as like in BSA. Based on sequence read count criteria, approximate copies of specific RNA present in a sample can be assessed. A

Bayesian model based on BSA can be able to identify the polymorphic SNP markers tightly linked to the target candidate gene.

Unlike RNA-seq studies, BSR-seq is unique as it doesn't demand replicated data to map the e-QTLs. BSR-Seq is both low-cost and efficient and very much appropriate for mapping genes and QTLs in species that possess huge genome sizes or lack whole genome information like onion, tree species, etc. BSR-Seq delivers the map position of a gene accountable for a mutant phenotype and the effects of a such mutant on comprehensive patterns of gene expression. Moreover, BSR-Seq provides a pool of SNPs strongly associated with the mutant and could be utilized for fine mapping of mutants or cloning of affected genes through chromosome walking.

BSR-Seq was performed on mutant and non-mutant pools of maize seedlings that were segregating for *glossy3* (*gl3*) gene, a recessive mutant that induces epicuticular wax deposition (Schnable et al., 1994). To map *gl3*, a Bayesian-based BSA technique was devised after measuring allele frequency by reading counts in RNA-Seq. The generated mapping data matched prior mapping results obtained using different methods. To make cloning of *gl3* easier, the mapping results were paired with transcriptional profiles from RNA-Seq data. The successful cloning of *gl3* confirms the efficacy of BSR-Seq (Liu et al., 2012). Also, BSR-seq has been successfully used in the mapping of stripe rust-resistant loci *YrMM58* and *YrHY1* on chr-2AS (Wang et al., 2018) and leaf senescence gene, *els1* on chr-2BS (Li et al., 2018) in wheat.

9.16 SLAF-SEQ

Specific-locus amplified fragment sequencing (SLAF-seq) technology, reported by Sun et al. (2013), is an efficient strategy for the *de novo* SNP discovery and genotyping of large populations based on an enhanced reduced representation library (RRL) sequencing method. SLAF-seq technique consists of three important phases (Figure 9.5).

9.16.1 CONDUCTING PRE-DESIGN SCHEME FOR SLAF SELECTION USING TRAINING DATA

During the phase, the restriction enzymes used in SLAF tag generation and the sizes of restriction fragments were optimized using training data. The efficiency of SLAF-seq can be improved by determining the three criteria: (i)

The number of SLAFs must be suitable for the specific needs of the research project; (ii) the SLAFs must be evenly distributed through the sequences to be examined; and (iii) repeated SLAFs must be avoided.

9.16.2 SLAF-SEQ LIBRARY CONSTRUCTION

Based on the predesign SLAF experiments, the genomic DNA obtained from genotypes was incubated with restriction enzymes (in sequential or in combination) designed for individuals. The enzyme-digested DNA fragments are subject to an A-tailing procedure. Then A-tailed DNA fragments then obtained were ligated with dual index sequence barcode adaptors to discriminate each individual and to facilitate the pooling and selection of a specific range of fragment size based on the results of the pilot study through gel-based separation. The qualified libraries were then enriched via PCR and then subjected to NGS-based sequencing. There are various SLAF library kits are available for conducting these steps.

9.16.3 DEEP SEQUENCING AND DATA ANALYSIS FOR THE POOLED RRLS

Raw reads were initially de-multiplexed by specific programs, followed by an analysis of average sequence quality per read and GC content. Adapter reads were searched by the Cutadapt program. Reads containing correct restriction sites were obtained by searching restriction site sequences in the raw reads, respectively. Clean data were produced by removing the adapter reads and reads with ambiguous or low quality (below a Phred score of Q30) bases. Later the clean reads obtained were mapped against the reference genome using burrows-wheeler aligner (BWA) software. Repetitive SLAFs should be removed. These can be identified by detecting the same SLAF if mapped on the same position with over 90% identity. SNP calls between genotypes or mutants were generated using the GATK software and Samtools.

This approach overwhelms numerous limitations of prior approaches (Huang et al., 2009; Xie et al., 2010; Kennedy et al., 2003). The major pros of SLAF-seq technology are: (i) Deep sequencing ensuring greater genotyping accuracy; (ii) a lower sequencing cost achieved through a predesign pilot study; (iii) a predesigned RRL scheme to optimize marker efficiency; (iv) and double barcode multiplexed sequencing system for a large population

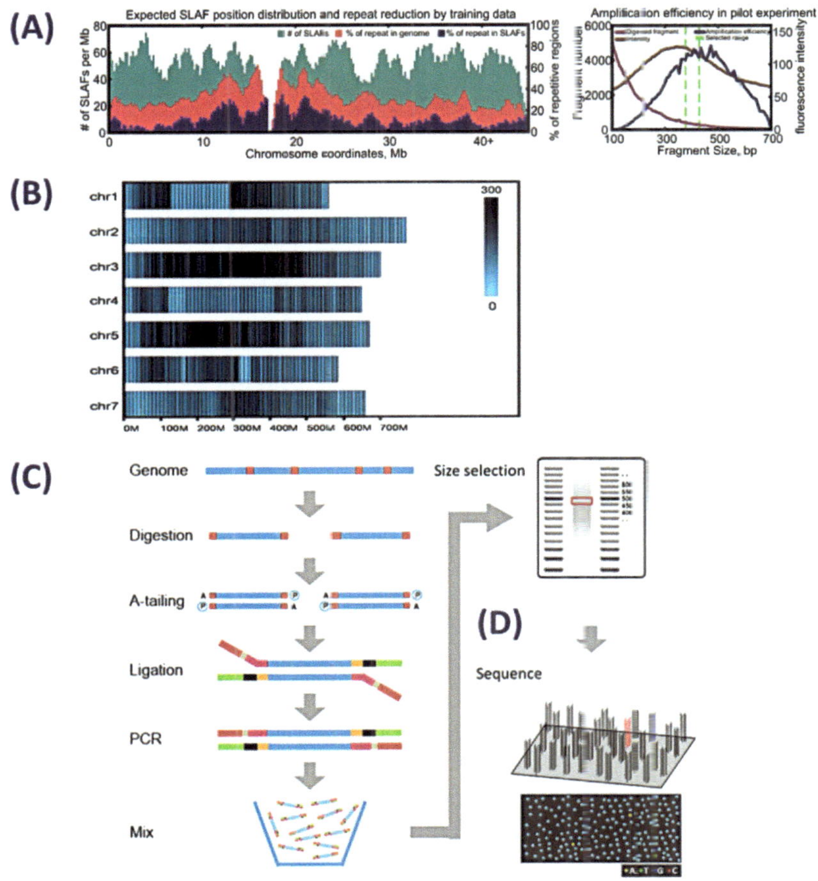

FIGURE 9.5 Schematic representation of SLAF-seq approach given by Zhu et al. (2021). (A) Predesign experiments related to SLAF selection using training data showing optimal fragment size and SLAF distribution throughout the genome. (B) SLAF distribution present among the genome of the pearl millet genome. (C) Flowchart related to SLAF library preparation. (D) DNA pools are subjected to NGS-based sequencing.

and large numbers of loci. However, one drawback of SLAF-seq is that it does not cover the whole genome.

9.17 CONCLUSION

The needs of mankind are changing every decade and to meet ever-changing human needs; the research in the field of agriculture should pick up double the pace. Traditional practices followed in agriculture research, particularly in plant breeding were often slow, resource expensive, and labor-intensive. However, the reward achieved so far was not up to the expectation. Integration of modern biotechnological principles along with conventional mutation breeding showed us the candlelight to shorten the breeding cycle thereby improving genetic gain. Popularity gained by next-generation sequencing (NGS) technologies for their fast, low-cost, and high throughput output marked the beginning of next-generation plant breeding. The MutMap and its variants integrate mutational breeding along with SNP marker technology can identify the genomic regions responsible for simply inherited traits with ease and precision and be demonstrated in a series of crops such as rice, sorghum, maize, soybean, chickpea, mustard, and cucumber. The limitation of the present technology is its dependency on the reference genome and there are progresses in several instincts to develop an algorithm that does not require the reference genome for mapping.

KEYWORDS

- **bulked segregant analysis**
- **ethyl methane sulphonate**
- **MutMap**
- **near-isogenic lines**
- **next generation mapping**
- **next-generation sequencing**
- **NGS techniques**
- **QTL-Seq**

REFERENCES

Abe, A., Kosugi, S., Yoshida, K., Natsume, S., Takagi, H., Kanzaki, H., et al., (2012). Genome sequencing reveals agronomically important loci in rice using MutMap. *Nat. Biotechnol., 30*(2), 174–178. https://doi.org/10.1038/nbt. 2095.

Arora, S., Steuernagel, B., Gaurav, K., Chandramohan, S., Long, Y., Matny, O., et al., (2019). Resistance gene cloning from a wild crop relative by sequence capture and association genetics. *Nat. Biotechnol., 37*(2), 139–143.

Ashelford, K., Eriksson, M. E., Allen, C. M., D'Amore, R., Johansson, M., Gould, P., et al., (2011). Full genome resequencing reveals a novel circadian clock mutation in *Arabidopsis*. *Genome Biol., 12*(3), 1–12.

Austin, R. S., Vidaurre, D., Stamatiou, G., Breit, R., Provart, N. J., Bonetta, D., et al., (2011). Next-generation mapping of *Arabidopsis* genes. *Plant J., 67*(4), 715–725. https://doi.org/10.1111/j.1365-313X.2011.04619.x.

Darvasi, A., & Soller, M., (1994). Selective DNA pooling for determination of linkage between a molecular marker and a quantitative trait locus. *Genetics, 138*, 1365–1373.

Fekih, R., Takagi, H., Tamiru, M., Abe, A., Natsume, S., Yaegashi, H., et al., (2013). MutMap+: Genetic mapping and mutant identification without crossing in rice. *PloS One, 8*(7), e68529.

Forster, B. P., & Shu, Q. Y., (2012). Plant mutagenesis in crop improvement: Basic terms and applications. In: Shu, Q. Y., Forster, B. P., Nakagawa, H., (eds.), *Plant Mutation Breeding and Biotechnology* (pp. 9–20). Wallingford: CABI.

Furey, T. S., (2012). ChIP-seq and beyond: New and improved methodologies to detect and characterize protein-DNA interactions. *Nat. Rev. Genet., 13*(12), 840–852.

Gbalint-Kurti, G. G., (2004). Wavepacket theory of photodissociation and reactive scattering. In: Rice, S. A., (ed.), *Advances in Chemical Physics* (Vol. 128, p. 257). Wiley: New York.

Han, Y. C., Lv, P., Hou, S. L., Li, S. Y., Ji, G. S., Ma, X., et al., (2015). Combining next generation sequencing with bulked segregant analysis to fine map a stem moisture locus in sorghum (*Sorghum bicolor* L. Moench). *PLoS One, 10*(5), e0127065.

Hu, T., Chitnis, N., Monos, D., & Dinh, A., (2021). Next-generation sequencing technologies: An overview. *Hum. Immunol.* https://doi.org/10.1016/j.humimm.2021.02.012.

Huang, X., Feng, Q., Qian, Q., Zhao, Q., Wang, L., Wang, A., et al., (2009). High-throughput genotyping by whole-genome resequencing. *Genome Res., 19*(6), 1068–1076.

Kawahara, Y., De La Bastide, M., Hamilton, J. P., Kanamori, H., McCombie, W. R., Ouyang, S., et al., (2013). Improvement of the *Oryza sativa* Nipponbare reference genome using next generation sequence and optical map data. *Rice, 6*(1), 1–10.

Kennedy, G. C., Matsuzaki, H., Dong, S., Liu, W. M., Huang, J., Liu, G., et al., (2003). Large-scale genotyping of complex DNA. *Nat. Biotechnol., 21*(10), 1233–1237.

Kharkwal, M. C., & Shu, Q. Y., (2009). The role of induced mutations in world food security. In: Shu, Q. Y., (ed.), *Induced Plant Mutations in the Genomics Era* (pp. 33–38). Rome: Food and Agriculture Organization of the United Nations.

Langmead, B., & Salzberg, S. L., (2012). Fast gapped-read alignment with Bowtie 2. *Nat. Methods., 9*, 357–359.

Leshchiner, I., Alexa, K., Kelsey, P., Adzhubei, I., Austin-Tse, C. A., Cooney, J. D., et al., (2012). Mutation mapping and identification by whole-genome sequencing. *Genome Res., 22*(8), 1541–1548.

Li, H., & Durbin, R., (2009). Fast and accurate short read alignment with Burrows–Wheeler transform. *Bioinformatics, 25*, 1754–1760.

Li, M., Li, B., Guo, G., Chen, Y., Xie, J., Lu, P., et al., (2018). Mapping a leaf senescence gene els1 by BSR-Seq in common wheat. *Crop J., 6*(3), 236–243.

Lister, R., Gregory, B. D., & Ecker, J. R., (2009). *Curr. Opin. Plant Biol., 12*, 107–118.

Liu, S., Yeh, C. T., Tang, H. M., Nettleton, D., & Schnable, P. S., (2012). Gene mapping via bulked segregant RNA-Seq (BSR-Seq). *PLoS One, 7*(5). e36406. doi: 10.1371/ journal.pone.0036406.

Livaja, M., Wang, Y., Wieckhorst, S., Haseneyer, G., Seidel, M., Hahn, V., et al., (2013). BSTA: A targeted approach combines bulked segregant analysis with next-generation sequencing and de novo transcriptome assembly for SNP discovery in sunflower. *BMC Genom., 14*, 628.

Mansur, L. M., Orf, J., & Lark, K. G., (1993). Determining the linkage of quantitative trait loci to RFLP markers using extreme phenotypes of recombinant inbreds of soybean (*Glycine max* L. Merr.). *Theor. Appl. Genet., 86*, 914–918.

Michelmore, R. W., Paran, I., & Kesseli, R. V., (1991). Identification of markers linked to disease-resistance genes by bulked segregant analysis: A rapid method to detect markers in specific genomic regions by using segregating populations. *PNAS, 88*, 9828–9832. doi: 10.1073/pnas.88.21. 9828.

Morin, R., Bainbridge, M., Fejes, A., Hirst, M., Krzywinski, M., Pugh, T., et al., (2008). Profiling the HeLa S3 transcriptome using randomly primed cDNA and massively parallel short-read sequencing. *Biotechniques, 45*(1), 81–94.

Nordström, K. J., Albani, M. C., James, G. V., Gutjahr, C., Hartwig, B., Turck, F., et al., (2013). Mutation identification by direct comparison of whole-genome sequencing data from mutant and wild-type individuals using k-mers. *Nat. Biotechnol., 31*(4), 325–330.

Nordström, K. J., Albani, M. C., James, G. V., Gutjahr, C., Hartwig, B., Turck, F., et al., (2013). Mutation identification by direct comparison of whole-genome sequencing data from mutant and wild-type individuals using k-mers. *Nat. Biotechnol., 31*(4), 325–330.

Novak, F. J., & Brunner, H., (1992). Plant breeding: Induced mutation technology for crop improvement. *IAEA Bulletin., 4*, 25–33.

Ossowski, S., Schneeberger, K., Clark, R. M., Lanz, C., Warthmann, N., & Weigel, D., (2008). Sequencing of natural strains of *Arabidopsis thaliana* with short reads. *Genome Res., 18*(12), 2024–2033.

Salvi, S., & Tuberosa, R., (2005). To clone or not to clone plant QTLs: Present and future challenges. *Trends Plant Sci., 10*, 297–304.

Sánchez-Martín, J., Steuernagel, B., Ghosh, S., Herren, G., Hurni, S., Adamski, et al., (2016). Rapid gene isolation in barley and wheat by mutant chromosome sequencing. *Genome Boil., 17*(1), 1–7.

Schnable, P. S., Stinard, P. S., Wen, T. J., Heinen, S., & Weber, D., (1994). The genetics of cuticular wax biosynthesis. *Maydica., 39*(4), 279–287.

Schneeberger, K., & Weigel, D., (2011). Fast-forward genetics enabled by new sequencing technologies. *Trends Plant Sci., 16*, 282–288.

Schneeberger, K., Ossowski, S., Lanz, C., Juul, T., Petersen, A. H., Nielsen, K. L., et al., (2009). SHOREmap: Simultaneous mapping and mutation identification by deep sequencing. *Nat. Methods, 6*(8), 550, 551. https://doi.org/10. 1038/nmeth0809-550.

Shu, Q. Y., Forster, B. P., & Nakagawa, H., (2012). Principles and applications of plant mutation breeding. In: Shu, Q. Y., Forster, B. P., & Nakagawa, H., (eds.), *Plant Mutation Breeding and Biotechnology* (pp. 301–325). Wallingford: CABI.

Singh, V. K., Khan, A. W., Saxena, R. K., Kumar, V., Kale, S. M., Sinha, P., et al., (2016). Next generation sequencing for identification of candidate genes for fusarium wilt and sterility mosaic disease in pigeonpea (*Cajanus cajan*). *Plant Biotechnol. J., 14*(5), 1183–1194.

Singh, V. K., Khan, A. W., Saxena, R. K., Sinha, P., Kale, S. M., Parupalli, S., et al., (2017). Indel-seq: A fast-forward genetics approach for identification of trait-associated putative candidate genomic regions and its application in pigeonpea (*Cajanus cajan*). *Plant Biotechnol. J., 15*(7), 906–914. doi: 10.1111/pbi.12685.

Song, J., Li, Z., Liu, Z., Guo, Y., & Qiu, L. J., (2017). Next-generation sequencing from bulked-segregant analysis accelerates the simultaneous identification of two qualitative genes in soybean. *Front. Plant Sci.*, 8. https://doi.org/10.3389/ fpls.2017.00919.

Steuernagel, B., Periyannan, S. K., Hernández-Pinzón, I., Witek, K., Rouse, M. N., Yu, G., et al., (2016). Rapid cloning of disease-resistance genes in plants using mutagenesis and sequence capture. *Nat. Biotechnol.*, *34*(6), 652–655.

Strope, C. L., Scott, S. D., & Moriyama, E. N., (2007). Indel-Seq-Gen A new protein family simulator incorporating domains, motifs, and indels. *Mol. Biol. Evol.*, *24*(3), 640–649.

Sun, X., Liu, D., Zhang, X., Li, W., Liu, H., Hong, W., et al., (2013). SLAF-seq: An efficient method of large-scale de novo SNP discovery and genotyping using high-throughput sequencing. *PloS One*, *8*(3), e58700. doi: 10.1371/journal.pone.0058700.

Takagi, H., Abe, A., Yoshida, K., Kosugi, S., Natsume, S., Mitsuoka, C., et al., (2013). QTL-seq: Rapid mapping of quantitative trait loci in rice by whole genome resequencing of DNA from two bulked populations. *Plant J.*, *74*(1), 174–183. https://doi.org/10.1111/ tpj.12105.

Takagi, H., Uemura, A., Yaegashi, H., Tamiru, M., Abe, A., Mitsuoka, C., et al., (2013). Mut map-gap: Whole-genome resequencing of mutant F_2 progeny bulk combined with de novo assembly of gap regions identifies the rice blast resistance gene Pii. *New Phytol.*, *200*(1), 276–283. https://doi.org/10.1111/nph.12369.

Tribhuvan, K. U., Kumar, K., Sevanthi, A. M., & Gaikwad, K., (2018). MutMap: A versatile tool for identification of mutant loci and mapping of genes. *Indian J. Plant Physiol.*, *23*(4), 612–621.

Trick, M., Adamski, N. M., Mugford, S. G., Jiang, C. C., Febrer, M., & Uauy, C., (2012). Combining SNP discovery from next-generation sequencing data with bulked segregant analysis (BSA) to fine-map genes in polyploid wheat. *BMC Plant Biol.*, *12*, 14.

Wadapurkar, R. M., & Vyas, R., (2018). Computational analysis of next generation sequencing data and its applications in clinical oncology. *Inform. Med. Unlocked*, *11*, 75–82.

Wambugu, P., Ndjiondjop, M. N., Furtado, A., & Henry, R., (2017). Sequencing of bulks of segregants allows dissection of genetic control of amylose content in rice. *Plant Biotechnol. Jl.* https://doi.org/10.1111/pbi.12752.

Wang, Y., Zhang, H., Xie, J., Guo, B., Chen, Y., Zhang, H., et al., (2018). Mapping stripe rust resistance genes by BSR-Seq: YrMM58 and YrHY1 on chromosome 2AS in Chinese wheat lines Mengmai 58 and Huaiyang 1 are Yr17. *Crop J.*, *6*(1), 91–98.

Wang, Z., Gerstein, M., & Snyder, M., (2009). RNA-Seq: A revolutionary tool for transcriptomics. *Nat. Rev. Genet.*, *10*(1), 57–63.

Wenger, W. J., Schwartz, K., & Sherlock, G., (2010). Bulk segregant analysis by high-throughput sequencing reveals a novel xylose utilization gene from *Saccharomyces cerevisiae*. *PLoS Genetics*, *6*, e1000942.

Williams, M. E., Torabinejad, J., Cohick, E., Parker, K., Drake E. J., Thompson, J. E., et al., (2005). Mutations in the *Arabidopsis* phosphoinositide phosphatase gene SAC9 lead to overaccumulation of PtdIns (4, 5) P2 and constitutive expression of the stress-response pathway. *Plant Physiol.*, *138*(2), 686–700.

Xie, W., Feng, Q., Yu, H., Huang, X., Zhao, Q., Xing, Y., et al., (2010). Parent-independent genotyping for constructing an ultrahigh-density linkage map based on population sequencing. *PNAS*, *107*(23), 10578–10583.

Yano, K., Ohmori, Y., & Fujiwara, T., (2017). Introduction to GWAS and MutMap for identification of genes/QTL using next-generation sequencing. In: *Plant Macronutrient Use Efficiency* (pp. 307–321). Academic Press.

Zhang, J., Chiodini, R., Badr, A., & Zhang, G., (2011). The impact of next-generation sequencing on genomics. *J. Genet. Genomics, 38*, 95–109. https://doi.org/10.1016/j.jgg.2011. 02.003.

Zheng, W., Wang, Y., Wang, L., Ma, Z. B., Zhao, J. M., Wang, P., et al., (2016). Genetic mapping and molecular marker development for Pi65(t),a novel broad-spectrum resistance gene to rice blast using next-generation sequencing. *Theor. Appl. Genet., 129*(5), 1035.

Zhu, Z., Sun, B., & Lei, J., (2021). Specific-locus amplified fragment sequencing (SLAF-Seq) as high-throughput SNP genotyping methods. In: *Crop Breeding Humana* (pp. 75–87). New York, NY.

Zuryn, S., Le Gras, S., Jamet, K., & Jarriault, S. A., (2010). Strategy for direct mapping and identification of mutations by whole-genome sequencing. *Genetics, 186*, 427–430.

CHAPTER 10

Allele Mining: Potential Tool for Accelerated Crop Breeding

VARSHA KUMARI,[1] PRIYANKA KUMAWAT,[2] SHYAM SINGH RAJPUT,[1] SHARDA CHOUDHARY,[3] RADHESHYAM SHARMA,[4] ATUL KUMAR UPADHYAY,[5] BHURI SINGH,[6] RAJDEEP MUNDIYARA,[7] and MAHESH SHARMA[1]

[1]*Department of Plant Breeding and Genetics, SKN Agriculture University, Jobner, Rajasthan, India*

[2]*Department of Agronomy, SKN Agriculture University, Jobner, Rajasthan, India*

[3]*Senior Scientist, ICAR–NRC on Seed Spices, Ajmer, Rajasthan, India*

[4]*Biotechnology Center, Jawaharlal Nehru Krishi Vishwa Vidyalaya, Jabalpur, Madhya Pradesh, India*

[5]*Department of Biotechnology, Thapar Institute of Engineering and Technology, Punjab, India*

[6]*Department of Plant Breeding and Genetics, Agriculture University, Kota, Rajasthan, India*

[7]*Department of Plant Breeding and Genetics, Agriculture University, Jodhpur, Rajasthan, India*

ABSTRACT

Allele mining is a promising approach used to identify allelic diversity of trait of interest within collections of enormous genetic resources. It could be a vital link between genomic resources and their efficient utilization due to its tremendous application potential. It is imperative to use genomic

Smart Breeding: Molecular Interventions and Advancements for Crop Improvement. Kailash Chandra, Subhash Chand, Ravi Prakash Saini, and Rohit Sharma (Eds.)

information for isolation and recognition of superior alleles of prime traits with available vast sequence information in public databases and crop gene banks through novel and efficient mining strategies. Allele mining approaches are practical, less cumbersome and comparatively cheaper due to rapid advancement in next-generation sequencing (NGS), recently discovered TILLING and eco-TILLING, association mapping (AM), etc. Allele mining helps in identification of allele/s, finding allelic synteny, evolution of alleles, haplotypes identification and diversity, similarity analysis, finding evolutionary relationship, and development of allele specific molecular markers for marker assisted selection (MAS) to improve crop productivity. This chapter examines the approaches, explores the concept and describes applications of allele mining, emphasizing the need of this approach for the discovery of novel alleles which could additionally be employed in molecular breeding practices.

10.1 INTRODUCTION

Allele mining is an enormous perspective utilized to dissect naturally existing allelic variations of candidate genes conferring key agronomic and productive traits which has promising applications in crop improvement. The creation of genetic diversity is the foundation to stand pillars of crop improvement. Wild relatives and traditional crop varieties harbor a large reservoir of valuable alleles that can be exploited to develop new varieties with ameliorated yield potential, nutritional quality, biotic and abiotic stress tolerance (McCouch et al., 2007; Sinha et al., 2020). Development of superior crop varieties with elevated yield was feasible by accumulation of favorable and positive gene(s) and allele(s) from existing genetic resources worldwide. A notable portion of supercilious alleles were left unutilized and underutilized during evolution and domestication. The existing untapped genetic variability in wild species, wild relatives and land races of crops could be gainfully exploited for evolution of cultivars superior agronomically and yield-wise (Kumar et al., 2010). Introgression of new alleles to cultivated varieties of crops from wild relatives has clearly demonstrated that certain allele combinations substantially alter trait expression when advanced to appropriate genomic background. By overcoming the genetic tailback which limits the introgression of beneficial traits to cultivars the vast genetic resources need to be relooked for novel allele variations and enhance the genetic aptitude of cultivars for numerous agronomic, yield, and yield contributing traits (Kumari et al., 2018).

Allelic and genetic heterogeneity are crucial to fulfill the requirement for crops that have high productivity, nutritional value, and are tolerant to stress conditions simultaneously. To explore allelic and genetic diversity and convey it to breeding programs an efficient system must be essentially devised. Access and identification of allelic diversity is most important for the exploitation and dissecting of germplasm resources in plant variety development. Germplasm collections and assembly appeared to harbor a store of undiscovered allelic variability. Unlocking of the existing variation from vast aggregate of accessions confined in gene banks is a big challenge. Identification of a tiny number of accessions and genome data from a wide collection formed on either phenotype or sequence is a major aim of allele mining. The genome data in a very efficient mean to pick out little accessions set for targeted phenotyping. The genomes dataset furnishes a digital filter for traits mapping and target sequences validation. Accessions with unique sequences and haplotypes can be identified and genome database can be searched through known sequences (Guo et al., 2019).

Extensive progress has been made in depositing sequence-related information into gene banks in current years. Sequence, as well as expression data assemblage in several genomic databases, accelerated annotation and disclosure of new alleles and allele-specific molecular markers (Chan, 2005; Spooner et al., 2005; Mardis, 2008a). Polymerase chain reaction (PCR) based approaches are concocted to isolate valuable alleles from an enormous radius of species based on gene and genome sequences. This enables direct access to major alleles responsible for enhanced yield and improved quality, conferring resistance to biotic and abiotic stresses and higher nutrient use efficiency (Latha et al., 2004).

Allele mining is a procedure aimed and objected to identify allelic variants of desired traits within gene pool clusters. It helps in tracing of novel alleles and haplotypes for the development and use of allele specific markers and their applications in marker-assisted selection (MAS). Allelic variants identification from germplasm entries helps in categorizing them for collection and conservation and imparts new genetic resources for transmission of novel alleles for focused character enhancement. Advance genomic gadgets could be utilized to identify homogeneous alleles conferring trait of interest and their variants in a greater number of genotypes. Genetic resources may be screened for allelic variation for identified genes of known function and DNA sequence by allele mining techniques *viz.,* TILLING, eco-TILLING, next-generation sequencing (NGS) and association mapping (AM) (Sharma et al., 2014; Pidon et al., 2020).

Allele mining utilizes DNA sequence of one genotype to discern functional alleles from related genotypes. It requires various bioinformatic tools for analyzing of data *viz.*, PlantCARE, JASPAR, MEME, DCPD, SCPD, ClustalW, etc. These ultra-modern tools are employed for sequence positioning and alignment in order to have a comparison between new sequence and reference genome sequence (Raddy et al., 2014). In this chapter, we have highlighted the concept of allele mining and its approaches perceiving the role, goal, and significance of allele mining in enhancing yield, nutrient efficiency, resistance to biotic and abiotic stresses, and improved nutritional quality for crop improvement.

10.2 VARIOUS APPROACHES OF ALLELE MINING

The major approaches for allele mining are modified TILLING procedure called eco-TILLING based allele mining, resequencing or sequencing-based allele mining, NGS for allele mining and AM-based allele mining.

10.2.1 ECO-TILLING-BASED ALLELE MINING

TILLING (Targeting Induced Local Lesions IN Genomes) is reverse genetics high-throughput (HT), low cost and non-transgenic method for identification of allele variation with respect to single base-pair/nucleotides in target gene (point mutations) and the discovery of induced mutations at particular point of DNA sequence by heteroduplex analysis (Sharma et al., 2014). The method that has been modified to detect natural mutation is called Eco-TILLING (Ecotype Targeting Induced Local Lesions IN Genomes).

It is economical and cheap because a single individual for each haplotype necessarily be sequenced lead to faster detection of variation in innumerable individuals concurrently. Eco-TILLING is a variant of TILLING involved in locating SNPs in old and elite cultivars regardless of genome size, chromosome number, ploidy level and pollinating system of plants (Barkley & Wang, 2008). These methods have been applied to many crop varieties and species of different crops.

TILLING comprises of various major steps: Generation of mutant population, DNA isolation, preparation, and pooling, PCR amplification, detection of heteroduplex, enzymatic cleavage of heteroduplex and phenotypic recovery. While eco-TILLING approaches are the same where base material are natural mutants. It is a fast, cost effective and well-established approach which is more efficient for extensive allele mining and high-efficient genotyping of natural allelic variations without previous recognition of SNP alleles of candidate genes of useful agronomic characters in divergent gene pools of different crops (McCallum et al., 2000; Till et al., 2006, 2007, 2010). Firstly, seeds are mutagenized by treatment with chemical mutagens *viz*., ethyl methane sulfonate (EMS), etc., to induce random mutations. Mutagenizing with EMS is done by soaking seeds in EMS solution of 50–100 mM concentration for about 14–18 hrs. Treated (M_1 seeds) seeds are planted in the field to obtain M_1 plants which further selfed to produce M_2 seeds to generate M_2 population. DNA is extracted from M_2 individuals, pooled and arrayed into 96 wells containing microtiter plates for mutational screening (Colbert et al., 2001). PCR amplification of a target fragment using gene-specific primers labeled with two different dyes for forward and reverse primer to find gene of interest (Tacele et al., 2010). Denaturing and re-annealing of PCR products leads to the formation of heteroduplex or mismatch represent single nucleotide polymorphisms (SNPs) or point mutation. Software are used to find out point mutation in gene such as COODLE which identify region where point mutation has a deleterious effect. Screening of mutant population compared with wild type at single nucleotide difference. After amplification of desired sequence of heteroduplex, single-strand specific nuclease *viz*., S1 nucleases are used to cleave it at the point of mismatch. Cleaved bands are visualized using denaturing polyacrylamide gel electrophoresis which represents point mutations or polymorphisms (Comai et al., 2004; Khan et al., 2018). Identification and confirmation of mutation are done by sequencing of band on gel (Figure 10.1).

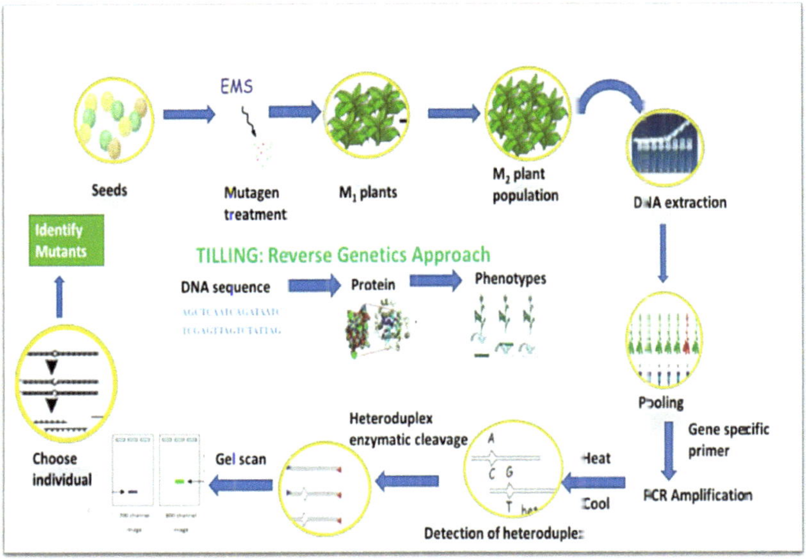

FIGURE 10.1 Outline of TILLING process.

10.2.2 RESEQUENCING OR SEQUENCING-BASED ALLELE MINING

Sequencing-based allele mining comprises PCR amplification and banding pattern of alleles of a gene in different genotypes and nucleotide variants in the alleles are recognized through DNA sequencing. This approach is used to recognize and analyze the effect and result of mutations on gene construct and sequence for the location of SNPs and insertions or deletions (InDels) mutations and point mutations for construction of haplotypes structure to infer genetic association studies in plants (Sharma et al., 2014). Unlike eco-tilling, sequencing-based allele mining does not involve much through-put laboratory, sophisticated equipment and tedious steps (Kumar et al., 2010). HT next-generation DNA sequencing technologies like 454 pyrosequencing, Illumina genome analyzer facilitate genotyping of huge genetic resources by sequencing of millions of different DNA contigs continuously at a manageable cost (Kumari et al., 2018). Procedure consists of: Collection of accession, DNA extraction, primer design, PCR amplification and sequencing and identification of allelic variation in the sample (Huang et al., 2009). Alleles are amplified primarily by PCR and polymorphism is detected by DNA sequencing.

10.2.3 NEXT-GENERATION SEQUENCING (NGS) FOR ALLELE MINING

Sequencing is a method to assess a precise nucleotide sequence of a DNA sample. Sanger dideoxy method of sequencing also called chain termination sequencing ruled the DNA sequencing research area for around 35 years where sequence read length has increased from 500 bases to 1 kb and above (Sanger et al., 1977; Madabhushi, 1998). With rapid advancement in the fields of biotechnology, bioinformatics, and nanotechnology, alternative strategies are developed and called next-generation or second-generation sequencing (NGS) methods. NGS expands the speed and throughput of DNA sequencing which overcome the high cost of Sanger sequencing as one of the major limitation. NGS technology is highly scalable and allows the entire genome to be sequenced at once in contrast to first generation sequencing. NGS is also known as high-throughput (HT) sequencing that could add substantial value to gene bank accession collections (Kilian & Graner, 2012). NGS technologies generate data faster and inexpensive way as compared with other methods. Thus, these technologies are becoming preferred choice in fields of population genetics, conservation genetics, and molecular biology to generate genetic data. Tremendous progress in sequencing technologies in recent years has made possible whole-genome resequencing (WGRS) by individual laboratories and increased feasibility of sequence capacity. Resequencing of candidate genes for thousands of individuals, entire transcriptome, whole plant genomes is quite possible, more efficient, economical, and in-depth knowledge can be obtained through this procedure (Hudson, 2008; Mardis, 2008b; Varshney et al., 2009). Hundreds to thousands of candidate genes can be traced in gene bank accession collections to represent useful genetic diversity among individuals in genetic resource pools. Harnessing information on candidate genes and candidate regions is essential to identify individual genes, locate quantitative trait loci (QTLs) and also to determine functional diversity (Kilian & Graner, 2012). Many modern sequencing platforms are available to sequence and screen large germplasm collection with greater accuracy *viz*., Illumina/GA, 454 pyrosequencing, applied biosystems/SOLiD, Roche/GS FLX and cPAL sequencing (Ashkani et al., 2015). Illumina/Solexa genome analyzer read 30–40 base sequence per run in the time period of 4–5 days and can sequence up to 1 billion bases (1 GB). Similarly, Roche/454 sequencing platform could make 100 Mb sequence and read length of around 250 bases with 98.9% or more accuracy. Likewise, applied biosystems/SOLiD can produce 2–3 GB of sequences per run instead of its shorter read length of 25–35 base sequence (Hutchison, 2007). Recently, single molecule sequencing (SMS) technology has been developed which is rapid and inexpensive than second-generation sequencing methodology of SOLiD and Illumina and considered as third-generation sequencing technology that gives platform for large scale SNPs development which can further deliver in genotyping and haplotype pattern (Gupta, 2008). NGS-based approach of TILLING is cost-effective, accurate measure for allele mining for detection of rare alleles in populations and useful for the faster identification of abiotic stress resistance genes (Tsai et al., 2011).

10.2.4 ASSOCIATION MAPPING (AM)-BASED ALLELE MINING

Association mapping (AM)-based allele mining is utilized to find out genomic regions associated with phenotypes by association studies through linkage-disequilibrium analysis. DNA markers are the means to identify beneficial alleles in wide reservoirs of genetic diversity. The approach works on the principle that alleles conferring phenotype accompanying with the flanking markers of the same locus are inherited as a block. AM uses either whole genome scanning (WGS) or candidate gene sequencing (CGS) of natural population (Rafalski, 2002). CGS-based approaches are time-consuming, costly, and utilize Sanger sequencing of PCR amplicons for candidate genes amidst thousands of the natural population genotypes. In general, Solexa approach of NGS sequence pools of PCR amplicons with the help of barcodes for a large number of candidate genes generated for 1,000 genotypes of the natural population. Therefore, Solexa sequence data for each run would be accessible within a short time and cost in the natural population for a sizeable count of candidate genes as compared to Sanger sequencing method. Unlikely, WGS approaches demands natural population screening with a broader set of genome-wide markers, which are quite impossible in many species of crop. Thus, NGS technologies can expedite the speedy enlargement of genome-wide markers that would be eventually availed for WGS approaches to AM.

10.3 APPLICATIONS OF ALLELE MINING FOR CROP IMPROVEMENT

Allele mining is a propitious method to unravel the diversity of tremendous gene pools in the world gene bank assemblies (Kaur et al., 2008a). It could be used functionally for uncovering of superior alleles from diverse germplasm via mining of alleles of gene and of target traits and identify the nucleotide sequence difference responsible for novel trait variations associated with superior alleles. It is the path of molecular discrimination between related traits and species. In addition, allele mining facilitates in ingression of rare alleles using MAS, development of allele-specific markers, allelic evolution, allelic variations and allelic synteny studies. The applications of allele mining for various crops are illustrated in Figure 10.2.

10.3.1 RICE

Allele mining could be potentially deployed in the formation of allele-specific markers, spotting of base pair and nucleotide variation associated with phenotypic variation, identification of new haplotypes and finding allelic synteny and evolutionary relationship in numerous crop plants (Table 10.1). Several case studies, in recent years demonstrated the utility of sequence variation at key genes through the development of allele specific molecular markers for MAS. As for example in rice, comparison of nucleotide sequences of waxy gene in 18 germplasm accessions divulge the presence of five different alleles associated with the alteration in phenotype (Mikami et al., 2008) and same have been deployed in cultivated rice and development of hybrid rice recently (Shao et al., 2020). Bao et al. (2006), validated SNPs for gelatinization temperature (GT) of starch and nucleotide diversity at starch synthase (SS) locus IIa in 30 rice varieties. Allele mining divulges 18 InDel variations and 91 nucleotide polymorphisms at *Pi-ta locus* (conferring blast disease resistance) among wild rice accessions of rice (Huang et al., 2008). Similarly, 28 InDels and 86 SNPs have been associated with for GS3 gene responsible for grain size in 54 rice accessions (Takanokai et al., 2009). Allelic variants of *Pi9* gene (nucleotide binding site (NBS)-leucine-rich repeat region) conferring blast resistance were observed with total of 136 polymorphic sites composed of 77 mutations and 59 InDels isolated from 338 Indian rice landraces (Imam et al., 2016; Zhou et al., 2020). Allele mining unlocks the genetic diversity of resistance to rice yellow mottle virus resistance (RYMV) genes in African-cultivated rice (*Oryza glaberrima* and *O. sativa*) (Pidon et al., 2020). Role of drought responsive gene, *DRO1* through allele mining was reported recently by Singh et al. (2021) in rice. Haplotypes-based allele mining in MAGIC rice population in Japan was also illustrated (Ogawa et al., 2018). These studies are evidence to support the existence of tremendous unexploited allelic diversity for various characters in the plant gene pool which impart opportunity to exploit and identify the allelic diversity for trait enhancement.

TABLE 10.1 Applications of Allele Mining in Important Crop Plants

Crops	Alleles	Traits	References
Rice	*Adh2*	Alcohol dehydrogenase	Yoshida & Miyashita, (2005)
Rice	*SSIIa*	Physicochemical properties	Bao et al. (2006)
Rice	*Alt4*	Aluminum tolerance	Fontecha et al. (2007)
Rice	*Wx*	Granule-bound starch synthase	Mikami et al. (2008); Shao et al. (2020)
Rice	*Pi-ta*	Blast resistance	Huang et al. (2008)
–	*Rc*	Pericarp color	Brooks et al. (2008)
Rice	*GS3*	Grain size	Takanokai et al. (2009)
Rice	*Badh2*	Fragrance	Sakthivel et al. (2009)
Rice	*Pi9*	Blast resistance	Imam et al. (2016); Zhou et al. (2020)
Rice	*RYMV2*	Rice yellow mottle virus	Pidon et al. (2020)

TABLE 10.1 *(Continued)*

Crops	Alleles	Traits	References
Rice	*DRO1*	Drought responsive gene	Singh et al. (2021)
Wheat	*Wx-A1*	Waxy gene and amylose biosynthesis	Saito & Nakamura, (2005)
Wheat	*Pm3*	Powdery mildew	Tommasini et al. (2006); Yahiaoui et al. (2006)
Wheat	*Pm3*	Powdery mildew	Kaur et al. (2008a, b)
Wheat	*TaALMT1*	Aluminum resistant	Raman et al. (2008)
Wheat	Viviparous-1 gene	Pre-harvest sprouting tolerance	Xia et al. (2008)
Wheat	Yield-related locus	Yield-related traits	Guo et al. (2019)
Barley	*rym4*	Yellow mosaic virus	Graner (2006)
Barley	*VRN-H1/H4*	Vernalization	Cockram et al. (2007)
Barley	Abiotic stress resistance locus	Abiotic stress resistance	Cseri et al. (2011)
Tomato	*Pto*	Resistance to *Pseudomonas syringae*	Rose et al. (2005)
Potato	*Rpi-blb1* and *Rpi-blb2*	Late blight resistant genes	Wang et al. (2008b)
Bean	*APA*	Lectin-related resistance genes	Lioi et al. (2007)
Apple	*Mal d 3*	Allergenicity	Gao et al. (2005)
Soybean	*GmDGAT*	Acylation of the Kennedy pathway	Wang et al. (2006, 2008a)
Pea	*TI1*	Trypsin inhibitors	Sonnante et al. (2005)
Sorghum	*AltSB*	Aluminum tolerance	Hufnagel et al. (2018)
Finger millet	*RGAs*	Blast resistance	Kumar et al. (2016)

10.3.2 WHEAT

Map-based cloning, haplotyping, and sequence analysis for *Pm3* locus responsible for powdery mildew (PM) disease resistance among the germplasm accessions led to the identification of nucleotide polymorphism and evolution of functional genetic markers in wheat (Tommasini et al., 2006; Yahiaoui et al., 2006). Allele mining approach was employed to utilize the naturally occurring diversity at *Pm3* locus that differ in LRR domain and set of resistant lines harboring potentially functional gene and serve as suitable candidates for relay to economically important but susceptible varieties of wheat to produce PM resistant varieties. The analysis of different allelic sequence variations helps in better practical utility in resistance gene function and evolution (Kaur et al., 2008a, b). Allele synteny could be explored using the sequence of allele mining among the identified loci across the species. Using syntenic allele sequence from wheat, homologs alleles were isolated in rye, grasses, and *Phaseolus vulgaris* superior for aluminum tolerance (Fontecha et al., 2007). Mutant alleles in wheat mutant resources carry phenotypic variation. Allele mining for favorable alleles to improve plant architecture and yield-related traits in such mutants and characterization of their effects will improve wheat characteristics (Guo et al., 2019). Furthermore, *Viviparous-1* gene linked with pre-harvest sprouting tolerance diversity was exploited through allele mining in European wheat varieties (Xia et al., 2008).

FIGURE 10.2 Applications of allele mining.

10.3.3 OTHER CROPS

Allele-specific markers have been generated in European barley based on haplotype diversity at vernalization locus (*VRN-H1/H4*) (Cockram et al., 2007). Many functional alleles were identified by demonstrating the use of plant genetic resources (PGRs) based on alleles which were further analyzed at phenotypic basis during the study of the sequence variation of *rym4* locus responsible for barley yellow mosaic virus resistance (Graner, 2006). QTLs controlling resistance to blast showed similarity in their sequence in rice and barley both supporting the theory of common evolutionary relationship of blast resistance gene (Chen et al., 2002). In tomato, alleles of the *Pto* gene controlling resistance to *Pseudomonas syringae* were amplified and sequenced in 49 genotypes and identified many nucleotides alteration liable for pathogen recognition and hypersensitive resistance (Rose et al., 2005). Late blight resistant genes, *Rpi-blb1* and *Rpi-blb2* in *Solanum* were compared with ancestral gene sequence for allelic frequency variation suggesting that *Rpi-blb1* has been derived from the recombination of genes, *RGA3-blb* and *RGA1-blb* where *RGA1-blb* has emerges before tuber formation and non-tuber divergence (Wang et al., 2008b). Characterization and identification of a resistance gene, *Rpi-blb1* in *Solanum bulbocastanum* (non-crossable species of *Solanum*) induces locating of *Rpi-blb1* homologs in *Solanum stoloniferum* (crossable species) (Song et al., 2003). Evolutionary review of lectin-related resistance genes in wild species as well as cultivated species of bean was done through allele mining approaches (Lioi et al., 2007). Allele mining and haplotype and sequence variation *viz.*, SNPs and INDELs were examined at nine loci involved in the abiotic stress response pathways in 96 barley cultivars and landraces comprising of drought tolerant and sensitive genotypes collected worldwide (Cseri et al., 2011).

10.4 CONCLUSION

Inadequacy of genetic diversity for novel alleles for important agronomic traits limits the crop improvement through conventional breeding. Allele mining through efficient data mining used for precise detection of variations created by natural mutations. It further confirms the basis of trait variations utilizing HT sequencing to identify candidate genes linked with major agronomic and productive traits.

10.5 FUTURE PROSPECTS

Allele mining could be a prime connection between the utilization of genome resources effectively and efficiently and modern crop breeding ascribed by genomics approaches. Sequence-based allele mining would appear as a method of choice in finding naturally occurring variation and revealing candidate genes associated with key agronomic traits that could be introgressed in the background of well-adapted variety to create a variety with novel traits for novel crop improvement activities in the future.

KEYWORDS

- **agronomic traits**
- **allele mining**
- **eco-TILLING**
- **ethyl methane sulfonate**
- **genetic diversity**
- **next generation sequencing**
- **TILLING**

REFERENCES

Ashkani, S., Yusop, M. F., Shabanimofrad, M., Azadi, A., Ghasemzadeh, A., Azizi, P., & Latif, M. A., (2015). Allele mining strategies: Principles and utilization for blast resistance genes in rice (*Oryza sativa* L.). *Current Issues in Molecular Biology, 17*, 57–74. doi.org/10.21775/cimb.017.057.

Bao, J. S., Corke, H., & Sun, M., (2006). Nucleotide diversity in starch synthase IIa and validation of single nucleotide polymorphisms in relation to starch gelatinization temperature and other physicochemical properties in rice (*Oryza sativa* L.). *Theoretical and Applied Genetics, 113*(7), 1171–1183. doi: 10.1007/s00122-006-0355-6.

Barkley, N. A., & Wang, M. L., (2008). Application of TILLING and EcoTILLING as reverse genetic approaches to elucidate the function of genes in plants and animals. *Current Genomics, 9*(4), 212–226. doi: 10.2174/138920208784533656.

Brooks, S. A., Yan, W., Jackson, A. K., & Deren, C. W., (2008). A natural mutation in rc reverts white rice pericarp to red and results in a new, dominant, wild-type allele: Rc-g. *Theoretical and Applied Genetics, 117*(4), 575–580. doi: 10.1007/s00122-008-0801-8.

Chan, E. Y., (2005). Advances in sequencing technology. *Mutation Research, 573*(1, 2), 13–40. doi: 10.1016/j.mrfmmm.2005.01.004.

Chen, H., Wang, S., Xing, Y., Xu, C., Hayes, P. M., & Zhang, Q., (2002). Comparative analysis of genomic locations and race specificities of loci for quantitative resistance to *Pyricularia grisea* in rice and barley. *Proceeding of National Academy of Sciences of the United States of America, 100*(5), 2544–2549. doi.org/10.1073/pnas.0437898100.

Cockram, J., Chiapparino, E., Taylor, S. A., Stamati, K., Donini, P., Laurie, D. A., & O'sullivan, D. M., (2007). Haplotype analysis of vernalization loci in European barley germplasm reveals novel VRN H1 alleles and a predominant winter VRN-H1/VRN-H2 multi-locus haplotypes. *Theoretical and Applied Genetics, 115*(7), 993–1001. doi: 10.1007/s00122-007-0626-x.

Colbert, T., Till, B. J., Tompa, R., Reynolds, S., Steine, M. N., Yeung, A. T., McCallum, C. M., Comai, L., & Henikoff, S., (2001). High throughput screening for induced point mutations. *Plant Physiology, 126*(2), 480–484. doi: 10.1104/pp.126.2.480.

Comai, L., Young, K., Till, B. J., Reynolds, S. H., Greene, E. A., & Codomo, C. A., (2004). Efficient discovery of DNA polymorphisms in natural populations by eco tilling. *Plant Journal, 37*(5), 778–786. doi: 10.1111/j.0960-7412.2003.01999.x.

Cseri, A., Cserhati, M., Korff, M. V., Nagy, B., Horvath, G. V., Palagyi, A., Pauk, J., Dudits, D., & Torjek, O., (2011). Allele mining and haplotype discovery in barley candidate genes for drought tolerance. *Euphytica, 181*(3), 341–356. doi: 10.1007/s10681-011-0445-7.

Fontecha, G., Silva-Navas, J., Benito, C., Mestres, M. A., Espino, F. J., Hernandez-Riquer, M. V., & Gallego, F. J., (2007). Candidate gene identification of an aluminum-activated organic acid transporter gene at the Alt4 locus for aluminum tolerance in rye (*Secale cereale* L.). *Theoretical and Applied Genetics, 114*(2), 249–260. doi: 10.1007/s00122-006-0427-7.

Gao, Z. S., Weg, W. E. V. D., Schaart, J. G., Meer, I. M. V. D., Kodde, L., Laimer, M., Breiteneder, H., Hoffmann-Sommergruber, K., & Gilissen, L. J. W. J., (2005). Linkage map positions and allelic diversity of two Mal d 3(non-specific lipid transfer protein) genes in the cultivated apple (Malus domestica). *Theoretical and Applied Genetics, 110*(3), 479–491. doi: 10.1007/s00122-004-1856-9.

Graner, A., (2006). Barley research at IPK. In: De Vicente, M. C., & Glaszmann, J. C., (eds.), *Molecular Markers for Allele Mining* (p. 25). IPGRI.

Guo, H., Xiong, H., Xie, Y., Zhao, L., Gu, J., Zhao, S., Ding, Y., & Liu, L., (2019). Functional mutation allele mining of plant architecture and yield-related agronomic traits and characterization of their effects in wheat. *BMC Genetics, 20*, 102. doi.org/10.1186/s12863-019-0804-2.

Gupta, P. K., (2008). Single-molecule DNA sequencing technologies for future genomics research. *Trends in Biotechnology, 26*(11), 602–611.

Huang, C. L., Hwang, S. Y., Chiang, Y. C., & Lin, T. P., (2008). Molecular evolution of the Pi-ta gene resistant to rice blast in wild rice (*Oryza rufipogon*). *Genetics, 179*(3), 1527–1538.

Huang, X., Feng, Q., Qian, Q., Zhao, Q., Wang, L., Wang, A., Guan, J., et al., (2009). High-throughput genotyping by whole-genome resequencing. *Genome Research, 19*, 1068–1076.

Hudson, M. E., (2008). Sequencing breakthroughs for genomic ecology and evolutionary biology. *Molecular Ecology Resources, 8*(1), 3–17. doi: 10.1111/j.1471-8286.2007.02019.x.

Hufnagel, B., Guimaraes, C. T., Craft, E. J., Shaff, J. E., Schaffert, R. E., Kochian, L. V. and V. Magalhaes, J. V., (2018). Exploiting sorghum genetic diversity for enhanced aluminum tolerance: Allele mining based on the AltSB locus. *Scientific Reports, 8*(1), 10094. doi: 10.1038/s41598-018-27817-z.

Hutchison, C. A., (2007). DNA sequencing: Bench to bedside and beyond. *Nucleic Acids Research, 35*(18), 6227–6237. doi: 10.1093/nar/gkm688.

Imam, J., Mandal, N. P., Variar, M., & Shukla, P., (2016). Allele mining and selective patterns of Pi9 gene in a set of rice landraces from India. *Frontiers in Plant Science, 7*, 1846. doi: 10.3389/fpls.2016.01846.

Kaur, N., Street, K., Mackay, M., Yahiaoui, N., & Keller, B. (2008a). Allele mining and sequence diversity at the wheat powdery mildew resistance locus Pm3. *Proceedings of the 11thInternational Wheat Genetics Symposium* (pp. 1–3). Sydney.

Kaur, N., Street, K., Mackay, M., Yahiaoui, N., & Keller, B., (2008b). Molecular approaches for characterization and use of natural disease in wheat. *European Journal of Plant Pathology, 121*, 387–397.

Khan, A., Abidi, I., Bhat, M. A., Dar, Z. A., Gowhar, A. G., Shikari, A. B., & Khan, M. A., (2018). Tilling and eco-tilling – a reverse genetic approach for crop improvement. *International Journal of Current Microbiology and Applied Sciences, 7*(6), 15–21. doi. org/10.20546/ijcmas.2018.706.003.

Kilian, B., & Graner, A., (2012). *NGS Technologies for Analyzing Germplasm Diversity in GeneBanks*. Briefings in functional genomics, elr046.

Kumar, A., Babu, B. K., Yadav, S., & Agrawal, P. K., (2016). Allele mining for resistance gene analogs (RGAs) in crop plants: Aspecial emphasis on blast resistance in finger millet (*Eleusine coracana* L.). *Indian Journal of Genetics, 76*(1), 1–9. doi 10.5958/0975-6906.2016.00001.8.

Kumar, G. R., Sakthivel, K., Sundaram, R. M., Neeraja, C. N., Balachandran, S. M., Rani, N. S., Viraktamath, B. C., & Madhav, M. S., (2010). Allele mining in crops: Prospects and potentials. *Biotechnology Advances, 28*, 451–461. doi: 10.1016/j.biotechadv.2010.02.007.

Kumari, R., Kumar, P., Sharma, V. K., & Kumar, H., (2018). Allele mining for crop improvement. *International Journal of Pure and Applied Bioscience, 6*(1), 1456–1465. doi: http://dx.doi.org/10.18782/2320-7051.6073.

Latha, R., Rubia, L., Bennett, J., & Swaminathan, M. S., (2004). Allele mining for stress tolerance genes in *Oryza* species and related germplasm. *Molecular Biotechnology, 27*(2), 101–108.doi.org/10.1385/MB:27:2:101.

Lioi, L., Galasso, I., Lanave, C., Daminati, M. G., Bollini, R., & Sparvoli, F., (2007). Evolutionary analysis of the APA genes in the *Phaseolus* genus: Wild and cultivated bean species as sources of lectin-related resistance factors. *Theoretical and Applied Genetics, 115*(7), 959–970. doi: 10.1007/s00122-007-0622-1.

Madabhushi, R. S., (1998). Separation of 4-color DNA sequencing extension products in noncovalently coated capillaries using low viscosity polymer solutions. *Electrophoresis, 19*, 224–230.

Mardis, E. R., (2008a). The impact of next-generation sequencing technology on genetics. *Trends in Genetics, 24*(3), 133–141. doi: 10.1016/j.tig.2007.12.007.

Mardis, E. R., (2008b). Next-generation DNA sequencing methods. *Annual Review of Genomics and Human Genetics, 9*, 387–402. doi.org/10.1146/annurev.genom.9.081307.164359.

McCallum, C. M., Comai, L., Greene, E. A., & Henikoff, S., (2000). Targeted screening for induced mutations. *Nature Biotechnology, 18*(4), 455–457. doi 10.1038/74542.

McCouch, S. R., Sweeney, M., Li, J., Jiang, H., Thomson, M., Septiningsih, E., Edwards, J., et al., (2007). Through the genetic bottleneck: O. rufipogon as a source of trait-enhancing alleles for O. sativa. *Euphytica, 154*, 317–339. doi: 10.1007/s10681-006-9210-8.

Mikami, I., Uwatoko, N., Ikeda, Y., Yamaguchi, J., Hirano, H. Y., Suzuki, Y., & Sano, Y., (2008). Allelic diversification at the wx locus in landraces of Asian rice. *Theoretical and Applied Genetics, 116*, 979–989.

Ogawa, D., Yamamoto, E., Ohtani, T., Kanno, N., Tsunematsu, H., Nonoue, Y., Yano, M., et al., (2018). Haplotype-based allele mining in the Japan-MAGIC rice population, *Scientific Reports, 8*, 4379. doi: 10.1038/s41598-018-22657-3.

Pidon, H., Chéron, S., Ghesquière, A., & Albar, L., (2020). Allele mining unlocks the identification of RYMV resistance genes and alleles in African cultivated rice. *BMC Plant Biology, 20*, 222. doi.org/10.1186/s12870-020-02433-0.

Raddy, A. M., Gambhire, V. B., & Raddy, R. T., (2014). Allele mining in crop improvement. *International Journal of Development Research, 4*, 300–305.

Rafalski, J. A., (2002). Novel genetic mapping tools in plants: SNPs and LD-based approaches. *Plant Science, 162*(3), 329–333.

Raman, H., Ryan, P. R., Raman, R., Stodart, B. J., Zhang, K., Martin, P., Wood, R., et al., (2008). Analysis of TaALMT1 traces the transmission of aluminum resistance in cultivated common wheat (*Triticum aestivum* L.). *Theoretical and Applied Genetics, 116*(3), 343–354. doi: 10.1007/s00122-007-0672-4.

Rose, L. E., Langley, C. H., Bernal, A. J., & Michelmore, R. W., (2005). Natural variation in the Pto pathogen resistance gene within species of wild tomato (Lycopersicon). I. Functional analysis of Pto alleles. *Genetics, 171*(1), 345–357. doi: 10.1534/genetics.104.039339.

Saito, M., & Nakamura, T., (2005). Two point mutations identified in emmer wheat generate null $Wx\text{-}A_1$ alleles. *Theoretical and Applied Genetics, 110*(2), 276–282. doi: 10.1007/s00122-004-1830-6.

Sakthivel, K., Shobha, R. N., Pandey, M. K., Sivaranjani, A. K. P., Neeraja, C. N., Balachandran, S. M., Madhav, M. S., et al., (2009). Development of a simple functional marker for fragrance in rice and its validation in Indian basmati and non-basmati fragrant rice varieties. *Molecular Breeding, 24*(2), 185–190. doi: 10.1007/s11032-009-9283-x.

Sanger, F., Nicklen, S., & Coulson, A. R., (1977). DNA sequencing with chain-terminating inhibitors. *Proceedings of the National Academy of Sciences of the United States of America, 74*(12), 5463–5467. doi: 10.1073/pnas.74.12.5463.

Shao, Y., Peng, Y., Mao, B., Lv, Q., Yuan, D., Liu, X., & Zhao, B., (2020). Allelic variations of the Wx locus in cultivated rice and their use in the development of hybrid rice in China. *PLoS ONE, 15*(5), e0232279. doi.org/10.1371/journal.pone.0232279.

Sharma, M. K., Singh, S., & Sengar, R. S., (2014). Allele mining and its applications. *Biotech Today: An International Journal of Biological Sciences, 4*(1), 34–37. doi: 10.5958/2322-0996.2014.00016.7.

Singh, B. K., Ramkumar, M. K., Dalal, M., Singh, A., Solanke, A. U., Singh, N. K., & Sevanthi, A. M., (2021). Allele mining for a drought responsive gene DRO1 determining root growth angle in donors of drought tolerance in rice (*Oryza sativa* L.). *Physiological and Molecular Biology of Plants, 27*(3), 523–534. doi.org/10.1007/s12298-021-00950-2.

Sinha, A., Chauhan, A., & Lohani, P., (2020). Potential of allele mining for improving drought tolerance in crops. *International Journal of Current Microbiology and Applied Sciences, 9*(5), 1098–1117. doi.org/10.20546/ijcmas.2020.905.121.

Song, J., Bradeen, J. M., Naess, S. K., Raasch, J. A., Wielgus, S. M., Haberlach, G. T., Liu, J., et al., (2003). Gene RB cloned from *Solanum bulbocastanum* confers broad spectrum resistance to potato late blight. *Proceeding of National Academy of Sciences of the United States of America, 100*(16), 9128–9133. doi: 10.1073/pnas.1533501100.

Sonnante, G., Paolis, A. D., & Pignone, D., (2005). Bowman–Birk. Inhibitors in lens: Identification and characterization of two paralogous gene classes in cultivated lentil and wild relatives. *Theoretical and Applied Genetics, 110*(3), 596–604. doi: 10.1007/s00122-004-1888-1.
Spooner, D., Van, T. R., & De Vicente, M. C., (2005). *Molecular Markers for Genebank Management.* International Plant Genetic Resources Institute, Rome, Italy, IPGRI Technical Bulletin No. 10.
Tadele, Z., Mba, C., & Till, B. J., (2010). TILLING for mutations in model plants and crops. In: Jain, S., & Brar, D., (eds.), *Molecular Techniques in Crop Improvement.* Springer, Dordrecht. doi.org/10.1007/978-90-481-2967-6_13.
Takanokai, N., Jiang, H., Kubo, T., Sweeney, M., Matsumoto T., Kanamori, H., Padhukasahasram, B., et al., (2009). Evolutionary history of GS3, a gene conferring grain length in rice. *Genetics, 182*(4), 1323–1334. doi.org/10.1534/genetics.109.103002.
Till, B. J., Comai, L., & Henikoff, S., (2007). Tilling and ecotilling for crop improvement. In: Varshney, R. K., & Tuberosa, R., (eds.), *Genomics-Assisted Crop Improvement.* Springer, Dordrecht. doi.org/10.1007/978-1-4020-6295-7_15.
Till, B. J., Jankowicz-Cieslak, J., Sagi, L., Huynh, O. A., Utsushi, H., & Swennen, R., (2010). Discovery of nucleotide polymorphisms in the *Musa* gene pool by eco tilling. *Theoretical Applied Genetics, 121*(7), 1381–1389. doi: 10.1007/s00122-010-1395-5.
Till, B. J., Zerr, T., Comai, L., & Henikoff, S., (2006). A protocol for TILLING and eco TILLING in plants and animals. *Nature Protocol, 1*(5), 2465–2477. doi: 10.1038/nprot.2006.329.
Tommasini, L., Yahiaoui, N., Srichumpa, P., & Keller, B., (2006). Development of functional markers specific for seven Pm3 resistance alleles and their validation in the bread wheat gene pool. *Theoretical Applied Genetics, 114*(1), 165–175. doi: 10.1007/s00122-006-0420-1.
Tsai, H., Howell, T., Nitcher, R., Missirian, V., Watson, B., Ngo, K. J., Lieberman, M., et al., (2011). Discovery of rare mutations in populations: TILLING by sequencing. *Plant Physiology, 156*(3), 1257–1268. doi: 10.1104/pp.110.169748.
Varshney, R. K., Nayak, S. N., May, G. D., & Jackson, S. A., (2009). Next-generation sequencing technologies and their implications for crop genetics and breeding. *Trends in Biotechnology, 27*(9), 522–530. doi: 10.1016/j.tibtech.2009.05.006.
Wang, H. W., Zhang, J. S., Gai, J. Y., & Chen, S. Y., (2006). Cloning and comparative analysis of the gene encoding diacylglycerol acyltransferase from wild type and cultivated soybean. *Theoretical and Applied Genetics, 112*(6), 1086–1097. doi: 10.1007/s00122-006-0210-9.
Wang, K. J., Takahata, Y., Kono, Y., & Kaizuma, N., (2008a). Allelic differentiation of Kunitz trypsin inhibitor in wild soybean (glycine soja). *Theoretical and Applied Genetics, 117*(4), 565–573. doi: 10.1007/s00122-008-0800-9.
Wang, M., Allefs, S., Berg, R. G. V. D., Vleeshouwers, V. G. A. A., Vossen, E. A. G. V. D., & Vosman, B., (2008b). Allele mining in solanum: Conserved homologues of Rpi-blb1 are identified in *Solanum stoloniferum*. *Theoretical and Applied Genetics, 116*(7), 933–943. doi: 10.1007/s00122-008-0725-3.
Xia, L. Q., Ganal, M. W., Shewry, P. R., He, Z. H., Yang, Y., & Roder, M. S., (2008). Exploiting the diversity of viviparous-1 gene associated with pre-harvest sprouting tolerance in European wheat varieties. *Euphytica, 159*(3), 411–417. doi: 10.1007/s10681-007-9576-2.
Yahiaoui, N., Brunner, S., & Keller, B., (2006). Rapid generation of new powdery mildew resistance genes after wheat domestication. *Plant Journal, 47*(1), 85–98. doi: 10.1111/j.1365-313X.2006.02772.x.

Yoshida, K., & Miyashita, N. T., (2005). Nucleotide polymorphism in the Adh2 region of the wild rice *Oryza rufipogon*. *Theoretical and Applied Genetics, 111*(6), 1215–1228. doi: 10.1007/s00122-005-0054-8.

Zhou, Y., Lei, F., Wang, Q., He, W., Yuan, B., & Yuan, W., (2020). Identification of novel alleles of the rice blast-resistance gene Pi9 through sequence-based allele mining. *Rice, 13*, 80. doi.org/10.1186/s12284-020-00442-z.

CHAPTER 11

Marker-Assisted Selection Approaches for Abiotic Stress Tolerance in Crop Plants

SUKHDEEP SINGH SIVIA,[1] SUNAYANA,[2] GURVINDER SINGH,[3] and REENA RANI[4]

[1]*Department of Genetics and Plant Breeding, CCSHAU, Hisar, Haryana, India*

[2]*Punjab Agricultural University, RRS, Abohar, Punjab, India*

[3]*Department of Plant Pathology, Dr. Y.S. Parmar University of Horticulture and Forestry, Nauni, Solan, Himachal Pradesh, India*

[4]*Division of Plant Improvement and Pest Management, ICAR–CAZRI, Jodhpur, Rajasthan, India*

ABSTRACT

Abiotic stresses such as drought, heat, salinity, and mineral toxicity adversely affect plant growth and development, resulting in the utmost threat to the global food production, nutritional security and safety. Being quantitative in nature, these stresses are controlled by genetic as well as environmental factors which impede the breeding strategies. Conventionally, efforts to develop abiotic stress tolerant varieties were completely based on phenotypic selection (PS) for crop improvement. However, there is an urgent need to implement new advanced molecular approaches to fulfill the demands of fast growing world population. The substantial knowledge about the genetic and molecular basis of abiotic stress tolerance will improve the capacity to appropriately manipulate the genes responsible for low yield in agricultural

Smart Breeding: Molecular Interventions and Advancements for Crop Improvement. Kailash Chandra, Subhash Chand, Ravi Prakash Saini, and Rohit Sharma (Eds.)

crops. With the introduction of marker-assisted selection (MAS), molecular markers have been established as the potential tool of choice for MAS in breeding programs to improve agriculturally important traits including abiotic stress tolerance. The biology of abiotic stress tolerance can be understood on the basis of molecular and genetic studies. The discovery of QTLs (quantitative trait loci) helps to deploy MAS to develop abiotic stress tolerance using sophisticated techniques, statistical analysis and marker modifications. Modern scientific advances and the agricultural challenges led to the discovery of high throughput tools which help to discover and develop genomes (genic and intergenic regions) to expand insights involved in molecular responses to provide accurate strategies for crop improvement. Further, gene identification and characterization followed by genetic transformation help to develop abiotic stress tolerant genotypes without yield penalty. In various cases, the identified genes provide the interesting insight into adaptive-evolution mechanisms in plants. In this chapter, we extensively highlighted the use of MAS to develop climate resilient crop cultivars by increased selection efficiency and reduced crop development period to combat abiotic stresses.

11.1 INTRODUCTION

Plants are highly affected by adverse environmental conditions including both biotic and abiotic stresses throughout various growth stages from seed sowing to harvesting (Zheng et al., 2010). Abiotic stresses predominantly cover drought stress, salinity stress, temperature stress, nutrient deficiency, hazardous metals and metalloids in soil. Among these abiotic stresses, drought, salinity, and temperature are the most common factors that decrease crop yield due to human-accelerated climate change (Younis et al., 2014). Limited water availability coupled with high atmospheric temperature leads to the accumulation of high salt concentration in the cultivated soil. Globally, saline soil covers about 22% of total agricultural land (FAO, 2008). It has been reported that only 10% global cultivated area is not threatened by abiotic stresses. Drought is considered as one of the major abiotic stresses and covers about 26% of global cultivated area. Abiotic stresses threaten global food and nutritional security, particularly in underdeveloped nations and affect plant diversity at different geographical and regional niches. India's 56% area is liable to cultivation but only 43% area is under cultivation and remaining area is exaggerated by abiotic stresses. Stress may be defined as environmental circumstance that restricts complete inherited expression of crop plants or an external cause that adversely affects plant growth is known as stress (Taiz et al., 1991). Abiotic stresses are persistent threats that are emphasized by modernization, mechanization, and global warming (Kang et al., 2017). Abiotic stresses lead to cellular, biochemical, and molecular changes in the plants that trigger the defense system of crop plants. The resulting plant responses are affected by the stage of the plant, genotype, severity, frequency, and duration of stress. The activated defense responses reduce the biological loss due to stress by signaling multiple defense pathways (Rejeb et al., 2014). Abiotic stresses lead to major changes from normal plant life cycles leading to senescence of plants (Figure 11.1). Affected plants show the response initially as alarming phase, then activation of defense system (resistance phase) and finally yield loss due to stress (Larcher et al., 2003). Salinity restricts the growth and productivity by inhibiting water uptake and altering the ion homeostasis (Parida et al., 2005). Further, drought will affect the survival of plants by restricting their proper growth and development due to lack of water availability in the soil and aggravate the condition by excessive evapotranspiration. It may also occur due to higher temperature under high humidity (Lipiec et al., 2013).

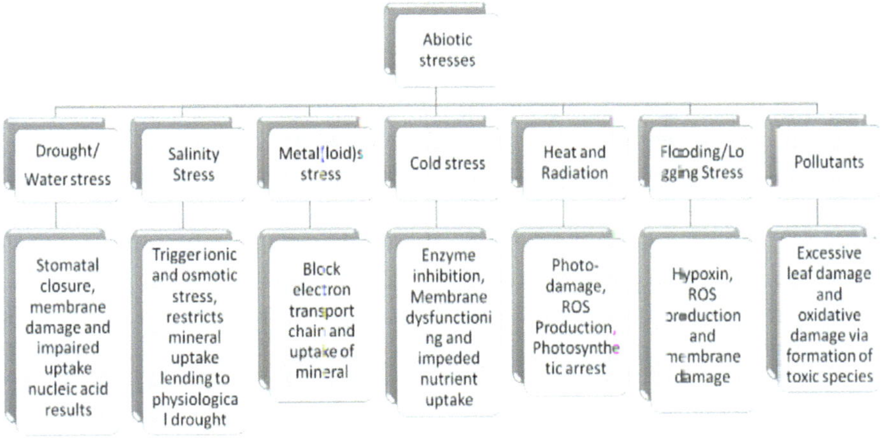

FIGURE 11.1 Classification of abiotic stresses.

It is well known that the conventional breeding methods alone are not adequate to develop stress resilient cultivars due to their high dependency on prevailing climatic conditions, quantitative nature of traits and diverse selection procedures. The knowledge of the molecular mechanisms of stress perception, signal transduction and response by plants may help to develop tolerant plants to multiple stresses. Recently several attempts have been made to identify the molecular basis of tolerance to abiotic stresses. Molecular markers are DNA fragments that are identified by allelic loci of genes for a specific trait (Xu & Crouch, 2008). Detection of parents, assessment of variation, recognition, and confirmation of linkage groups (LGs) on the genetic level are possible with the help of molecular/DNA marker.

Markers are classified as DNA markers and traditional markers. Advancement in marker assisted selection (MAS) through various polymorphism-searching methods has the scope for overcoming the limitations of conventional breeding (Xu, 2010). Introduction of molecular markers led to the mapping of genes and different genetic approaches for improvement of crop plants (Kage et al., 2016). Improvement and advancement in molecular markers played a significant role in the research of genomics. Polymerase chain reaction (PCR) based markers are less time consuming than the hybridization based for genetic mapping. Heterozygosity, maternal homozygosity and paternal homozygosity of every genotype are studied completely with the help of molecular markers. The construction of linkage maps requires the genes on the same chromosome and their recombination frequency (Ramzan et al., 2018). MAS have been successfully applied in various crops like rice, wheat, maize, and sorghum, etc. (Jena & Mackill, 2008; Prasanna et al., 2010; Miedaner & Korzun, 2012; Mohamed et al., 2014) to enhance the effectiveness of conventional breeding.

Plant breeding have a significant role to develop stress (abiotic and biotic) tolerant crop varieties with high yield and better quality (Simova-Stoilova et al., 2016). Taking into consideration this fact, there is an urgent need to develop stress tolerance and water-use efficient genotypes (Pandey et al., 2017). The advancement in DNA markers led to an important encroachment to remove the main limitations of classical breeding programs based on plant morphology leading to linkage drag (Reddy, 2017). To fight against the problems of recent era to develop tolerance to stresses, higher yield and better quality crops; there is need of MAS with advance DNA/molecular markers (Oliveira et al., 2008; Rumanti et al., 2016). It helps in accurate selection of stress-tolerant crop plants (Kiriga et al., 2016). Similarly, waterlogged-tolerant lines have been developed in crops using MAS (Devi et al., 2017). Genetic mapping of quantitative traits is escalating the incorporation of biotechnology and conventional breeding. Therefore, the objective of this chapter is to describe the consequences of abiotic stresses, classification of molecular markers and MAS approaches for the development of abiotic stress tolerant genotypes in different crops.

11.2 CLASSIFICATION OF MARKERS

1. **Genetic Markers:** These are defined as the DNA sequence with a specific gene location on chromosome and work as a marker for the targeted genes responsible for desired traits (Collard et al., 2005). These are classified as: traditional markers and molecular markers (Jiang, 2013) as explained below (Figure 11.2):

   i. **Traditional Markers:** These markers constitute the morphological, cytological, and biochemical markers:

      a. **Morphological Markers:** These are the simple markers that help in the identification of desired genes on the basis of morphological traits like shape and size of seed, color of flower, germination, and other visibly recognized traits. Morphological markers do not need any specific phenomenon or mechanism to work properly. These are extensively used by the plant breeders since long back. Limitations: they are limited in quantity, affected by environmental conditions and different stages of germination (Eagles et al., 2001; Karaköy et al., 2014).

      b. **Cytological Markers:** These markers are resultant due to variation in chromosome number, size, shape, and banding pattern of euchromatin and heterochromatin. These can be further used to differentiate typical and evolved chromosomes. These markers were also used for identification of linkage and actual mapping (Jiang, 2013).

      c. **Biochemical Markers:** These are defined as biochemical molecules like antigens, antibodies to detect any changes from the normal condition of plants. As biochemical markers are the variations of enzymes which can be used for estimation of gene and genotypic frequencies as well as analysis of genetic divergence. These are co-dominant, cost-effective, and manipulative markers. However, reduced polymorphism, restricted quantity and sensitive to isolation techniques and germination phases are the limitations of biochemical markers (Mondini et al., 2009).

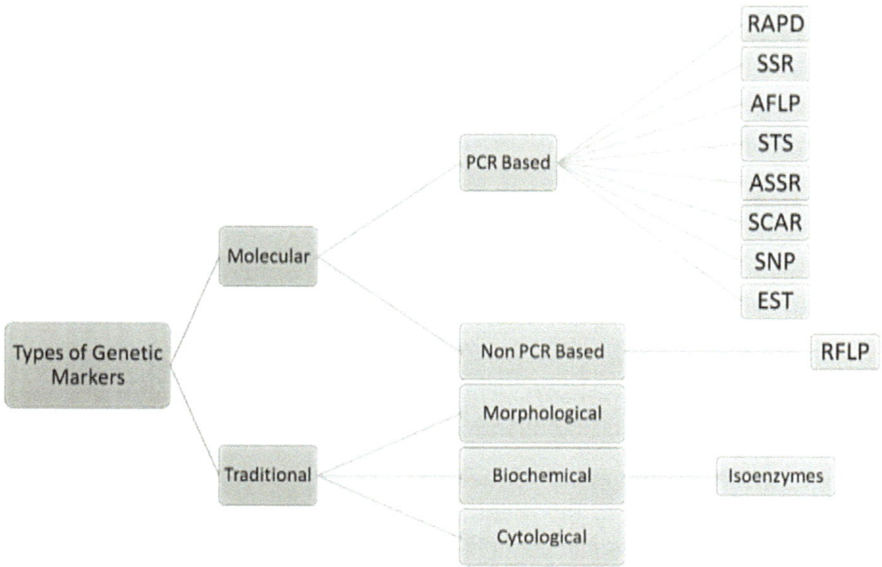

FIGURE 11.2 Types of genetic markers.

ii. **Molecular Markers:** These genetic markers are small piece of DNA that reveals sites of polymorphisms within the whole genome. The advantages of molecular markers include independence on environmental conditions, co-dominant nature, abundance, reproducible, and absence of epistasis (Raza et al., 2018).

   Types of molecular/DNA markers:
   a. Co-dominant or dominant markers;
   b. Hybridization or PCR-based.

Comparison among indispensable properties of DNA markers is described in Table 11.1.

11.3 ABIOTIC STRESSES: TYPES AND MECHANISMS

Abiotic stresses occur at various crop growth stages that adversely affect the global agricultural production and contributing yield loss nearly 50% of important crops like wheat, rice, corn, etc. (Mittler & Blumwald, 2010; Rosenzweig et al., 2014). To overcome the prevailing consequence of abiotic

TABLE 11.1 Essential Features of Different Molecular Markers

Characteristics	RFLP	RAPD	AFLP	ISSR	SSR	SNP
Dominance	Co-dominant	Dominant	Dominant	Dominant	Co-dominant	Co-dominant
Reproducibility	High	High	Intermediate	Medium-high	High	High
Polymorphism level	Medium	Very high	High	High	High	High
DNA quality/quantity	High/high	High/medium	High/low	Low/low	Low/low	High/low
DNA required (ng)	10,000	20	500–1,000	50	50	50
Genome abundance	High	Very high	Very high	Medium	Medium	Very high
Cost	High	Less	High	High	High	Variable
Sequencing	Yes	No	No	No	Yes	Yes
PCR based	No	Yes	Yes	Yes	Yes	Yes
Visualization	Radioactive	Agarose gel	Agarose gel	Agarose gel	Agarose gel	SNP

stresses, it is imperative for the breeders to develop new cultivars having high photosynthetic rate and per day productivity with tolerance to major abiotic stresses like flooding, droughts, freezing, and high temperatures (Hirabayashi et al., 2013; Pryor et al., 2013; Iizumi et al., 2013; Mickelbart et al., 2015).

11.3.1 DROUGHT

It is a period of time when the shortage of water accessibility including precipitation and soil moisture storage capacity in an area or region is experienced, hampering the expression of genetic potential of crop plants. Burgeoning human population is also increasing demand for water availability which results in water scarcity for crops. There is need to develop drought tolerant genotypes to overcome water scarcity. The duration of drought and its rigorousness determine the intensity and effect of drought stress on the yield potential of crop (Sorresls et al., 2000).

1. **Drought Resistance:** It is the phenomenon resulting in minimum yield loss in a drought stress environment compared to the maximum yield in the control environment of the crop.
2. **Drought Escape (DE):** Drought susceptible variety gives higher yield under drought stress conditions by avoiding drought period is known as DE.
3. **Dehydration Avoidance:** Avoids the dehydration by adopting phenomenon like osmotic adjustment, reduced transpiration, cuticular wax, leaf pubescence, leaf angle, leaf rolling, deep root system, root length, and root hydraulic resistance.
4. **Dehydration Tolerance:** This phenomenon occurs by maintenance of membrane integrity, plant growth, seedling survival, seed germination, stem reserve translocation and proline accumulation.
5. **Source of Drought Resistance:** Sources for drought resistance are cultivated varieties, land races, related wild species and genetic engineering. A number of identified and isolated transgenes proved to be a potential source of resistance to abiotic stresses like *hsf* for drought resistance in tobacco (Table 11.2).

TABLE 11.2 Transgene: For Protection to Abiotic Stresses

Transgene	Isolated From	Function	Transferred Into	Abiotic Stress Protection
mlt1D	*E. coli*	Mannitol-1-phosphate dehydrogenase	*Arabidopsis*	Salt
hsf	–	Heat shock factor	Tobacco	Drought
sacB	*Bacillus subtilis*	Fructan accumulation	Tobacco	Heat
ala3	Chemically synthesized	Antifreeze protein	Tobacco	Freeze
hva1	Barley	Lea protein	Rice	Salt
Sal1	*Arabidopsis*	Sulfur assimilation	Yeast	Salt tolerance
codA	*Arthobacter globiformis*	Choline oxidase	*Arabidopsis*	Salt and cold

6. **Selection Criteria for Drought Tolerance (DT):** It should be easy to estimate, high heritability, huge genetic variability and significant association with drought resistance.

11.3.2 SALINITY

It is another important abiotic stress affecting germination, tillering, yield, and quality of crops. The occurrence of metal ions (magnesium and calcium) as bicarbonates, sulfates, and chlorides contribute in varying degrees of soil salinity.

11.3.2.1 SALT STRESS RESISTANCE

It is plant's ability to stop, reduce or prevail over the injurious effects caused directly or indirectly by the presence of excessive soluble salts in the root zone. On the basis of anatomical, morphological, and developmental features, different species develop different strategies to cope up with the stress. Resistant plants maintain the ionic composition to grow and reproduce under salt stress conditions through avoidance, exclusion of ions or osmoregulation.

11.3.2.2 SOURCE OF SALINITY RESISTANCE

Sources for salinity resistance are cultivated varieties, land races, related wild species and genetic engineering. A number of identified and isolated transgenes proved to be a potential source of resistance to abiotic stresses like *mlt1D* for salt resistance in *Arabidopsis* (Table 11.2).

11.3.3 TEMPERATURE STRESS

Every genotype has an optimal temperature range for proper growth and development. When the temperature range exceeds above optimal range of the genotype then it negatively affects the stress. The temperature stress is classified as follows:

1. **Heat Stress:** When the temperature higher than optimal adversely affects the plants is called heat stress. It affects plant growth and development and survival. The degree of the effects is based on the temperature range, plant species and their interrelationship.
2. **Chilling Stress:** When plants are subjected to temperature above freezing, i.e., >0°C called chilling stress.
3. **Freezing Stress:** When plants are subjected to subzero temperature is called freezing stress.
4. **Heat Stress Tolerance:** It is the ability of a genotype to perform better than other genotypes under the same level of heat stress. Additive as well as dominance gene actions are responsible for heat tolerance. The genetic variability for heat tolerance occurs in many ways like germination, growth, and growth recovery, photosynthesis, translocation, and stability of cellular membranes. Therefore, recurrent selection in variable population under heat stress environment will lead to genetic gain for the trait of interest.

11.4 STRATEGIES FOR ABIOTIC STRESS TOLERANCE

The quantitative or qualitative nature of characters, number of concerned genes and nature of gene action will help to choose the breeding procedures for abiotic stress resistance.

11.4.1 SELECTION INDEPENDENT OF ENVIRONMENTS

Selection in traditional breeding for abiotic stress like drought, salinity, etc., resistance can only be done under specific stress environments that are either present at specific locations or created at well-controlled environments. Selection for abiotic stresses is considered most difficult for breeding programs (Xu, 2010).

11.4.2 GENOTYPE-BY-ENVIRONMENT INTERACTIONS (GEIS)

Statistical methods are of utmost importance to distinguish genetic effects from the environmental effects and their interaction and are also helpful for MAS systems for tolerance to abiotic stress. There is an enormous need for incorporated support tools for genotype-by-environment interaction (GEI) analysis like field experimental designs to define target population of environments (TPE) and genotypes, to maximize responses to selection under field conditions, QTL mapping and QTL-environment interaction, selection index for phenotypic and molecular marker to select the best genotypes for next cycle of selection, incorporating environmental or genotypic variables into statistical models to explain the causes of GEI and LD mapping for desired traits.

11.5 MARKER-ASSISTED SELECTION (MAS)

DNA marker maps are developed with the help of molecular markers. The MAS helps to improve targeted traits in crops using information obtained from the map position of desired traits linked with molecular markers. However, the accomplishment of MAS depends on the genetic base of the desired trait and its association with molecular marker. It also depends upon the number of individuals and the genetic background of the recipient genotype. The MAS involves certain steps which are discussed in subsections (Figure 11.3).

11.5.1 DEVELOPMENT OF POPULATION

Selection of parents with desired traits (high yield, better quality, and stress resistance) is the basis for any breeding program. Contrasting desired parents are selected after extensive germplasm screening to combine the desirable genes into either a new variety or improvement from the previous one. Homozygous parents are to be selected for MAS. More number of plant genotypes are being screened for recognition of exact gene combinations if large no of genes are segregating (Witcombe & Virk, 2001). It includes the selection of parents-crossing of parents (F_1 progeny) – F_1 screening. Hybrids bearing the markers are selected and used to develop F_2 generation for segregation pattern of RFLPs.

11.5.2 QTL MAPPING AND QTL VALIDATION

Isolate the DNA from every plant of F_2 with standard protocols. Linkage map is constructed using phenotypic evaluation for trait and QTL analysis. After QTL mapping, it is validated by position confirmation and QTL effect, then verify QTL in independent population and testing in different genetic backgrounds then fine mapping.

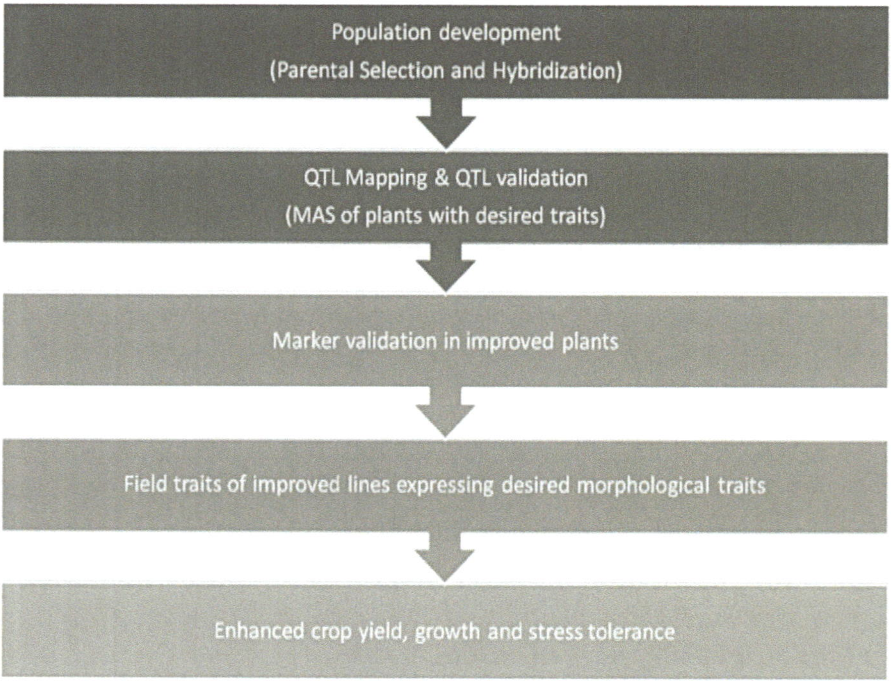

FIGURE 11.3 Depiction of different procedures of marker-assisted selection.

11.5.3 MARKER VALIDATION

DNA probes that are corresponding to the molecular marker sequence of desired trait help to detect the variation of the DNA/molecular markers among the parents and the occurrence of marker alleles in each plant of F_2. Generally, for detection of similar fragments phosphorous (32P) is used. The DNA probes hybridize with particular complementary segments. It helps in the identification of the individuals having the marker/indicator alleles.

11.5.4 ASSOCIATION WITH MORPHOLOGICAL TRAITS

For the confirmation of indirect selection through DNA markers is correlated with the field traits. After recognition of the association of DNA markers and morphological characters, MAS will help to improve the crops on genetic background. Selected lines are evaluated for stress resistance, desired quality traits and yield potential on the field.

MAS help to increase the speed of breeding. Marker assisted breeding (MAB) helps to examine the genomic regions of germplasm of different crops under water limited conditions. Conventional markers (RFLPs, RAPDs, AFLPs, and SSRs) do not originate from the genes oneself due to convoluted gene cloning in polyploidy and large genome size species. Polymorphism/variation in functional genes within transcribed regions helps to design functional markers (FMs). These markers are associated with the function of genes responsible for precise selection of aimed genes (Ivandic et al., 2003). MAS have been used for monogenic, oligogenic, and polygenic traits (Dwivedi et al., 2007). Genotypic selection with high precision of complex traits like abiotic stress tolerance requires closely linked markers or functional nucleotide polymorphic markers in combination with precision phenotyping to increase detection power (Andersen & Lübberstedt, 2003). For efficient MAS, improvement of suitable selection criteria is vital for both marker trait association and the following MAS, particularly for abiotic stresses. Marker-assisted recurrent selection (MARS) schemes have been developed for 'forward breeding' of inhabitant genes and QTL for quantitative traits such as abiotic stress tolerance and grain yield (Eathington, 2005). MARS has helped in genetic gain in maize since long back. Marker genotypes are produced for QTL flanking markers which are incorporated in selection indices. Genotyping of plants is done at each cycle for selection of specific combinations (Xu, 2010). QTL can be studied under adverse environments, near iso-environments or a uniform environment by replicating doubled haploid (DH) or recombinant inbred line (RIL) populations and a segregating population.

11.6 MAS FOR ABIOTIC STRESS TOLERANCE IN CROP PLANTS

11.6.1 RICE

Drought stress is a matter of primary concern in rice production. To enhance production of rice in drought susceptible areas; identification and introgression of drought resistance QTLs would be effective. Progress on MAB of the identified QTLs is partial. The studies on MAS in rice are depicted in Table 11.3 (Wani et al., 2018). Under drought stress; 3 QTLs (chr. 1, 4, and 6) were mapped using SSR markers for physiological and yield traits for introgression into elite rice genotypes for superior yield (Prince et al., 2015). Genotyping by using SNP helped in mapping of 6 QTLs on different chromosomes (1, 2, 4, 7, and 10) for deep rooting, root length and root dry weight for imparting DT (Lou et al., 2015; Lang et al., 2013). While using SSR genotyping 10 QTLs were identified for seed characters, yield related traits and physiological parameters using backcross inbred lines (BILs) for drought conditions (Sangodele et al., 2014) (Table 11.3). Heat stress in changing climatic conditions is a potential threat for rice production and productivity. Using SSR and SNP markers 19 QTLs were mapped for fertility of spikelet, flowering duration and shedding of pollen grains on chromosomes 1, 2, 3, 4, 5, 7, 8, 10, and 11 to impart heat tolerance (Zhao et al., 2016; Ye et al., 2015). To survive under submergence SUB1 gene along with recent QTLs proved to be better for superior yield in rice. Around 41 QTLs on chromosomes 1, 2, 3, 4, 6, 7, 8, 9, 10, and 12 have been identified using SNP, SSR, and RFLP markers for submergence tolerance (Septiningsih et al., 2012; Manangkil et al., 2013; Gonzaga et al., 2016). Salinity is also affecting the rice production. QTL mapping using 131 SSR and SNP mapped 85 QTLs on chromosome 1, 2, 3, 4, 6, 7, 8, 10, 11, and 12 for diverse traits like pollen fertility, sodium ion concentration, sodium, and potassium ratio, shoot K^+ concentration, score of salt injury, plant height, and shoot dry weight (Hossain et al., 2015; Leon et al., 2016). The identified QTLs are used efficiently for the development of high yielding and better quality rice using MAB for salinity tolerance. In reduced soil, excessive ferrous iron (Fe^{2+}) created problem in rice. Higher adaptability of *Oryza glaberrima* proved to be a potential source of resistance to Fe^{2+} toxicity. Therefore, SSR and SNP based QTL mapping on chromosomes 1, 2, 3, 4, 5, 7, 8, 10, and 12 for morphological, physiological traits and leaf bronzing identified 7 QTLs (Dufey et al., 2015; Wu et al., 2014).

11.6.2 WHEAT

Introgression of drought-tolerant QTLs can be effective to combat the moisture stress in wheat. Using recombinant inbred lines (RILs) three QTLs were identified for higher yield and biomass on the chromosomes 7 AS, 1 BS and 2BS while in F_4 population 13 QTLs were identified for abscisic acid (ABA) on chromosomes 5A, 4A, and 3B for development drought-tolerant cultivars (Merchuk-Ovnat et al., 2016; Barakat et al., 2015). Around 22

QTLs on chromosomes 7D, 1B, 2A, 7B, 2B, 3B, 6D, 4A, 5D, and 6A for root and shoot parameters, seedling characters, stem, and leaf parameters were reported by Zhang et al. (2013a). Around 26 QTLs were identified for seminal root angle and root number, grain weight (GW), spike number, weight, and harvest index through DArT and SSR markers in DH, F_3 and F_4 populations for DT (Christopher et al., 2013; Golabadi et al., 2011). Globally, high temperature is severely affecting wheat production and adversely affecting food and nutritional security. Azam et al. (2015) identified the 37 QTLs in a DH population using SSR and AFLP markers on chromosomes 7D, 1A, 4A, 1B, and 2B for fluorescence and maximum quantum efficiency of photosystem II for heat stress tolerant genotypes. Around 10 QTLs were recognized on chromosome 1D, 7D, 6B, 7B, 2D, 2B, and 7A for grain filling duration, GW, canopy temperature and grain yield for tolerance to heat stress through SSR markers (Tiwari et al., 2013; Paliwal et al., 2012). Salinity is a leading cause of worry for wheat production in the global market. SSR-based studies in 119 wheat genotypes identified single QTL on chromosome 4B for percent proportions of dead leaves (Turki et al., 2015) Around 18 additives and 16 epistatic QTLs (E-QTLs) for shoot potassium and sodium ion concentration while for total dry weight on chromosome 5A and for salinity tolerance on chromosome 4B (Xu et al., 2013). On the basis of DArT and SSR mapping 36 and 10 QTLs for root dry weight index, shoot dry weight index, and total dry weight index for water logging tolerance were identified (Yu & Chen, 2013).

11.6.3 MAIZE

Drastic yield reduction in maize occurs due to drought stress. Assessment of biparental populations led to identification of 203 QTLs using SNPs and SSR markers responsible for different drought-related morphological and physiological characters like anthesis-silking interval, ears per plant, stay green and plant-to-ear height ratio, grain yield, kernel numbers and weight on chromosomes 1, 3, 4, 5, 7, and 10 (Almeida et al., 2014; Li et al., 2016) Water logging is also a major constraint that leads to considerable yield loss in high rainfall areas. Incorporation of QTLs on the basis of MAS for water logging tolerance could be a better strategy to tackle this problem. Zaidi et al. (2015); and Zhang et al. (2013b) reported 31QTLs on chromosomes 1, 2, 3, 4, 5, 6, 7, 8, 9, and 10 using SNPs and SSRs for different morpho-physiological traits such as brace roots, root length, root fresh and dry weight, stem, and root lodging, chlorophyll content and seedling length. Low temperature negatively affects the growth and yield of maize. Allam et al. (2016); and Revilla et al. (2016) used SNPs to identify 33 QTLs on chromosome 1, 2, 3, 4, 5, 6, 7, 8, 9, and 10 for days to emergence, germination, chlorophyll content and early growth for development of cold tolerance genotypes. These results for cold tolerance will unlock the pathways for the genetic improvement of maize genotypes by using MAS. Salinity also hampers the maize production. Hoque et al. (2015) identified 15 QTLs for salt tolerance traits like root and shoot parameters and plant dry weight on chromosomes 1, 2, 4, 5, 6, 7, 8, 9, and 10.

11.6.4 BARLEY

Relationship between molecular markers and traits related to DT in 52 wild barley lines were studied and 14 QTLs were identified on chromosome 4H using SNP and SSR markers (Forster et al., 2000). These identified QTLs were used to develop salt and drought tolerant genotypes in barley.

11.6.5 COTTON

Drought and salt stress adversely affects the growth and development of cotton, resulting in yield reduction. MAS could be a better solution to develop drought and salt tolerant cotton genotypes for stable yield production under abiotic stress conditions. Saeed et al. (2014); and Abdelraheem et al. (2015) identified 14 QTLs using SSR and SSCP markers for a number of bolls per plant, boll weight and chlorophyll content to develop stress tolerant genotypes.

11.6.6 SOYBEAN

Soybean is an essential source of oil and protein for human health and livestock feeding. It is very sensitive to aluminum (Al) toxicity and breeding for Al-tolerant varieties could be a way out for Al toxicity in soybean. Root architecture such as root extension is an important trait to alleviate Al toxicity and would help to develop Al tolerant genotypes. Under Al stress, six QTLs were identified for root extension and root length mapped in RILs (Abdel-Haleem et al., 2014; Sharma et al., 2011). To develop salt tolerant soybean cultivars; a QTL in RIL population using SSR and SNPs for leaf scorching was identified (Ha et al., 2013). Abdel-Haleem et al. (2014) identified 7 QTLs for DT under different regimes. The linked markers can be effectively

TABLE 11.3 Identified QTLs for Different Abiotic Stresses in Agriculturally Important Field Crops

Crop	QTLs	Traits	Mapping Population	Markers	Chromosome	Stress	References
Rice	3 QTLS	physiological and yield traits	RILs	SSRs	1, 4, and 6	Drought	Prince et al. (2015)
	6 QTLS	Ratio of deep rooting	RILs	SNP	1, 2, 4, 7, and 10	Drought	Lou et al. (2015)
	15 QTLS	Physiological, productivity, and related traits	BIL	SSR	1, 2, 3, 7, 8, and 9	Drought	Sangodele et al. (2014)
	11 QTLS and 8 QTLS	Spikelet fertility, daily flowering time and pollen shedding level	CSSLs, 3 ways cross	SSRs and SNP	2, 4, 3, 8, 10, 11, 5, and 7	Heat	Zhao et al. (2016), Ye et al. (2015)
	41 QTLS	Submergence, shoot length, root length and shoot fresh weight	RILs, BILs, $F_{2:3}$	SSR, RFLPs, and SNP	1, 2, 3, 4, 6, 7, 8, 9, 10, and 12	Water logging/ submergence	Gonzaga et al. (2016), Manangkil et al. (2013), Septiningsih et al. (2012)
	85QTLS	Shoot k conc., sodium–potassium ratio, salt injury score, plant height and shoot dry weight, pollen fertility, Na^+ concentration, and Na/K ratio in the flag leaf	RILs, F_2	SSR and SNP	1, 2, 3, 4, 6, 7, 8, 10, 11, and 12	Salinity	Leon et al. (2016); Hossain et al. (2015)
	38 QTLS	Different morphological and physiological traits, leaf bronzing	BC3DH RILs	SSR and SNP	1, 2, 3, 4, 5, 7, 8, 10, and 12	Iron (Fe) toxicity	Dufey et al. (2015), Wu et al. (2015)
Wheat	3 QTLS	Yield and biomass	NILs	SNP	1BL, 2BS, and 7AS	Drought	Merchuk-Ovnat et al. (2016)
	13 main QTLs	ABA content	F_4	TRAP, SRAP, and SSR	3B, 4A, and 5A	Drought	Barakat et al. (2015)

TABLE 11.3 *(Continued)*

Crop	QTLs	Traits	Mapping Population	Markers	Chromosome	Stress	References
	22 QTLs	Coleoptiles length, root traits, seedling traits, root, and shoot fresh and dry weight	RILs	SSR, ISSR, STS, SRAP, and RAPD	1B, 2A, 2B, 3B, 4A, 5D, 6A, 6D, 7B, and 7D	Drought	Zhang et al. (2013a)
	26 QTLs	1,000 grain weight, grain weight per spike, number of grains per spike, spike weight, spike harvest index and harvest index, seminal root angle and seminal root number	DHs, F_3 and F_4	DArT and SSR	3B, 7B, 1B, 2B, 1B, 2B 2A, 3D, 6A, 5D, 4A, 1B, 3A, 3B, and 6B	Drought	Christopher et al. (2013); Golabadi et al. (2011)
	37 QTLs	Parameters of chlorophyll fluorescence kinetics	DH	AFLP and SSR	1A, 1B, 2B, 4A, and 7D	Heat	Azam et al. (2015)
	10 stable QTLs	Grain yield, 1,000 grain weight, grain filling duration, and canopy temperature	DH, RILs	SSR	1D, 6B, 2D and 7A, 2B, 7B, and 7D	Heat	Tiwari et al. (2013); Paliwal et al. (2012)
	1 QTL	Proportion of dead leaves	Varieties	SSR	4B	Salinity	Turki et al. (2015)
	18 additive and 16 epistatic QTLs	Root, shoot, and total dry weight, K^+, Na^+ concentration and K^+/Na^+ ratio	RILs	SSR	1A, 2A, 4B, 5D, 1B, 3A, 6D, 7B, 1D, 2B, 5A, 5B, 7A, 4A, 6A, and 6B	Salinity	Xu et al. (2013)
	36 QTLs and 10 QTLs	Root dry weight index, shoot dry weight index, total dry weight index	RILs	DArT and SSR	18 chromosomes	Water logging	Yu & Chen (2013)

TABLE 11.3 *(Continued)*

Crop	QTLs	Traits	Mapping Population	Markers	Chromosome	Stress	References
Maize	203 QTLs	ASI, ears per plant, stay green and plant-to-ear height ratio, grain yield, kernel numbers and weight	RILs $F_{2:3}$ $F_{2:3}$	SNPs and SSRs	1, 3, 4, 5, 7, and 10	Drought	Almeida et al. (2014); Li et al. (2016)
	31 QTLs	Yield brace roots, chlorophyll content, % stem and root lodging, seedling height, root length, shoot fresh weight, root fresh weight, shoot dry weight and root dry weight	RILs, BC2F2	SNP markers using KASP platform and SSRs	1, 2, 3, 4, 5, 6, 7, 8, 9, and 10	Water logging	Zaidi et al. (2015); Zhang et al. (2013)
	23 QTLs	Days to emergence, chlorophyll content germination and early growth	Inbred population	SNP	1, 2, 3, 4, 5, 6, 7, 8, 9, and 10	Cold	Allam et al. (2016); Revilla et al. (2016)
	15 QTLs	Shoot length, root length, ratio of root length and shoot length, shoot fresh weight, root fresh weight, plant fresh weight, plant dry weight, shoot dry weight, root dry weight, ratio root dry weight and shoot dry weight	$F_{2:3}$	SSRs	1, 2, 4, 5, 6, 7, 8, 9, and 10	Salinity	Hoque et al. (2015)
Soybean	6 QTLs	Root length and root extension	RILs	BARC, RFLP, and SSR	Gm08 and LGs B2, C2, F, and G Gm16	Al tolerance	Abdel-Haleem et al. (2014); Sharma et al. (2011)
–	1 QTL	Salt tolerance	RILs	SSR and SNP	3	Salinity	Ha et al. (2013)

TABLE 11.3 *(Continued)*

Crop	QTLs	Traits	Mapping Population	Markers	Chromosome	Stress	References
–	7 QTLs	Canopy wilting	RILs	SSR	Gm12	Drought	Abdel-Haleem et al. (2012)
Barley	15 QTL	Yield and physiological traits Na^+, K^+ content, and Na^+:K^+ ratio	52 wild barleys	SSR and SNP	4H	Drought and salt tolerance	Forster et al. (2000)
Cotton	14 QTLS	Morpho-physiological traits	Inbred population	SSR, SSCP	11	Drought and salt tolerance	Saeed et al. (2014); Abdelraheem et al. (2015)
Sorghum	7 QTLs	Yield traits	RILS	RFLP	1 and 2	Drought tolerance	Kabede et al. (2001)

utilized for recognition of drought tolerant genotypes having enviable QTLs for accelerating the speed of soybean molecular-breeding program.

11.7 CONCLUSION

Ever increasing human population necessitates increased crop productivity with better nutritional quality globally. The amalgamation of conventional breeding with cutting-edge technologies could meet the present and future demands. In this context, MAS helped the plant breeders to develop high yielding varieties with better quality. MAS has been extensively used for drought, salinity, temperature, and freezing stress tolerance cultivation in major agricultural crops such as wheat, rice, maize, barley, soybean, and cotton. In addition, MAS is extremely helpful to discover QTLs linked with abiotic stress tolerance. Molecular genetics provide abundant DNA markers that discover genetic modification (GM), genotypic resistance and stress tolerant lines to different abiotic stresses. Although, recent sophisticated marker applications help to discover specific genes accountable for abiotic stress tolerance. Molecular markers with QTL mapping reveal stress tolerance genes on particular chromosomal loci. However, demand for continuous improvement in molecular marker technology will permit complete scrutiny of abiotic stress tolerance under changing climatic conditions. The major shortcoming of MAS is high cost; otherwise, it is helpful for the development of high yielding superior cultivars being tolerant to abiotic stresses.

KEYWORDS

- **abiotic stress**
- **functional markers**
- **genotype-by-environment interaction**
- **marker-assisted selection**
- **MAS and molecular markers**
- **polymerase chain reaction**

REFERENCES

Abdel-Haleem, H., Carter, T. E., & Rufty, T. W., (2014). Quantitative trait loci controlling aluminium tolerance in soybean: Candidate gene and single nucleotide polymorphism marker discovery. *Molecular Breeding, 33*, 851.

Abdelraheem, A., Mahdy, Z., & Zhang, J. F., (2015). The first linkage map for a recombinant inbred line population in cotton (*Gossypium barbadense*) and its use in studies of PEG-induced dehydration tolerance. *Euphytica, 205*, 941–958.

Allam, M., Revilla, P., & Djemel, A., (2016). Identification of QTLs involved in cold tolerance in sweet x field corn. *Euphytica, 208*, 353.

Almeida, G. D., Makumbi, D., & Magorokosho, C., (2013). QTL mapping in three tropical maize populations reveals a set of constitutive and adaptive genomic regions for drought tolerance. *Theoretical and Applied Genetics, 126*, 583.

Andersen, J. R., & Lubberstedt, T., (2003). Functional markers in plants. *Trends in Plant Science, 8*, 554–560.

Azam, F., Chang, X., & Jing, R., (2015). Mapping QTL for chlorophyll fluorescence kinetics parameters at seedling stage as indicators of heat tolerance in wheat. *Euphytica, 202*, 245.

Barakat, M. N., Saleh, M. S., & Al-Doss, A. A., (2015). Mapping of QTLs associated with abscisic acid and water stress in wheat. *Biologia Plantarum, 59*, 291.

Christopher, J., Christopher, M., & Jennings, R., (2013). QTL for root angle and number in a population developed from bread wheats (*Triticum aestivum*) with contrasting adaptation to water-limited environments. *Theoretical and Applied Genetics, 126*, 1563.

Collard, B. C. Y., Jahufer, M. Z. Z., Brouwer, J. B., & Pang, E. C. K., (2005). An introduction to markers, quantitative trait loci (QTL) mapping and marker-assisted selection for crop improvement: The basic concepts. *Euphytica. 142*(1, 2), 169–196.

Devi, E. L., Devi, C. P., Kumar, S., Sharma, S., Beemrote, A., Chongtham, S. K., & Akoijam, R., (2017). Marker assisted selection (MAS) towards generating stress tolerant crop plants. *Plant Gene., 11*, 205–218.

Dufey, I., Draye, X., Lutts, S., Lorieux, M., Martinez, C., & Bertin, P., (2015). Novel QTLs in an interspecific backcross *Oryza sativa* x *Oryza glaberrima* for resistance to iron toxicity in rice. *Euphytica, 204*, 609.

Dwivedi, S. L., Crouch, J. H., Mackill, D. J., Xu, Y., Blair, M. W., Ragot, M., Upadhyaya, H. D., & Ortiz, R., (2007). The molecularization of public sector crop breeding: Progress, problems and prospects. *Advances in Agronomy, 95*, 163–318.

Eagles, H. A., Bariana, H. S., Ogbonnaya, F. C., Rebetzke, G. J., Hollamby, G., Henry, R. J., Henschke, P., & Carter, M., (2001). Implementation of markers in Australian wheat breeding. *Australian Journal of Agricultural Research, 52*, 1349–1356.

Eathington, S. R., Crosbie, T. M., Edwards, M. D., Reiter, R. S., & Bull, J. K., (2007). Molecular markers in a commercial breeding program. *Crop Science, 47*, 154–163.

FAO (Food and Agriculture Organization of the United Nations), (2008). *FAO Production Yearbook*. FAO, Rome.

Forster, B. P., & Thomas, W. T. B., (2004). Doubled haploids in genetics and plant breeding. *Plant Breeding Reviews, 25*, 57–88.

Golabadi, M., Arzani, A., & Mirmohammadi, S. A. M., (2011). Identification of microsatellite markers linked with yield components under drought stress at terminal growth stages in durum wheat. *Euphytica, 177*, 207.

Gonzaga, Z. J. C., Carandang, J., & Sanchez, D. L., (2016). Mapping additional QTLs from FR13A to increase submergence tolerance in rice beyond *SUB1*. *Euphytica, 209*, 627.

Ha, B. K., Vuong, T. D., & Velusamy, V., (2013). Genetic mapping of quantitative trait loci conditioning salt tolerance in wild soybean (*Glycine soja*) PI 483463. *Euphytica, 193*, 79.

Hoque, M. M. I., Jun, Z., & Guoying, W., (2015). Mapping QTLs associated with salinity tolerance in maize at seedling stage. *International Journal of Current Microbiology, 3*(10), 1–23.

Hossain, H., Rahman, M. A., Alam, M. S., & Singh, R. K., (2015). Mapping of quantitative trait loci associated with reproductive-stage salt tolerance in rice. *Journal of Agronomy and Crop Science, 201*(1), 17–31.

Iizumi, T., Sakuma, H., Yokozawa, M., Luo, J. J., Challinor, A. J., Brown, M. E., Sakurai, G., & Yamagata, T., (2013). Prediction of seasonal climate-induced variations in global food production. *Nature Climate Change, 3*, 904.

Ivandic, V., Thomas, W. T. B., Nevo, E., Zhang, Z., & Forster, B. P., (2003). Association of SSRs with quantitative trait variation including biotic and abiotic stress tolerance in *Hordeum spontaneum*. *Plant Breeding, 122*, 300–304.

Jena, K., & Mackill, D., (2008). Molecular markers and their use in marker-assisted selection in rice. *Crop Science, 48*, 1266–1276.

Jiang, G. L., (2013). Molecular markers and marker-assisted breeding in plants. In: *Plant Breeding from Laboratories to Fields* (pp. 45–83). Intech Croatia.

Kage, U., Kumar, A., Dhokane, D., Karre, S., & Kushalappa, A. C., (2016). Functional molecular markers for crop improvement. *Critical Reviews in Biotechnology, 36*(5), 917–930.

Kang, J. S., Singh, H., Singh, G., Kang, H., Kalra, V. P., & Kaur, J., (2017). Abiotic stress and its amelioration in cereals and pulses: A review. *International Journal of Current Microbiology and Applied Sciences, 6*, 1019–1045.

Karaköy, T., Baloch, F. S., Toklu, F., & Özkan, H., (2014). Variation for selected morphological and quality related traits among 178 faba bean landraces collected from Turkey. *Plant Genetic Resources, 12*, 5–13.

Kiriga, W. J., Yu, Q., & Bill, R., (2016). Breeding and genetic engineering of drought-resistant crops. *International Journal of Agriculture and Crop Science, 9*(1), 7–12.

Lang, N. T., Nha, C. T., HA, P. T. T., & Buu, B. C., (2013). Quantitative trait loci (QTLs) associated with drought tolerance in rice (*Oryza sativa* L.). *Journal of Breeding and Genetics, 45*(3), 409–421.

Larcher, W., (2020). Physiological plant ecology: Ecophysiology and stress physiology of functional groups. *Plants, 9*, 1374.

Leon, T. B. D., Linscombe, S., & Subudhi, P. K., (2016). Molecular dissection of seedling salinity tolerance in rice (*Oryza sativa* L.) using a high-density GBS-based SNP linkage map. *Rice, 9*, 52.

Li, C., Sun, B., Li, Y., Liu, C., Wu, X., Zhang, D., & Wang, T., (2016). Numerous genetic loci identified for drought tolerance in the maize nested association mapping populations. *BMC Genomics, 17*(1), 894.

Lipiec, J., Doussan, C., Nosalewicz, A., & Kondracka, K., (2013). Effect of drought and heat stresses on plant growth and yield: A review. *International Agrophysics, 27*, 463–477.

Lou, Q., Chen, L., Mei, H., Wei, H., Feng, F., Wang, P., Xia, H., Li, T., & Luo, L., (2015). Quantitative trait locus mapping of deep rooting by linkage and association analysis in rice. *Journal of Experimental Botany, 66*(15), 4749–4757.

Manangkil, O. E., Vu, H. T. T., Mori, N., Yoshida, S., & Nakamura, C., (2013). Mapping of quantitative trait loci controlling seedling vigor in rice (*Oryza sativa* L.) under submergence. *Euphytica, 192*, 63–75.

Mateu-Andres, I., & DePaco, L., (2004). Allozymic differentiation of the *Antirrhinum majus* and *A. siculum* species groups. *Annals of Botany, 95*, 465–473.

Merchuk-Ovnat, L., Barak, V., Fahima, T., Ordon, F., Lidzbarsky, G. A., Krugman, T., & Saranga, Y., (2016). Ancestral QTL alleles from wild emmer wheat improve drought resistance and productivity in modern wheat cultivars. *Frontiers in Plant Science, 7*, 452.

Mickelbart, M. V., Hasegawa, P. M., & Bailey-Serres, J., (2015). Genetic mechanisms of abiotic stress tolerance that translate to crop yield stability. *Nature Reviews Genetics, 16*, 237–251.

Miedaner, T., & Korzun, V., (2012). Marker-assisted selection for disease resistance in wheat and barley breeding. *Phytopathology, 102*, 560–566.

Mittler, R., & Blumwald, E., (2010). Genetic engineering for modern agriculture: Challenges and perspectives. *Annual Review of Plant Biology, 61*, 443–462.

Mohamed, A., Ali, R., & Elhassan, O., (2014). First products of DNA marker-assisted selection in sorghum released for cultivation by farmers in sub-Saharan. *African Journal of Plant Science and Molecular Breeding, 3*, 1–10.

Mondini, L., Noorani, A., & Pagnotta, M. A., (2009). Assessing plant genetic diversity by molecular tools. *Diversity, 1*, 19–35.

Oliveira, L. K., Melo, L. C., Brondani, C., Peloso, M. J. D., & Brondani, R. P. V., (2008). Backcross assisted by microsatellite markers in common bean. *Genetics and Molecular Research, 7*(4), 1000–1010.

Paliwal, R., Roder, M. S., Kumar, U., Srivastava, J. P., & Joshi, A. K., (2012). QTL mapping of terminal heat tolerance in hexaploid wheat (*T. aestivum* L.). *Theoretical and Applied Genetics, 125*, 561–575.

Pandey, P., Irulappan, V., Bagavathiannan, M. V., & Senthil-Kumar, M., (2017). Impact of combined abiotic and biotic stresses on plant growth and avenues for crop improvement by exploiting physio-morphological traits. *Frontiers in Plant Science, 8*, 1–15.

Parida, A. K., & Das, A. B., (2005). Salt tolerance and salinity effects on plants: A review. *Ecotoxicology and Environmental Safety, 60*, 324–349.

Prasanna, B., Pixley, K., Warburton, M. L., & Xie, C. X., (2010). Molecular marker-assisted breeding options for maize improvement in Asia. *Molecular Breeding, 26*, 339–356.

Prince, S. J., Beena, R., Michael, G. S., Senthivel, S., & Chandra, B. R., (2015). Mapping consistent rice (*Oryza sativa* L.) yield QTLs under drought stress in target rainfed environments. *Rice, 8*, 25.

Pryor, S., Barthelmie, R., & Schoof, J., (2013). High-resolution projections of climate-related risks for the Midwestern USA. *Climate Research, 56*, 61–79.

Ramzan, F., Kim, H. T., Shim, K. K., Choi, Y. H., Younis, A., & Lim, K. B., (2018). Genetic diversity and relationship assessment of *Lilium lancifolium* Asiatic hybrid 'chianti' progeny by ISSR markers. *European Journal of Horticultural Sciences, 83*, 142–150.

Raza, A., Shaukat, H., Ali, Q., & Habib, M., (2018). Assessment of RAPD markers to analyze the genetic diversity among sunflower (*Helianthus annuus* L.) genotypes. *Turkish Journal of Agriculture – Food Science and Technology, 6*, 107–111.

Reddy, V. R. P., (2017). New concepts in plant breeding and genetics. *Advances in Plants and Agricultural Research, 7*(1), 00245.

Rejeb, I., Pastor, V., & Mauch-Mani, B., (2014). Plant responses to simultaneous biotic and abiotic stress molecular mechanisms. *Plants, 3*, 458–475.

Revilla, P., Rodríguez, V. M., Ordás, A., Rincent, R., Charcosset, A., Giauffret, C., Melchinger, A. E., et al., (2016). Association mapping for cold tolerance in two large maize inbred panels. *BMC Plant Biology, 16*(1), 1–10.

Rosenzweig, C., Elliott, J., Deryng, D., Ruane, A. C., Müller, C., Arneth, A., Boote, K. J., et al., (2014). Assessing agricultural risks of climate change in the 21st century in a global gridded crop model intercomparison. *Proceedings of National Academy of Sciences, 111*, 3268–3273.

Rumanti, I. A., Nugraha, Y., Wening, R. H., Gonzaga, Z. J. C., Nasution, A., Kusdiaman, D., & Septiningsih, E. M., (2016). Development of high-yielding rice varieties suitable for swampy lands in Indonesia. *Plant Breeding and Biotechnology, 4*(4), 413–425.

Saeed, M., Guo, W., & Zhang, T., (2014). Association mapping for salinity tolerance in cotton (*Gossypium hirsutum* L) germplasm from US and diverse regions of China. *Australian Journal of Crop Science, 8*, 338–346.

Sangodele, E. A., Hanchinal, R. R., Hanamaratti, N. G., Shenoy, V., & Kumar, M. V., (2014). Analysis of drought tolerant QTL linked to physiological and productivity component traits under water-stress and non-stress in rice (*Oryza sativa* L.). *International Journal of Current Research and Academic Review, 2*(5), 108–113.

Septiningsih, E. M., Sanchez, D. L., & Singh, N., (2012). Identifying novel QTLs for submergence tolerance in rice cultivars IR72 and Madabaru. *Theoretical and Applied Genetics, 124*, 867.

Sharma, A. D., Sharma, H., & Lightfoot, D. A., (2011). The genetic control of tolerance to aluminum toxicity in the 'Essex' by 'Forrest' recombinant inbred line population. *Theoretical and Applied Genetics, 122*, 687–694.

Simova-Stoilova, L., Vassileva, V., & Feller, U., (2016). Selection and breeding of suitable crop genotypes for drought and heat periods in a changing climate: Which morphological and physiological properties should be considered? *Agriculture, 6*(2), 1–19.

Taiz, L., & Zeiger, E., (1991). *Plant Physiology* (pp. 100–119). The Benjamin Cummings Publishing Co., Inc.: Redwood City, CA, USA.

Tiwari, C., Wallwork, H., Kumar, U., Dhari, R., Arun, B., Mishra, V. K., Reynolds, M. P., & Joshi, A. K., (2013). Molecular mapping of high temperature tolerance in bread wheat adapted to the Eastern Gangetic plain region of India. *Field Crop Research, 154*, 201–210.

Turki, N., Shehzad, T., & Harrabi, M., (2015). Detection of QTLs associated with salinity tolerance in durum wheat based on association analysis. *Euphytica, 201*, 29.

Wani, S. H., Choudhary, M., Kumar, P., Akram, N. A., Surekha, C., Ahmad, P., & Gosal, S., (2018). Marker-assisted breeding for abiotic stress tolerance in crop plants. *Biotechnologies of Crop Improvement, 3*, 1–23.

Wani, S. H., Choudhary, M., Kumar, P., Akram, N. A., Surekha, C., Ahmad, P., & Gosal, S., (2018). Marker-assisted breeding for abiotic stress tolerance in crop plants. *Biotechnologies of Crop Improvement, 3*, 1–23.

Witcombe, J. R., & Virk, D. S., (2001). Number of crosses and population size for participatory and classical plant breeding. *Euphytica, 122*, 451–462.

Wu, L. B., Shhadi, M. Y., Gregorio, G., Matthus, E., Becker, M., & Frei, M., (2014). Genetic and physiological analysis of tolerance to acute iron toxicity in rice. *Rice, 7*, 8.

Xu, Y., & Crouch, J. H., (2008). Marker-assisted selection in plant breeding: From publications to practice. *Crop Science, 48*(2), 391–407.

Xu, Y., (2010). Marker assisted selection: Practice. In: *Molecular Plant Breeding.* CAB International, U.K.

Xu, Y., (2010). *Molecular Plant Breeding*. Cabi. Wallingford, U.K.

Xu, Y., Li, S., Li, L., Zhang, X., Xu, H., & An, D., (2013). Mapping QTLs for salt tolerance with additive, epistatic and QTL 3 treatment interaction effects at seedling stage in wheat. *Plant Breeding, 132*, 276–283.

Younis, A., Siddique, M. I., Kim, C. K., & Lim, K. B., (2014). RNA interference (RNAi) induced gene silencing: A promising approach of hi-tech plant breeding. *International Journal of Biological Sciences, 10*, 1150–1158.

Yu, M., & Chen, G., (2013). Conditional QTL mapping for waterlogging tolerance in two RILs populations of wheat. *Springer Plus, 2*, 245.

Zaidi, P. H., Rashid, Z., Vinayan, M. T., Almeida, G. D., Phagna, R. K., & Babu, R., (2015). QTL mapping of agronomic waterlogging tolerance using recombinant inbred lines derived from tropical maize (*Zea mays* L.) germplasm. *PLoS One, 10*(4), e0124350.

Zhang, H., Cui, F., Wang, L., Li, J., Ding, A., Zhao, C., Bao, Y., Yang, Q., & Wang, H., (2013a). Conditional and unconditional QTL mapping of drought-tolerance-related traits of wheat seedling using two related RIL populations. *Journal of Genetics, 92*, 213–231.

Zhang, X., Tang, B., Yu, F., Li, L., Wang, M., Xue, Y., & Qiu, F., (2013). Identification of major QTL for waterlogging tolerance using genome-wide association and linkage mapping of maize seedlings. *Plant Molecular Biology Report, 31*(3), 594–606.

Zheng, J., Fu, J., Gou, M., Huai, J., Liu, Y., Jian, M., Huang, Q., et al., (2010). Genome-wide transcriptome analysis of two maize inbred lines under drought stress. *Plant Molecular Biology, 72*, 407–423.

CHAPTER 12

Marker-Assisted Selection Approaches for Improving Quality Traits

SUNIL KUMAR NAIR[1] and MOHAMMAD TAQI RABBANI[2]

[1]*Department of Genetics and Plant Breeding, College of Agriculture, IGKV, Raipur, Chhattisgarh, India*

[2]*Department of Agronomy and Plant Breeding, Faculty of Agriculture, Balkh University, Mazar I Sharif, Balkh, Afghanistan*

ABSTRACT

Plant breeders are almost concerned with the quality improvement of agriculturally important crops to meet consumers' desires and demands. Quality traits, including flavor, color, shape, size, storability, and nutritional values, are of main challenges for most plant breeders to improve. Integration of marker-assisted selection (MAS) into plant genetic improvement could significantly enhance the reliability and efficiency of breeding programs of plants of interest by fast-tracking and precise monitoring of desirable gene(s) within breeding materials. Indeed, the application of reliable, easy-to-use, and cost-effective DNA markers derived from the fine mapping studies of desirable gene(s) for quality characteristics and MAS approaches will provide the breeders the chance of developing well-performing and high-quality varieties. The adoption of MAS in plant genetic improvement programs remains a challenge due to the uncertainty of combining MAS approaches with conventional breeding techniques, unreliability, and relatively high expenses of MAS application. However, novel achievements in omics studies, development of modern genomic tools and marker validation methods, and recent models for exploitation of quantitative trait loci (QTLs) in plant genetic improvement programs stand to triumph over these

Smart Breeding: Molecular Interventions and Advancements for Crop Improvement. Kailash Chandra, Subhash Chand, Ravi Prakash Saini, and Rohit Sharma (Eds.)

obstacles. In this chapter, the recent advances in the applications of MAS in quality improvement of crop species will be briefly reviewed.

12.1 INTRODUCTION

Nowadays, the majority of consumers prefer to purchase agricultural products with improved quality attributes, including flavor, color, shape, size, storage ability, nutritional values (protein content, amino acid constituents, minerals, and vitamins concentration) and industrial characteristics (oil content, carbohydrates content, sprouting, fiber quality, etc.) (Table 12.1). A successful plant-breeding program should be aimed to efficiently change the crop genetic makeup to meet minimal criteria for different morphological, organoleptic, and nutritional characteristics so that their products become much more desirable regarding the crop of interest (Debnath & Guha, 2015; Mishra & Singh, 2015). Crop product quality is determined by polygenes that usually interact with each other (epistasis) and their expression substantially influenced by prevailing environmental factors and could be harnessed by effective use of plant breeding methods (Figure 12.1). The simultaneous segregation of individual regions with small effects are not easy to characterize, and consequently, to get tangible genetic impact, numerous genomic regions should be synchronously manipulated. Furthermore, multiple field trials are required to precisely identify QTLs effects and assess their stability and interactions under various environments (Q×E) that confer a resolute superiority across a variety of environments. Determination of Q×E interactions and genotype-by-environment (G×E) may lead to the characterization of individuals with better adaptation potential for a particular environment (Fox et al., 1997; Francia et al., 2005; Mishra & Singh, 2015). However, the quality traits may be controlled by the involvement of independent complementary genes whose expressions are regulated by other genes (Costantini et al., 2008; Bergamini et al., 2013). Developing crop varieties with improved quality in field pea, strawberry, soybean, tomato, potato, and wheat have had a positive influence on consumers' preference and resulted into increased demand (Luby & Shaw, 2009). Marker-assisted selection (MAS) refers to the application of molecular (DNA or RNA variations), biochemical, or morphological (visual) traits in breeding programs where markers are tightly linked to the trait(s) of interest and consequently used for indirect selection of individual traits expressing better performance, productivity, and quality (Ashaf et al., 2012; Kumawat et al., 2020). This approach has been widely explored and suggested for crop breeding schemes.

TABLE 12.1 Distinct Categories of Quality Traits in Crop Plants

Type of Quality Trait	Description	Example	Crop Plants
Morphological appearance	Better morphological attributes of plant products	Fruit shape, color, and size	Strawberry, tomato, field pea
Organoleptic properties	Enhanced playability of plant products	Flavor, softness, aroma, and taste	Strawberry, tomato, soybean
Nutritional values	Improved value of the plant products for human consumption or animal feeding	Protein content, fatty acid composition, micronutrients content, vitamin content	Soybean, wheat, sorghum, rice, cotton
Industrial properties	Improved quality if plant products for industrial purposes	Fiber quality, physical attributes of the dough, carbohydrates content, sprouting, soluble solid content	Potato, tomato, cotton, wheat, sorghum
Consumer/market preference	Better marketability and/or storability of plant products	Shelf life, storability, grain properties	Tomato, rice

To accelerate the process of improved varietal development, conventional breeding techniques should be coupled with advanced biological tools to improve the reliability, selection, effectiveness, and simultaneously decreases the breeding cycle durations in breeding programs (Luby & Shaw, 2009). In fact, a molecular marker is served as a "genetic tag" to characterize a particular DNA segment (or gene/s) within a plant genome. Smith & Simpson, (1986); and Soller & Beckmann, (1983) were the first who proposed the concept of MAS in genetic improvement programs Prior to DNA markers, biochemical markers (isozymes) have been widely used to speed-up morphological selection in tomatoes (Tanksley et al., 1981). However, DNA markers have been earlier suggested to be exploited for breeding purposes (Soller, 1978). Integration of reliable, easy to use and low-cost molecular markers have been developed from the fine mapping studies of desirable gene(s) to improve quality characteristics and would help to the breeders to develop well-performing and high-quality varieties (Collard & Mackill, 2008; Jena & Mackill, 2008; Cobb et al., 2019) after their extensive field trials at various stages (Chand et al., 2020). The major purposes for which plant breeders employ MAS approaches are: (i) fast tracking and monitoring of desirable alleles to make a combination of desirable genes/QTLs; (ii) characterizing elite genotypes among segregating populations on the basis of target allelic constitution for the whole or a small part the genome; and (iii) breaking the genetic linkage between favorable and unfavorable genes (Francia et al., 2005).

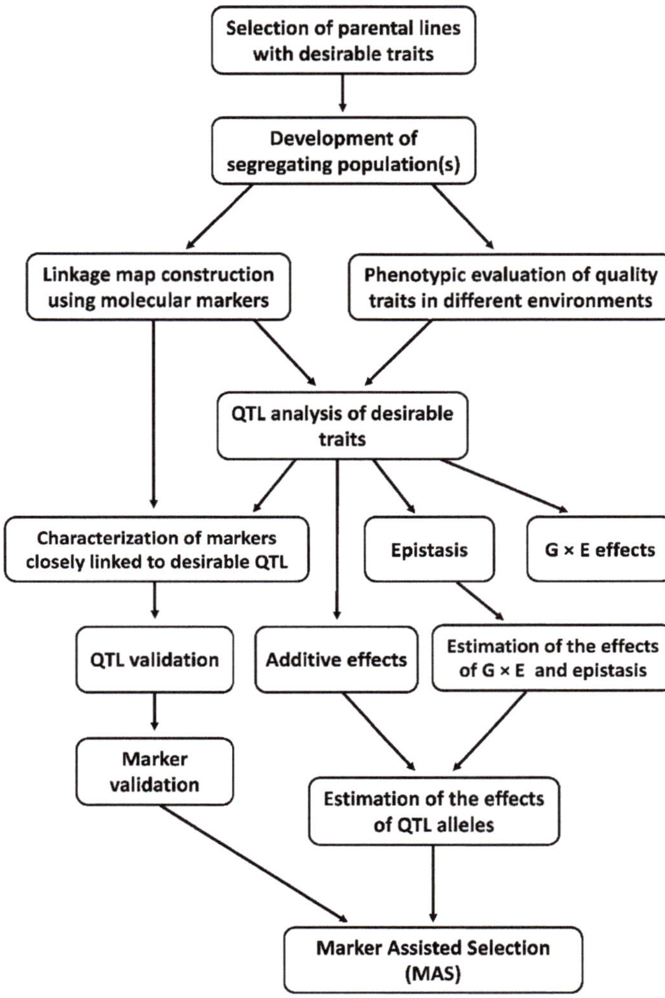

FIGURE 12.1 A simplified diagram of various steps involved in MAS for quality improvement.

Source: Modified from Francia et al. (2005); Collard & Mackill (2008).

Although MAS is highly beneficial to enhance the efficiency of breeding programs, it is essential to ensure the effectiveness of markers linked with genes/QTLs in various genetic backgrounds and environmental conditions. Therefore, the markers should be first mapped in early generations of breeding populations and then confirmed/validated in advanced generations (Holland, 2007; Wickneswari et al., 2012; Bergamini et al., 2013; Gelli et al., 2017).

However, traditional plant breeding methods alone are not adequate to satisfy a rise in demand for food around the world. Indeed, the application of MAS technology is mostly limited to developed countries where MAS is routinely combined with conventional selection methods. It is noteworthy that the MAS efficiency could be significantly influenced by factors such as population size, trait heritability, the density of genomic markers, the number of associated markers to the index, and the statistical procedure used to detect linkage or association between a DNA marker and target genes/QTLs (Gimelfarb & Lande, 1994). Compared to purely/sole phenotypic selection (PS), breeders could benefit the MAS efficiency while working on quite large populations and selecting for traits with relatively low heritability (Gimelfarb & Lande, 1994; Wittaker et al., 1995). Therefore, the efficiency of MAS appertains to the genetic nature of the characteristic. Interestingly, purely PS will be more efficient than MAS under several successive selection cycles that is mostly due to the higher proportion of fixation of undesirable alleles at small effect QTLs in advance generations that could equate the higher efficiency of MAS on large effect QTLs in early generations (Hospital et al., 1997). On the other hand, phenotypic evaluations coupled with molecular markers data could result in maximum selection efficiency than each technique alone (Lande & Thompson, 1990; Wittaker et al., 1995). The application of DNA markers could result into relatively fast and highly reliable characterization of well-performing individuals in early segregating generations by employing a small number of evaluating lines (Table 12.2) (Eathington et al., 1997).

TABLE 12.2 Comparison of Phenotypic Selection and MAS for Improvement of Quality Traits

Selection Criteria	Phenotypic Selection	MAS
Evaluation time	Latter stages of growth and development	Seeds or early stages of growth and development
Cost of procedure	Low	High
Equipment required	Simple	Laboratory
Scoring method	Fast morphological scoring	Time consuming molecular markers scoring
Number of required individuals	Large	Small
Linkage drags	Large	Small
Progeny test for recessive traits	Required	Not required
Accuracy of selection	Low	Relatively high

12.2 MAS FOR QUALITY IMPROVEMENT

Although most plant quality attributes exhibit quantitative variation and consequently, their expressions are highly influenced by environments, however, MAS approaches have been reliably exploited for selection of elite individuals in several agronomically and horticulturally important crops (Table 12.3).

12.2.1 FIELD PEA

Field pea (*Pisum sativum* L.) is one of the most important cool-season legumes whose seeds contain high protein (~25% w/w) content, compound sugars, essential vitamins, macro- and micro-elements, and phytochemicals (Burstin et al., 2011). Advanced molecular approaches have been successfully used to find genes/QTLs associated with seed quality attributes in field pea that are advantageous for rapid individuals' selection and development of varieties with improved seed quality. Mineral concentrations and contents are amongst the most important quality traits of pea to be improved. Ma et al. (2017) found single nucleotide polymorphisms (SNPs) by employing genotyping by sequencing (GBS) technique in a pea recombinant inbred line (RIL) population to identify QTLs responsible for seed mineral concentration. They ultimately characterized a total of 46 QTLs for seed mineral concentration and 37 QTLs for seed mineral contents such as iron (Fe), calcium (Ca), magnesium (Mg), molybdenum (Mo), and manganese (Mn). Highly positive correlations between different element content, i.e., Ca and Mn, Mg and Mn, and Fe, and sulfur (S) indicates the possibility of simultaneous breeding for two different elements meaning enhancing one of these mineral elements could concurrently increase the concentration of the other mineral element.

Another major breeding objective of pea is to develop varieties with improved visual seed quality traits. The genetic improvement of seed quality attributes involves working with quantitative genetic variability and environment sensitive gene expressions that makes it highly challenging. Therefore, well understanding of genetic pattern, characterization of QTLs associated with morphological/visual quality characteristics and identifying/developing tightly linked molecular markers could be extremely useful for accelerating pea-breeding programs. Ubayasena et al. (2011) characterized several QTLs associated with seed perceptible traits including seed coat color, seed shape and seed dimpling with green and yellow seeds. They simultaneously

TABLE 12.3 Selected Examples of Improved Quality Traits in Crop Plants

Crop Species	Trait Improved	Genes/QTL	Molecular Marker	References
Field pea	Mineral content Boron content	*[B]-Ps1.1, [B]-Ps5.1, [B]-Ps6.1, [B]-Ps7.1, [B]-Ps7.2*	SNPs	Ma et al. (2017)
	Calcium content	*[Ca]-Ps4.1, [Ca]-Ps5.1, [Ca]-Ps7.1, [Ca]-Ps7.2, [Ca]-Ps7.3*		
	Iron content	*[Fe]-Ps2.1, [Fe]-Ps2.2, [Fe]-Ps5.1, [Fe]-Ps5.2*		
	Magnesium content	*[Mg]-Ps3.1, [Mg]-Ps4.1, [Mg]-Ps5.1, [Mg]-Ps5.2*		
	Seed visual quality traits Seed lightness	*CCTC200, C19_308, A102.214, AGTG202, CAAT170, CAAC286*	–	Ubayasena et al. (2011)
	Seed greenness	*CCTT186, CCTG187, AGAC198*		
	Seed dimpling	*AGTG138, AGTC139*		
	Seed shape	*CCTG187, AGTG138, CGAG219*		
Potato	Dry matter content (DMC)	*StCDF1* gene (*Solanum tuberosum* CDF gene 1	SNPs	Massa et al. (2015)
	Cooking quality	*SGT02-2, SGT03-2, STG02-5*, and *SGT03-3* Plastidic phosphorylase A (*StPho1a*) and cytosolic phosphorylase (*StPho2*)	SSR, AFLP	Werij et al. (2012)
		Pain-1 *AuxRP* and *Hsp90*	RT-PCR technique	Sowokinos (2001); Sołtys-Kalina et al. (2015)
	Gelatinization properties	*SGS02-8, SGS03-8, SGS02-3*, and *SGS03-9*	DArT	Śliwka et al. (2016)
	Sugar content	*Sut1, Sut2, Sug6a, Sug6b*, and *Sug6c*	RFLP, AFLP, and candidate gene markers	Menéndez et al. (2002)
		AGPaseS, AGPaseB, SbeI, GapC, Invap, Ppa1	SNP	Li et al. (2013)

TABLE 12.3 *(Continued)*

Crop Species	Trait Improved	Genes/QTL	Molecular Marker	References
		InvCD141 (invertases), *SSIV* (starch synthase), *StCDF1* (transcription factor), *LapN* (leucine aminopeptidase)	SNPs	Schönhals et al. (2016)
	Enzymatic discoloration (ED)	Chlorogenic acid and tyrosine levels PPO gene	SSR, AFLP	Werij et al. (2007)
Tomato	Fruit weight	*fw1.1*, *fw2.2*, *fw3.1*, *fw4.1*, fascinated and locule-number	Simple sequence repeats (SSR)	Grandillo et al. (1999)
	Fruit shape	*ovate*, *sun*, and *fs8.1*	RFLP, SCAR, CAPS	Liu et al. (2003)
	Total soluble solids (sugar and acids)	*Brix9-2-5*	RFLP, SCAR, CAPS	Fridman et al. (2000)
		Brix TA1150	RFLP and SCAR	Frary et al. (2003)
Strawberry	Total anthocyanins	*Ant*	SNP, SSR, SSCP, and AFLP markers	Labadie et al. (2020)
	Anthocyanins (colorimetry)	*ANTHc*		
	fruit color	*COLOR*		
	Pelargonidin-3-glucoside-malonate	*PgGsM*		
	Soluble solid content (SSC; sugar biosynthesis)	*UDP*-glucose 4-epimerase *GEPI48*-like gene (*FvH4_5g04740.1*), UDP-glucose 6-dehydrogenase family protein (*FvH4_5g14230.1*), Glucose-6-phosphate/phosphate translocator-related protein (*FvH4_5g05430.1*)	ddRAD-seq technique	Natarajan et al. (2020)

TABLE 12.3 *(Continued)*

Crop Species	Trait Improved	Genes/QTL	Molecular Marker	References
	Sugar related traits (SSC, fructose, glucose, and Sucrose and raffinose content)	*SSC-VIam2*, *GLU-VIa-f1*, *SUCR-VIa-m1* *qSuc-V-4*, *qRaf-V-4*	SSR	Lerceteau-Kőhler et al. (2012); Vallarino et al. (2019)
	Acid-related traits (pH, titratable acidity, citrate content, malate acid)	*pH-f1*, *pH-f_3. TA-f_3. CIT-f_2*, *MAL-f_2*	SSR	Lerceteau-Kőhler et al. (2012)
Wheat	Grain protein content (GPC)	*QGpc.ccsu-2D.1* *QGpc.ccsu-5A.1* *Gpc-B1* QGpc-3B-3.1 and QGpc-1B-2	STMS STMS	Prasad et al. (1999); Harjit-Singh et al. (2001)
		QGpc-6A-2, *QGpc-4B-1.4*, and *QGpc-1B-2*	CAPS, DArTs, SNPs, SSR, and EST-SSRs	Guo et al. (2020); Goel et al. (2019)
		QGpc.2B-yume	SSR	Terasawa et al. (2016)
	Dough strength (HMW-GC)	*ω-secalin*, *Glu-B1-2a* (*By8*), *Glu-D1-1d* (*Dx5*), *Glu-A3d*, *Glu-B3*, *Glu-1*	Multiplex PCR SCAR	Mohammadi et al. (2013); Radovanovic & Cloutier, (2003); Ma et al. (2003)

TABLE 12.3 *(Continued)*

Crop Species	Trait Improved	Genes/QTL	Molecular Marker	References
Soybean	Vitamin content (VE)	*QTVEC2_1*, *TVEC2_2*, *QTVE D1b_1*, *QTVEN _1*, and *QTVEO _1*	SSR	Li et al. (2010)
	Soluble protein content (SPC)	*qsp8-4, and qsp8-5*	SSR	Lu et al. (2013)
	Total isoflavonoids content (genistein, daidzein, and glycitein)	*TGly*, *Daid*, *Gen*	SSR	Kassem et al. (2004); Primomo et al. (2005)
	Linolenic acid content	*LN* *QLNB2_1*, *QLNB2_2*, *QLNB2_3*, *QLNB2_4*, *QLND1b_1*	SSR, STS SSR, SNPs	Reiprecht et al. (2006); Xie et al. (2012)
Sorghum	Endosperm texture	ET	SSR, SNPs	Guindo et al. (2019)
	Protein content	PR		
	Amylose content	AMY		
	Lipid content	LIP		
	Color of the testa layer	TESTA		
	Presence of anthocyanins	ANTHO		
	Fiber-related traits for biofuel production Acidic detergent fiber	ADF	SSR, EST-SSR, AFLP	Shiringani et al. (2010)
	Neutral detergent fiber	NDF		
	Acidic detergent lignin	ADL		
	Cellulose content	CC		
	Hemicellulose content	HC		
	Protein digestibility	HD, LD	SSR	Winn et al. (2009)

TABLE 12.3 *(Continued)*

Crop Species	Trait Improved	Genes/QTL	Molecular Marker	References
Cotton	Fiber quality	*qFL-D02-1, qFL-D02-4, qFS-D11-1*	SNP	Gu et al. (2020)
	FL	*qFL01.1, qFL14.1, qFL17.1, qFL26.1*		
	FM	*qFM05.1, qFM07.1, qFM20.2, qFM23.1, qFM25.1*	SSR	Liu et al. (2017)
	FS	*qFS07*.1, *qFS14.2*	SNPs	Fang et al. (2107)
	FE	*qFE-chr15-2, qFE-chr20-2, qFE-chr24-1, qFE-chr26-1*, and *qFE-chr26-2*	SNPs	Zhu et al. (2021)
	Protein content	*qPP-D9-1, qKPCA01.1, qKPCA01.2, qKPCA01.3, qKPCA08.1, qKPCA08.2, qKPCA09.1, qKPCA11.1, qKPCA13.1, qKPCD02.1, qKPCD02.2, qKPCD03.1*, and *qKPCD09.1*	SSR	Song et al. (2007)
	Oil content	*qOP-D8-1* *qKOC-A01.2, qKOC-D05.1, qKOC-9.1*	SSR SNPs	Song et al. (2007); Wang et al. (2019)
	Amino acid content Aspartic acid	*D03.1qAsp-A11-1*	SSR	Song et al. (2007)
	Serine	*qSer-A8-1*		
	Glycine	*qGly-A11-1, qGly-A8-1*		
	Isoleucine	*qIle-D3-1*		
	Leucine	*qLeu-D2-1*		
	Phenylalanine	*qPhe-A8-1*		
	Argentine	*qArg-A5-1*		

investigated two RIL populations and found that all visual seed quality traits inherited quantitatively and segregated transgressively. Finally, a total of 36 QTLs for yellow seed lightness, yellow seed greenness, seed feature, and seed dimpling have been successfully characterized. However, only five QTLs showed consistency over at least two different environments.

12.2.2 POTATO

Potato (*Solanum tuberosum* L.) is an extremely important vegetable plant to human being and widely cultivated globally. The crop produces underground tubers that are rich in starch and other valuable vitamins and minerals. Various quality traits should be improved to meet both individual and industrial purposes/demands. Dry matter content (DMC), an important quality trait, plays a crucial role in the processing industries for various purposes. The potato production efficiency depends, in part, upon the improvement of DMC in potato tubers (Ahmed et al., 2018). DMC is mostly composed of tuber starch content (TSC) and determined by the specific gravity that is directly influenced by the length of the life cycle. This trait is controlled by polygenes and several QTLs were found to be associated with it in both diploid and tetraploid potato species (Freyre & Douches, 1994; Werij et al., 2012; McCord et al., 2011). Different biochemical (isozymes) and molecular markers (RAPD, RFLP, AFLP, DArT, SNPs, etc.,) have been exploited to identify QTLs responsible for improved TSC. Freyre & Douches, (1994) characterized 11 QTLs by employing biochemical markers (isozymes), RFLP, and RAPD markers as well as 10 putative QTLs on various diploid potato chromosomes. Furthermore, another 18 putative QTLs associated with TSC on all chromosomes of diploid potato have been characterized (Schäfer-Pregl et al., 1998). Although DMC and specific gravity are important processing attributes and show strong correlation, However, McCord & co-workers (2011) found independent QTLs for these characteristics that were not co-located with each other. Genetic map constructed based on SNP polymorphism had been employed to identify a major QTL on chromosome 5 for associated with plant maturity that was located ~0.25 Mb from the *StCDF1* gene (*Solanum tuberosum* CDF gene 1 (Massa et al., 2015). A total of 25 major QTLs associated with carbohydrate storage in potato tubers have been identified on all 12 potato chromosomes where most QTLs have been found on the same loci for both glucose and fructose content (Werij et al., 2012).

Cooking quality of potato tubers, a key quality trait, depends upon gelatinization temperature (GT). Various small and large effect QTLs associated with GT were detected on chromosomes 2, 3, and 5 of diploid potato species. The two major QTLs, *SGT02*-2 and *SGT03*-2, are mapped on chromosome 2 showed significant consistency over two years than only one-year significant effect of the two small effects QTLs, *SGT02*-5 and *SGT03*-3 on chromosomes 2 and 3, respectively. The coincidence of QTLs for starch GT on chromosome 2 with two genes playing important role in starch biosynthesis, *plastidic phosphorylase* A (*StPho1a*) and cytosolic phosphorylase (*StPho2*) have also been reported (Werij et al., 2012). Amylose content (AC) is another important quality attribute in potato breeding. Amylose biosynthesis is modulated by an enzyme, i.e., granule-bound starch synthase (SS) or GBSS, encoded by the *Waxy* gene on chromosome 8 (Van de Wal et al., 2001). Five QTLs linked with starch concentration in diploid potato species on chromosomes 2, 3, 5, 7, and 10 have also been characterized (Werij et al., 2012). They proposed that these QTLs have likely regulatory role in amylose biosynthesis.

Enzymatic discoloration (ED) is another important quality trait for consumers to keep product longtime as such without any color change. It is an oxidation reaction catalyzed by the enzyme polyphenol oxidase (PPO) that transforms phenolic compounds to quinones. Later on, quinones are converted to dark color pigments (melanin). The ED severely affects tuber nutritional quality, flavor, and color that results into poor processing quality and marketing industry. Werij & co-workers (2007) identified four QTLs associated with ED were detected on chromosomes C3, C8, E1, and E8. The QTLs for chlorogenic acid and tyrosine content were identified on chromosomes C2 and C8, respectively. They also illustrated that chromosome 8 contains a candidate gene *PPO* for ED. However, the QTLs had no overlap that indicates the absence of genetic correlations between these QTLs.

12.2.3 TOMATO

Tomato (*Solanum lycopersicum*) is an extremely important vegetable crop cultivated worldwide, and its quality improvement is directly related to consumer preference and health benefits. Tomato quality improvement is challenging and involves both conventional and biotechnological approaches. Soluble-solids content, a key parameter for tomato processing industry, is directly correlated with production duration. The concentration process requires much less energy requirement when the tomato variety has abundant soluble solids as opposed to when it has. Introgression of desired QTLs derived from tomato gene pool is an applied strategy to enhance tomato quality traits. Frary et al. (2003) introgressed a QTL (*BrixTA1150*) for soluble-solids content from *L. chmielewskii* and developed a near-isogenic line (NIL) carrying a 56 cM introgressed DNA segments from chromosome 1 of wild tomato species, *L. chmielewskii*, namely NILTA1150. A F_2 segregating population was subsequently developed by crossing between introgressed NIL and a cultivated tomato genotype to generate sub-NILs with a small size introgressed DNA segment of the wild species, *L. chmielewskii*. To identify recombinant individuals, RFLP analysis was employed to characterize markers flanking to the introgressed QTL. Transferring a QTL associated with total soluble solids (carbohydrates and acids), namely *Brix 9-2-5*, derived from a wild relative of cultivated tomato, *L. pennellii*, has been resulted into improved glucose (>28%), fructose (>18%), and soluble solids content in different genetic background across various environmental conditions (Fridman et al., 2000). This QTL contains the *Lin5* gene, encoding a fruit-specific apoplastic invertase, responsible for regulation of fruit sink strength/capacity. Genetic studies had illustrated that QTL had incomplete dominance pattern and not associated with fresh fruit weight.

Fruit size is also an important trait for tomato possessing industry and shows quantitative inheritance pattern. The QTLs *fw1.1*, *fw2.2*, *fw3.1*, and *fw4.1* were reported to be tightly associated with only increased fruit weight, except the QTL *fw3*.1 (Grandillo et al., 1999; van der Knaap et al., 2002; van der Knaap & Tanksley, 2003). Among these QTLs, *fw2*.2 is the best-known QTL for fruit size at the molecular level. The natural genetic variation ensues at the locus *fw2.2* and could alone enhance the fruit weight by maximum 30% (Frary et al., 2000). This QTL could be found in two different forms, small-fruit alleles and large-fruit alleles in wild and cultivated tomato species, respectively. Investigations confirmed that mutations in the promoter of *fw2.2* are not only responsible for knock-down the gene transcription during the mitotic cell divisions at fruit growth and developmental stages but also alter gene expression timing (Cong et al., 2002). Furthermore, investigations employing isogenic lines had indicated that the changes in *fw2.2* gene expression were mostly happened during the late phase of fruit enlargement. However, histological findings in the same isogenic lines illustrated that earlier occurrence of fruit size changed within tomato fruit development (Lippman & Tanksley, 2001; van der Knaap & Tanksley, 2003). The genetic variations in this QTL is mostly confined with fruit size whereas, other plant structural organizations, fruit appearance or seed size/seed number or plant productivity are not associated with genetic variation at this QTL (Nesbitt & Tanksley, 2001; Liu et al., 2003). It has been observed that if plants have more number of locules, then it would produce larger fruit size and more fruit number (Lippman & Tanksley, 2001; van der Knaap & Tanksley, 2003). Two other genes, *fasciated* on chromosome 11 and *locule-number* on chromosome 2 could influentially determine the number of pistils in a flower and subsequently define fruit size. The occurrence of mutations at one or both of these loci is resulted into multilocular tomatoes with very large fruit size. However, natural mutations occurred at *fasciated* locus play a more important role than *locule-number* locus. Varieties carrying mutations at this locus can develop more than 15 locules. The presence of simultaneous mutations at both genes' loci, *fasciated* and *locule*-number, could extremely increase fruit size (>500 g). The epistatic interaction between these two loci could produce fruits with an unusually high number of locules and subsequently large fruit size (Lippman & Tanksley, 2001).

Tomato fruit shape is another important quality trait to be improved for industrial use and consumer preference. Tomato fruits could be found into different shapes such as wild and semi-wild tomato genotypes possess round shape fruits, whereas domesticated tomatoes have wide range of fruit shapes *viz.,* orbicular, pear-shaped, torpedo-shaped, and bell pepper shaped (Tanksley, 2004). *Ovate*, *sun*, and *fs8.1* are three major QTLs associated with this trait and have no significant effect on fruit size. The *ovate* gene carries a mutation that produces elongated or pear-shaped fruits rather than round shaped. It is due to induction of an early stop codon in the second exon formed a nonfunctional allele with recessive inheritance pattern (Liu et al., 2002). Although the allele is recessive, its phenotypic expression mostly depends upon the genetic background. On the other hand, the tomato fruit may be elongated shape with highly constricted neck growth to little or no neck constriction. The second QTL, *sun* that is mapped on chromosome 7, could also produce tomatoes with elongated fruit shape. Although both *sun* and *ovate* alleles affect tomato fruit shape, they illustrate different important genetic, morphological, and developmental attributes (Tanksley, 2004). The third QTL, *fs8.1*, is also associated with blocky and slightly elongated fruit shape in processing tomatoes (Grandillo et al., 1996). The QTL is mostly functional in early floral/carpel development, a common feature this QTL shares with QTLs, ovate, *fasciated*, and *locule number* (Ku et al., 2000).

Tomato fruits are rich in lycopene that could be served as an important source of lipid-soluble antioxidants in our daily diet. This phytochemical could protect human cells against the induction or multiplication of oxidative stresses (Wu & Kubota, 2008; Ilahy et al., 2011; Sun et al., 2012). Finding QTLs associated with lycopene content (LYC) is very beneficial in developing tomato varieties with improved quality. Sun & co-workers (2012) identified QTLs controlling LYC and soluble solid content (SSC) in a segregating population derived from an interspecific cross of *S. lycopersicum* and *S. pimpinellifolium*. Total, nine QTLs associated with LYC and SSC at three different fruit ripening steps were identified. Two, one, and three QTLs for LYC have been detected in the color changing period (CP), half-ripe period (HP), and full-ripe period (FP), respectively. It was also observed that the QTLs for SSC were co-located with those for LYC on chromosome 4. It could be mostly attributed due to pleiotropic effects of a single gene alone or a group of genes through shared physiological pathways between SSC and LYC contents (Sun et al., 2012). Notable that among numerous QTLs identified for LYC and SSC, a few QTLs for LYC have been practically employed for MAS in tomato breeding programs (Thorup et al., 2000; Saliba-Colombani et al., 2001; Heather et al., 2004; Foolad, 2007; Causse et al., 2007).

12.2.4 STRAWBERRY

Strawberry (*Fragaria* × *ananassa*) is one of the important fruit crops with high esthetic property, flavor, and enriched nutritional values (Alvarez-Suarez et al., 2014; Henning et al., 2010). The popularity of strawberry is being significantly increased since long back, and it would be better to improve fruit quality with increased production to meet the consumer demand and preferences (Natarajan et al., 2020). Various fruit sensorial quality traits including fruit color, flavor, aroma, and taste are the major breeding targets of cultivated strawberry (Mezzetti et al., 2018). The fruit redness color is the consequence of anthocyanin accumulation in the strawberry fruits. Investigations using pseudo F_1 progeny have paved a new paradigm to find QTLs corresponding for anthocyanin, flavonols, and flavan-3-ols content and fruit color. Recently, it has been observed that metabolite QTLs (mQTLs) associated with anthocyanin and other fruit colors are mostly specific to sub-genome of *F. vesca* but the contribution of other strawberry sub-genomes to color variation was also tangible. Variation in fruit color could be attributed due to occurrence of several insertion/deletions (INDELS) in the cis-regulatory element of a *MYB102*-like *ODORANT* gene and deletion of a putative *MADS* box binding motif in the 5′ untranslated region (UTR) upstream area of an *anthocyanidin reductase* (*ANR*) gene in strawberry (Labadie et al., 2020).

SSC is another important quality trait in strawberry. Glucose and fructose are the two major sugars representing over 80% of SSC and approximately 40% of total fruit dry weight (Wrolstad & Shallen-Berger, 1981). Several QTLs associated with sugar-related traits including SSC, glucose, fructose, and sucrose content have yet been identified (Lerceteau-Kőhler et al., 2012; Verma et al., 2017; Natarajan et al., 2020). Lerceteau-Kőhler & co-workers (2012) employed a pseudo full-sib F_1 population obtained from a cross between the varieties 'Capitola × CF1116' and 10 QTLs associated with sugar-related traits were detected. Among them, seven QTLs were reported for fruit firmness, and 19 QTLs for fruit color on different linkage groups (LGs). Their findings showed that QTL associated with fruit quality were mostly located on two LGs (LGIII-a and LGVI-a). Furthermore, by employing double digest restriction-associated DNA sequencing (ddRAD-Seq technique), Natarajan & associates (2020) characterized four QTLs for SSC that described 23–50% of all phenotypic variations. In addition, 14 sugar biosynthesis associated genes including Glucose-6-phosphate/phosphate translocator-related protein (*FvH4_5g05430.1*), UDP-glucose 6-dehydrogenase family protein (*FvH4_5g14230.1*), UDP-glucose 4-epimerase GEPI48-like gene (*FvH4_5g04740.1*), and *FAN_iscf00049825.1* were also identified. The latter gene contains the flanking DNA marker *MF472* of QTL *qSSC-2B* encoding a *UDP-glucosyltransferase* gene probably playing potential roles in sugar biosynthesis (Natarajan et al., 2020). Totally, four SNPs-derived ddRAD-seq markers (*MF3293-MF3696*, *MF994-MF1321-MF472*, *MF1180-MF154-MF518*, and *MF811-MF1169*) tightly linked to QTLs for SSC have been identified (Natarajan et al., 2020).

Acid-related traits including pH, titratable acidity (TA), citrate content (CIT), and malate acid (MAL) are also important quality traits in strawberry as keeping quality and consumer's preferences are concerned. Citric acid is the major constituent of strawberry fruit and accounts for about 88% of total acid content. Lerceteau-Kőhler & associates (2012) characterized 21 QTLs for acid-related traits-6, 5, 6, 4 QTLs for pH, TA, CIT, and MAL, respectively. In addition, strawberry fruit is a rich source of antioxidants essential for reducing the threat of chronic diseases (Demmig-Adams & Adams, 2002; Battino et al., 2009). The anthocyanin pelargonidin-3-glucoside is amongst the major strawberry antioxidants (Wrolstad et al., 1970; Tulipani et al., 2009; Lerceteau-Kőhler et al., 2012). Several QTLs controlling the traits have been identified and used for enhancing strawberry fruit quality. In strawberry, QTLs associated with both agronomical traits and fruit quality traits are found as a cluster of QTLs that have been non-randomly distributed across the chromosomes. Clustering of strawberry for agronomical and quality traits revealed the existence of either pleiotropic effects or closely coordinated gene expression (Lerceteau-Kőhler et al., 2012; Natarajan et al., 2020). Noteworthy, various markers have recently been characterized that are tightly linked to quality traits in strawberry. SSR *EMF006* (allele *v006205c*) marker, closely flanked AFLP markers *tgta383* and *gata170*, and the SSR *Fvi020* (allele *g020175*) marker, tightly linked to the AFLP *ccta303* marker, could be effectively used for screening and selecting individuals for improved sweetness, color, and anthocyanin content (Labadie et al., 2020). On the other hand, integration of flavonoids mQTL with quality traits QTLs would be highly beneficial to identify genomic regions linked to both sensorial and nutritional quality traits in strawberry.

12.2.5 WHEAT

Grain protein and starch are the functional chemicals determining the end-use quality of wheat. Protein and starch-related quality characteristics consist of various traits including grain protein content (GPC), physical (rheological) attributes of the dough during bread making, sedimentation index, falling number (FN), and starch pasting characteristics that have significant influence on both nutritional and marketing values (Guo et al., 2020; Kumar et al., 2021). Exploiting a set of RILs derived from a cross between two bread wheat varieties with different GPC, Prasad & co-workers (1999) characterized a QTL, *QGpc.ccsu-2D.1*, for high GPC on chromosome 2D tightly flanked by a sequence-tagged microsatellite site (STMS) marker, *Xwmc41*. Another QTL for high GPC, *QGpc.ccsu-5A.1*, with a tightly flanked markers, *Xwmc415*, was identified on chromosome 5A (Harjit-Singh et al., 2001). The allelic relationship within other QTLs for GPC from different wheat species had also been reported and validated earlier (Snape et al., 1995; Perretant et al., 2000; Turner et al., 2004; Lafiandra et al., 2007). Although several major QTLs for GPC with significant phenotypic variation (PV) on chromosome 2AS, 2BS, 3AS, 4AL, 6BS, 7AS, and 7BL have been identified, however the inconsistency of the aforementioned QTLs across the environments is a serious issue and hampers applicability of MAS in GPC improvement (Lafiandra et al., 2007). *GPC-B1* is the most important GPC QTL that was first identified in tetraploid wheat, *Triticum turgidum* var. *dicoccoides*

(Avivi, 1978). Notably, the gene *Yr36* conferring resistance to stripe rust is tightly linked to this locus and the same marker used for tagging *GPC-B1* can be used for *Yr36*, and *vice-versa*. Although the QTL have been extensively introgressed into both tetra- and hexaploid breeding lines and is highly consistent across environments, however its effect is mostly influenced by genetic backgrounds. Hexaploid wheat carries two *GPC-B1* homoeologous genes, on wheat chromosome 6A and 6D, *GPC-A1* and *GPC*-D1, respectively that could significantly improve both GPC, zinc and iron content (Uauy et al., 2006). GPC and grain yield in most cereals including wheat have been reported to be negatively correlated with each other (Blanco et al., 2012; Bogard et al., 2013). Employing a set of RILs derived from a cross between 'Tainong 18 × Linmai 6' (TL-RILs), Guo & co-workers (2020) characterized 10 QTLs for GPC on chromosomes 1B, 3B, 4A, 4B, 4D, and 6A. However, three GPC QTLs, *QGpc-6A-2*, *QGpc-4B-1.4*, and *QGpc-1B-2* were among the most frequent QTLs (RHF-QTLs) characterized across different environments. Finding genes/QTLs potentially associated with GPC without compromising the overall yield would be promising. *QGpc.2B-yume* is a novel major GPC QTL that has been recently identified on chromosome 2BS. It is tightly flanked by the markers *Xgwm37* and *Xwmc179* making it possible to incorporate the QTL into different genetic background to improve GPC without possible grain yield loss (Terasawa et al., 2016; Kumar et al., 2018).

Grain starch accounts for over 70% of dry grain weight (GW) and predominantly consist of two storage carbohydrates-amylose and amylopectin. The relative content and structure of starch is a critical determinant of functional attributes of flour or semolina parameters used for food and industrial purposes (Zeng et al., 1997; Yoo & Jane, 2002). Several enzymes, namely SS, are involved in starch biosynthesis with minimum five well-known isoforms. The granule bound starch synthases (GBSSI or waxy proteins) are the SS uniquely catalyzing amylose biosynthesis in storage tissues (Nakamura et al., 1993). Wx-A1, Wx-D1, and Wx-B1 are the three waxy proteins found in bread wheat corresponding to three genes *Wx-A1*, *Wx-D1* and *Wx-B1* located on chromosomes 7A, 7D, and 4A, respectively (Miura & Tanii, 1994; Yamamori et al., 2000). The endosperm gluten proteins comprising mostly two fractions, gliadins, and glutenins, play an important role in defining the quality parameters in wheat. Dough rheological properties during mechanical treatment mostly depend upon high molecular weight glutenin subunits (HMW-GS), especially the alleles *Glu-A1b* and *Glu-D1d*, then gliadins and low molecular weight glutenin subunits (LMW-GS)

(Ma et al., 2003; Francia et al., 2005). The genes encoding HMW-GS are located at the *Glu-1* loci on the long arms of the homoeologous group 1 chromosomes. The genes that encode gliadin components, the loci *Gli-1* and *Gli-2*, are located on the short arm of the homoeologous groups 1 and 6 chromosomes of wheat genomes. Electrophoretic studies of grain proteins, which permit the investigation of glutenin polypeptide profiles of wheat genotypes, are capable of discriminating allelic variations at multiple loci (Yamamori et al., 1994; Monari et al., 2005). In addition, multiplex PCR assay has been developed to study wheat lines in order to investigate waxy allelic variations and trace their origin (Nakamura et al., 2002). A multiplex PCR assay established by Zhang & co-workers (2008) to study allelic variations of genes/loci *ω-secalin*, *Glu-B1-2a* (*By8*), *Glu-D1-1d* (*Dx5*), *Glu-A3d*, *Glu-B3*. For comprehensive use in different genetic background, the assay has been validated using 70 wheat varieties and advanced breeding lines. The genes could be targeted to enhance gluten parameters. It was concluded that the developed multiplex PCR assays are capable of exploiting QTLS in MAS targeting enhanced pan bread-making. Similarly, three molecular markers, *Glu-A1b* (2*), *Glu-B1i* (17 + 18), and *Glu-D1d* (5 + 10), have been utilized for allelic selection of loci *Glu-A1*, *Glu-B1*, and *Glu-D1* attributed to improved gluten content (Mohammadi et al., 2013).

12.2.6 SOYBEAN

Soybean (*Glycine max* [L.] Merr) is one of the most popular legume crops grown around the globe that provides high-quality edible oil as well as the cheapest source of protein meal. The crop has several quality traits targeted in soybean breeding programs. Vitamin E (VE) in soybean seed is a highly valuable phytochemical used for both human purposes (foods, medicines, cosmetics) and animal feeding. VE not only plays a key role in protecting plant organs, i.e., chloroplasts from photo-oxidative burst (Munne-Bosch & Alegre, 2002), but also improves the functionality of human immune system to prevent a range of serious diseases (Adachi et al., 1997; Prasad et al., 1999; Pryor, 2000; Li et al., 2010). To detect QTLs for high VE content, Li, & co-workers (2010) developed a mapping RIL population derived from a cross between two parental lines 'Hefeng 25' and 'OAC Bayfield. Totally, five QTLs associated with seed VE concentration, *QTVEC2_1* (Satt376), *TVEC2_2* (Satt286), *QTVE D1b_1* (Satt172), *QTVEN_1* (Sat_125), and *QTVEO _1* (Satt592), were identified. The phenotypic variation explained with QTLs was between 2.9% and 10.9%. Furthermore, three simple sequence repeat markers (SSR), *Satt376*, *Satt286*, and *Satt266*, were found to be closely linked to total seed VE content across different environments. The markers were potentially capable of using MAS for the development of varieties with improved VE content.

Proteins derived from soybean are nutritionally functional ingredients widely used for human consumption (Rhee, 1994; Qi et al., 1997). It is a major quality trait targeted in soybean breeding programs. The percentage of soluble protein content (SPC) from soybean seed is the conventional index illustrating soybean product quality (Walsh et al., 2003; Malhotra & Coupland, 2004). From both scientific and industrial prospective, solubility characteristic is also an important quality trait, since the trait directly affects the efficiency of industrial protein isolation methods. Identification of QTLs governing higher SPC could pave the way for MAS to improve it. Lu & co-workers (2013) identified two QTLs for high SPC, *qsp8-4*, and *qsp8-5*, with high consistency across environments that were flanked by makers, *Sat_115-GMENOD2B*, *GMENOD2B*, and *Sat_215*, respectively. They concluded that SPC QTLs and linked markers could be promisingly useful in soybean breeding programs.

Soybean seed is a rich source of various bioactive phytochemicals including isoflavones, saponins, phytic acids, phytosterols, trypsin inhibitors, phenolic acids, etc. The isoflavonoids and other phytoestrogenic compounds are potentially capable of preventing and treating numerous chronic diseases (Munro et al., 2003). Therefore, the phytoestrogens such as genistein, daidzein, and glycitein have attracted considerable attention and are the focus of various medical research. However, the biological effects of isoflavones may seem to be depended upon various genetic and physiological factors including organism, genetic makeup, age, purity, concentration, and cofactors (Knight & Eden, 1996). Therefore, soybean breeding programs would be targeted for both increased or decreased isoflavone and other phytoestrogen compounds in soybean seed depending upon consumers need and preference. Various QTLs associated with genistein (five regions), daidzein (four regions) and glycitein (two regions) have been characterized (Kassem et al., 2004). Interestingly, two significantly consistent loci controlling seed genistein, daidzein, and glycitein content were identified in mapping populations. The genomic region closely associated with the marker *BARC_Satt063*, mapped on linkage group 2, controlled the two traits of seed, i.e., genistein and seed daidzein. Additionally, the QTL mapped on chromosome is potentially capable of modulating both seed glycitein and total glycitein contents; and tightly flanked by SSR markers, *BARC Satt197*, *Satt251* and *Satt415*

(Kassem et al., 2004). Similarly, nine genomic regions in different LGs A1, C2, D1a, F, G, H, J, K, and M associated with isoflavone contents have also been identified (Primomo et al., 2005). It was concluded that close association between QTLs for isoflavone content and molecular markers could be potentially used to increase isoflavonoid content using MAS in soybean breeding programs.

Oil quality is significantly reduced by oxidation of polyunsaturated fatty acid (PUFA) including linolenic acid (18:3, LN) containing a *cis*, *cis*-1, 4-pentadiene structure catalyzed by lipoxygenase. The best strategies to avoid oxidation are to develop varieties either with lower LN content or without seed lipoxygenase activity. To increase oil flavor stability and improve oil storage-ability, the linolenic acid content should be 1% or less (Khanna & Singh, 1991). The less linolenic acid content the oil contains, the more flavor stability the oil has. A major QTL for LN was identified in a RIL population derived from a cross between a line with low LA content (RG10), and a seed lipoxygenase free line (OX948). The QTL explained approximately 78% of phenotypic variations for LN content and tightly flanked by SSR marker *Satt534* and STS marker *Fad3i6* (Reiprecht et al., 2006). Among six QTLs, *QLNB2_1*, *QLNB2_2*, *QLNB2_3*, *QLNB2_4*, *QLND1b_1*, and *QLNG_1* associated with LA content were identified by Xie & co-workers (2012), however, only two LN QTL, *QLNB2_1* and *QLNB2_2* showed considerable potential for MAS across different environments. Since phenotypic evaluation of individuals for linolenic acid content is difficult to perform, the selection efficiency could be significantly enhanced by MAS. However, the inconsistency of QTLs across environments could be a serious issue while employing QTLs for MAS of elite individuals in environmental conditions differing from the environment(s) where the QTLs were first detected.

12.2.7 SORGHUM

Sorghum (*Sorghum bicolor* (L.) Moench) is an important multi-purpose staple crop grown not only for human consumption, but also for animal feeding, and biofuel production. Nowadays, the crop is receiving more attention to satisfy a sharp rise in demand for gluten-free and health-promoting grains around the globe. Sorghum breeding programs involve improving various traits including yield performance, yield stability, grain quality traits, and biofuel production related traits. Biochemical and physical attributes of sorghum grain are determinant for the quality of the ultimate sorghum products playing critical roles in different steps of the preparation process.

QTLs related to sorghum quality traits have been identified (Rhodes et al., 2017; Boyles et al., 2017; Guindo et al., 2019). The latter studied the quality traits of the crop grains associated with a sorghum meal known as *tô*. They totally detected 23 QTLs for various grain quality traits including endosperm texture (*ET*), protein content (*PR*), amylose content (*AMY*), lipid content (*LIP*), color of the testa layer (*TESTA*), and anthocyanin contents (*ANTHO*) on different sorghum chromosomes. These QTLs could contribute to developing high-yielding sorghum varieties with improved grain quality desirable alleles through the application of MAS.

Sorghum grain has the highest antioxidant content among major cereal crops including wheat, barley, corn, rice, and oats (Alfieri et al., 2017). However, the antioxidant attributes of sorghum grain mostly depend upon grain polyphenols categorized as flavonoids and phenolic acids that principally protect human body from damages of oxidative bursts and reduce the risk of chronic diseases such as cardiovascular diseases, type II diabetes, and cancers (Habyarimana et al., 2019; Habyarimana and Lopez-Cruz, 2019). Though tannin, an astringent grain flavonoid in sorghum, have almost always a negative effect on feed efficiency in human and animals due to tightly binding ability to proteins and minerals, the grain rich in tannins has shown the potential for fighting against human obesity (Santos-Buelga & Scalbert, 2000). Consequently, fortifying sorghum grain with higher concentrations of such health-promoting phytochemicals has been the major breeding objectives in several genetic improvement programs. Habyarimana et al. (2019) conducted a GWAS of grain antioxidants in sorghum accessions and *S. bicolor* × *S. halepense* RIL populations and identified a total of 55 QTLs for antioxidant contents explaining 15 to 31% of phenotypic variations. Additionally on chromosome 4, they have localized several major effect markers closely linked to a putative *MYB-bHLH-WD40* complex genes associated with flavonoid biosynthesis. Rhodes et al. (2014) also established a GWAS to detect novel QTLs related to polyphenol contents in sorghum. Some QTLs mapped for polyphenols were shown to be co-localized with ortho- or homologous genes of flavonoid biosynthesis pathway such as an Arabidopsis *TRANSPARETN TESTA 16* and a corn *Purple aleurone 1* (*Pr1*, flavonoid 3′-hydroxylase).

In sorghum, protein digestibility is another important quality trait to be improved. The crop contains a relatively high proportion of proteins; however, its digestibility is lower as compared with other major cereal crops. Development of lines with higher digestibility (HD lines, i.e., P721 Opaque, a high-lysine sorghum mutant), has resulted in significantly improved sorghum protein digestibility (Massafaro et al., 2016). Although HD trait is inherited as a single recessive gene, two opposed and tightly linked QTLs for protein digestibility, one QTL undesirably-(*LD*) and one QTL desirably affects digestibility (*HD*) have been characterized on chromosome 1 (Winn et al., 2009).

Today world is facing with serious issues including climate change, rapid population growth, and increased demand for fuels, and hence, renewable energy sources (bioenergy) is gaining considerable attraction, particularly in developed countries. Among the crop plants including sugarcane, corn, and switchgrass, sorghum has attracted considerable attention as a potentially renewable source of bioenergy. It is a high-yielding C4 cereal crop species producing high concentration sugar syrup stored in the stem (Tarpley & Vietor, 2007). To develop sorghum varieties capable of producing in biofuel, the best strategy is to enhance juice yield instead of increasing in SSCs that involves a physiological limit (Ritter et al., 2008; Guan et al., 2011; Felderhoff et al., 2012). Guan et al. (2011) detected several QTLs for bioenergy related traits in a $F_{2:3}$ population. A total of four QTLs for stem sugar content (Brix) and six QTLs for juice weight (JW) were detected. Only two JW QTLs located on chromosomes SBI-07 and SBI-09 were stably expressed (SE) across different environments. Felderhoff et al. (2012) characterized a major-effect QTL for SSCs on chromosome 3 in a set of $F_{3:4}$ lines. However, the co-localization of this QTL with those related to unfavorable biofuel production traits implies that selection for this trait may result into simultaneous decrease in grain yield and/or biomass production. Identification of QTLs associated with fiber-related traits is another effective approach for the development of sorghum varieties with improved biofuel capacity. Shiringani et al. (2010) characterized QTLs for acidic detergent fiber (ADF), neutral detergent fiber (NDF), acidic detergent lignin (ADL), cellulose content (CC), and hemicellulose content (HC). They detected a total of 67 QTLs for ADF, NDF, ADL, CC, and HC. Co-localization and overlapping of these QTLs strongly suggest that the genes encoding for cell wall components might be closely linked and tend to be coordinately regulated. However, the QTLs for such traits should be validated/confirmed in different genetic backgrounds to get the potential of employing in MAS.

12.2.8 COTTON

Upland Cotton (*Gossypium hirsutum* L.) is the most widely grown cotton species due to exhibiting better adaptation potential to the adverse environments and higher yield performance accounting for over 90% of total global cotton production (Zhang et al., 2015). Fiber length (FL) and fiber strength (FS) are the most important fiber quality attributes determining yarn quality (Yang et al., 2016). However, the quality of textile products also depends upon other fiber characteristics such as fiber elongation (FE), fiber uniformity (FU), and fiber micronaire (FM) that are of highly importance fiber quality traits for textile industries (Ijaz et al., 2019). Fiber quality attributes are complex traits controlled by QTLs/minor genes that are highly affected by environmental conditions (Lacape et al., 2010; Wang et al., 2015). In addition, linkage drags are common phenomenon between fiber quality traits and fiber yield in cotton (Si et al., 2017). Currently, a large number of QTLs related to major fiber quality traits, FL, FU, FS, FE, and FM in RIL populations (Wu et al., 2009; Tang et al., 2015; Ma et al., 2017; Ijaz et al., 2019), paternal backcross population (Ma et al., 2020), and MAGIC population (Fang et al., 2021) have been characterized. However, the majority of these QTLs show low stability or transferability across different environments or plant populations and only a little number of QTLs for fiber quality traits have been widely used in cotton MAS breeding (Feng et al., 2020; Fang et al., 2021). Tan et al. (2015) characterized several fiber quality QTLs, *qFL07.1*, *qFS07.1*, *qFE07.1*, and *qFM07.1* that closely flanked by SSR markers (*DPL0757* and *BNL1604*) and displayed high stability across different environments. Ling et al. (2020) identified a total of 35 QTLs for fiber quality traits in a paternal backcross population of upland cotton. However, only six QTLs for fiber quality traits, *qFL-Chr1–2*, *qFS-Chr5–1*, *qFS-Chr9–1*, *qFS-Chr21–1*, *qFM-Chr9–1* and *qFM-Chr9–2,* showed consistency across different environments. Additionally, their findings suggested that backcrossing male parents to F_1 hybrids is an efficient approach for characterizing novel QTLs dutiable for MAS.

Gene introgression is an effective approach for introducing favorable quality genes/QTLs in cotton. *G. mustelinum* that is a distant relative to *G. hirsutum* has the genetic potential for broadening the genetic base of upland cotton for fiber quality improvement. In an interspecific advanced backcross (BC) population derived from a cross between *G. mustelinum* and *G. hirsutum* more than 100 QTLs associated with fiber quality attributes have been identified. However, only two FL QTLs, *qUHM-1-1* and *qUHM-4-1*, and one FS QTL, *qSTR-23-1*, were SE across different environment. Identification of background-independent (BI) and SE QTLs would be of special importance for rapid genetic improvement of fiber quality traits in cotton. Consequently, introducing favorable SE-QTLs from *G. mustelinum* into the genetic background of Upland cotton would play a crucial role in further cotton breeding programs (Chen et al., 2020). The QTL for FL, *qFL-c10-1*, mapped in immortalized BC populations derived from RIL population has been reported to SE in different environments explaining over 35% of phenotypic variations (Wang et al., 2016). One QTL for FS (*qFS-Chr.D02*), has been successfully introgressed from *G. hirsutum* race *yucatanense* into upland cotton. This QTL fine mapped in a ≈551 kb-interval closely flanked by two markers, *INTR1027* and *JESPR-231* and has significantly improved FS compared with the recipient parent (Feng et al., 2020). The FS QTL (*qFS-c7-1*), characterized using a multiparent advanced generation intercross (MAGIC) population is highly transferable and significantly stable across adverse environments. The QTL affects not only FS, but also FM, FL, and FU and solely explain over 21.9% of phenotypic variance (Fang et al., 2021).

QTLs related to fiber quality traits are highly susceptible to environmental factors making it crucial to study the interactions of introgressed QTL with both recipient genetic background and different environments. Shi et al. (2020) assessed two introgressed lines (ILs) derived from crosses between *G. hirsutum* × *G. barbadense* across multiple environments. They detected a set of 61 novel SE-QTLs, consisting of two BI-QTLs. The genetic background was found to be more effective on fiber quality traits than the environments. Furthermore, QTLs for fiber quality traits that SE under stress conditions, i.e., salinity could provide highly valued insights for the rapid and effective genetic improvement of cotton fiber quality traits. Employing a RIL population, Guo et al. (2021) identified three major QTLs related to FL, *qFL-Chr1-1*, *qFL-Chr5-5*, and *qFL-Chr24-4*, SE under saline stress. Interestingly, several genes located within intervals flanked by two FL QTLs, *qFL-Chr1-1* and *qFL-Chr5-1*, were highly overexpressed. The stable QTLs for FL such as *qFL06.1*, *qFL16.1*, *qFL21.1*, and FM, *qFM07.1* have been used to identify candidate genes in cotton (Ali et al., 2018). Two QTLs for FS, *qFS-chr25-4* and *qFS-C9-1* were reported to be SE across different environments (Zhang et al., 2015; Yang et al., 2016). Liu et al. (2017) detected 50 QTLs for fiber quality traits SE in multiple environments. However, only four QTLs for FL, *qFL01.1*, *qFL14.1*, *qFL17.1*, and *qFL26.1*, five QTLs for FM, *qFM05.1*, *qFM07.1*, *qFM20.2*, *qFM23.1*, and *qFM25.1*, and two QTLs for FS, *qFS07.1* and *qFS14.2*, are capable of using for MAS. Gu et al. (2020) identified a large number of QTLs associated with fiber quality traits in large RIL population. A total of 66 QTLs, including 10 FL QTLs, 14 FS QTLs, 12 FU QTLs, and 19 FE QTLs were identified. However, only two FL QTLs, *qFL-D02-1* and *qFL-D02-4*, and one FS QTL, *qFS-D11-1*, were SE across different environments. Zhou et al. (2021) detected 32 unique QTLs for different fiber quality traits in the D sub-genome of upland cotton, implying that functional mutations in this sub-genome plays a key role for fiber quality improvement in cotton. Additionally, SE-QTLs associated with fiber quality traits enhance the effectiveness and reliability of the QTL mapping for MAS in cotton breeding programs and used for fine mapping and to identify candidate genes for a particular trait.

Sea Island cotton (*G. barbadense*) is the second most important cotton species exceptionally produces high quality fibers (Shi et al., 2015). Although this cotton species could produce markedly high-quality fibers suitable for the textile industry and be used for broadening the genetic base of upland cotton for fiber quality, it has been less studied to identify QTLs for fiber quality traits found in natural populations. Su et al. (2020) investigated 279 sea Island cotton accessions collected from various origins under saline and normal stress conditions and detected a total of 34 stable QTLs associated with fiber quality traits. Furthermore, in QTL *TM6004*, they have characterized an *E3 ubiquitin-protein ligase* gene playing a key role in fiber quality traits.

Cotton is a rich source of protein (27.83–45.6%) that could be used principally for animal feeding. Protein content in cotton is a polygenic trait even might be simultaneously regulated by nuclear and/or maternal genes, and cytoplasmic or environmental factors (Ye et al., 2003). Attempts have been made to characterize QTLs related to protein content in cotton (Yu et al., 2012; Alfred et al., 2012; Liu et al., 2015, 2017; Shang et al., 2016; Wang et al., 2019). Song & Zhang (2007) detected 10 QTLs for oil content, protein content, and amino acid concentration, including aspartic acid, serine, glycine, isoleucine, leucine, phenylalanine, and argentine, *qOP-D8-1*, *qPP-D9-1*, *qAsp-A11-1*, *qSer-A8-1*, *qGly-A11-1*, *qGly-A8-1*, *qIle-D3-1*, *qLeu-D2-1*, *qPhe-A8-1*, *qArg-A5*-1, in a BC_1S_1 population. They characterized two epistatic QTLs (E-QTLs) for leucine and cysteine that are of special importance in developing cotton varieties with improved amino acid contents. Wang et al. (2019) identified 58 QTLs related to seed size, and oil and protein contents in a RIL population using specific locus amplified fragment sequencing (SLAF-seq) technique. A total of two QTLs for oil content, *qKOC-D03.1,* and protein content, *qKPC-D03.1*, were SE across three different environments. These SE-QTLs for nutritional properties have the potential for characterizing candidate genes, better understanding of molecular mechanisms responsible for protein and oil synthesis, and MAS in cotton genetic improvement programs.

12.3 MAS APPLICATIONS

The major applications of MAS are marker-assisted gene pyramiding (MAGP), marker-assisted introgression (MAI), marker-assisted recurrent selection (MARS), marker-assisted backcrossing (MABC), and F_2 enrichment (Collard & Mackill, 2008; Xu & Crouch, 2008). Gene pyramiding, also known as multi-trait introgression refers to the technique of stacking multiple genes/QTLs into a single genotype to combine the benefits of multiple favorable gene/QTLs through conventional breeding schemes (Joshi & Nayak, 2010). The method has the potential to be implemented through recurrent selection, backcrossing, and multi-parent crossing. Gene stacking and its effects have been summarized in different plants, including rice, soybean, wheat, corn, barley, lentil, and tomato (Tarán et al., 2003; Lecomte et al., 2004; Richardson et al., 2006; Jiang et al., 2007; Emebiri et al., 2009; Sacco et al., 2013; Luo et al., 2016; Chukwu et al., 2019; Dormatey et al., 2020; Jamaloddin et al., 2020; Ramalingam et al., 2020). The success of gene pyramiding depends upon several factors, such as heritability of favorable genes/QTLs, number of genes/QTLs to be transferred, the number of parental lines donating the genes/QTLs of interest, association between molecular markers and target genes/QTLs, and relative cost of procedure. However, the method has proven to be an effective approach for quality improvement in major crop species including tomato, corn, rice, wheat, and cotton (Guo et al., 2005; Sacco et al., 2013; Sarika et al., 2018; Wang et al., 2012; Gautam et al., 2020). Gene pyramiding has been successfully used to improve the antioxidant contents of tomato fruit. QTLs associated with increased ascorbic acid content (AsA), phenols and SSC were pyramided from *S. pennellii* into the recipient parent M82. The F_3 individuals produced higher AsA, total phenolic compounds and SSC compared with recurrent parent M82 (Sacco et al., 2013). *Opaque* mutants in corn are rich in lysine and tryptophan playing an important role in human nutrition. MAS has significantly expedited the stacking of two major *opaque* genes, *opaque2* (*o2*) and *opaque16* (*016*), in sub-tropical corn and improved corn nutritional quality (Sarika et al., 2018). MAGP has been also used for pyramiding desirable quality QTLs in upland cotton. Guo et al. (2005) stacked two major effect QTLs for FS, *qtlfs-1* and *qtlfs-2*, and found that coupling QTL pyramiding with MAS could significantly improve the efficiency of selection for FS in upland cotton.

MABC that is a simple form of MAS could help breeders to target desirable genes/QTLs introgressed into another superior line or genotype to improve a characteristic. The technique relies on molecular markers tightly linked with favorable genes/QTLs instead of phenotypic performance. MABC ensures the quick recovery of recurrent parent genome (RPG) (background selection) with minimal linkage drag. It could optimistically shorten the duration of BC breeding programs by minimizing the numbers of BC generations and a number of individuals necessary in each generation. MABC is best suited for introgression of desirable genes/QTLs such as gene pyramiding where PS is not efficient enough or phenotype may be masked epistatically and accelerate fixation of desirable gene/QTLs (Dormatey et al., 2020). Notably, a properly established MABC could ensure the stable expression of desirable traits. In tomato, MABC has been successfully employed to transfer QTLs for fruit quality traits into tomato lines. However, the efficiency of gene pyramiding was proposed to be dependent upon the genetic background of the recurrent parent demonstrating significant interactions between QTLs and genetic backgrounds (Lecomte et al., 2004).

MARS is another improved system of genotype selection allowing breeders to integrate multiple desirable genes/QTLs in obtaining a "designed" individual containing the most desired combinations of favorable genes/QTLs (Bohra, 2013). The technique has been used for gene pyramiding-based quality improvement in rice and corn (Xie et al., 2010; Bankole et al., 2017). Compared with PS, MARS has even almost doubled the genetic improvement in corn (Ribaut et al., 2006). MARS may demonstrate to be a highly effective strategy in breeding programs as it can facilitate genes/QTLs transferring controlling the expression of quantitative traits that is a limitation in case of MAS and MABC (Gokidi et al., 2016). Integration of these MAS based techniques into conventional breeding programs will expedite the process of evaluating, transferring, and combining target genes/QTLs. However, to maximize the rate and precision of MAS based gene/QTL introgression, there is a need for proper adaptation of new strategies, including improved scoring techniques, appropriate procedure for quantitative genetic analysis, large-sized population, and novel marker validation approaches.

12.4 CONCLUSION

Quality improvement is the major objective of numerous breeding programs, the process that inevitably requires huge financial and human resources to be implemented. During the last decades, the use of molecular markers as a complementary biological toolkit, so called MAS, has gained considerable importance for breeding of both simple and complex traits. However, several factors including number of transferring target gene/QTL(s), proximity between target gene/QTL and neighboring marker(s), number of individuals selected in each round of selection, germplasm nature, and facilities available at the level of molecular markers are contributed to successfully adoption of MAS. Recent progress in plant molecular biology, genomics, genetics, and bioinformatics have resulted in better understanding of both molecular markers and the nature of QTLs controlling major agronomic and quality traits allowing selection of individuals carrying genes/QTLs of interest with greater speed and precision. The integration of MAS into plant breeding schemes such as MAGP, MAI, MARS, MABC, and F_2 enrichment is dramatically beneficial for plant breeder to fast track and precisely monitor the transfer of desirable quality genes/QTLs (foreground selection) within breeding materials. However, it still bears its challenges to be indefinitely feasibly used in combination with conventional breeding techniques. Lack of confidence in MAS protocols, uncertainty in how to combine MAS approaches with PS methods, and relatively high cost of MAS are the factors limited comprehensive application of MAS in breeding programs, particularly in the less developed countries. Nonetheless, the development of new genomic tools and locus validation approaches, and the introduction of novel models for the exploitation of QTLs in crop improvement programs could potentially help breeders overcome these issues. Promisingly, with the development of next-generation sequencing (NGS) approaches, novel molecular markers such as SNPs, and genome-wide association studies (GWAS), the practicality, efficiency, and effectiveness of MAS are expected to be tangibly enhanced in the near future.

KEYWORDS

- **dry matter content**
- **genotyping by sequencing**
- **marker-assisted backcross**
- **marker-assisted selection**
- **MAS efficiency**
- **phenotypic selection**
- **quality improvement**
- **quantitative trait loci**

REFERENCES

Adachi, N., Migita, M., Ohta, T., Higashi, A., & Matsuda, I., (1997). Depressed natural killer cell activity due to decreased natural killer cell population in a vitamin E-deficient patient with Shwachman syndrome, reversible natural killer cell abnormality by a-tocopherol supplementation. *European Journal of Pediatrics, 156*, 444–448.

Ahmed, S., Zhou, X., Pang, Y., Jin, L., & Bao, J., (2018). Improving starch-related traits in potato crops, Achievements and future challenges. *Starch-Stärke., 70*, 1700113

Alfieri, M., Balconi, C., Cabassi, G., Habyarimana, E., & Redaelli, R., (2017). Antioxidant activity in a set of sorghum landraces and breeding lines. *Maydica, 62*, 1–7.

Alfred, Q., Liu, H. Y., Xu, H. M., Li, J. R., Wu, J. G., Zhu, S. J., & Shi, C. H., (2012). Mapping of quantitative trait loci for oil content in cottonseed kernel. *Journal of Genetics, 91*(3), 289–295.

Ali, I., Teng, Z., Bai, Y., Yang, Q., Hao, Y., Hou, J., Jia, Y., et al., (2018). A high density SLAF-SNP genetic map and QTL detection for fiber quality traits in *Gossypium hirsutum*. *BMC Genomics, 19*, 879.

Avivi, L., (1978). High protein content in wild tetraploid *Triticum dicoccoides* Korn. In: Ramanujam, S., (ed.), *Proceedings of the 5th International Wheat Genetics Symposium* (pp. 372–380). Indian Society of Genetics and Plant Breeding (ISGPB): New Delhi

Bankole, F., Menkir, A., Olaoye, G., Crossa, J., Hearne, S., Unachukwu, N., & Gedil, M., (2017). Genetic gains in yield and yield related traits under drought stress and favorable environments in a maize population improved using marker assisted recurrent selection. *Frontiers in Plant Science,* 8.

Battino, M., Beekwilder, J., Denoyes-Rothan, B., Laimer, M., McDougall, G. J., & Mezzetti, B., (2009). Bioactive compounds in berries relevant to human health. *Nutrition Review, 67*, 145–150.

Bergamini, C., Cardone, M. F., Anaclerio, A., Perniola, R., Pichierri, A., Genghi, R., Alba, V., et al., (2013). Validation assay of *p3_VvAGL11* marker in a wide range of genetic background for early selection of stenospermocarpy in *Vitis vinifera* L. *Molecular Biotechnology, 54*, 1021–1030.

Blanco, A., Mangini, G., Giancaspro, A., Giove, S., Colasuonno, P., Simeone, R., Signorile, A., et al., (2012). Relationship between grain protein content and grain yield components through quantitative trait locus analyses in a recombinant inbred line population derived from two elite durum wheat cultivars. *Molecular Breeding, 30*, 79–92.

Bogard, M., Allard, V., Martre, P., Heumez, E., Snape, J. W., Orford, S., Griffiths, S., et al., (2013). Identifying wheat genomic regions for improving grain protein concentration independently of grain yield using multiple inter-related populations. *Molecular Breeding, 31*, 587–599.

Bohra, A., (2013). Emerging paradigms in genomics-based crop improvement. *The Scientific World Journal, 2013,* 17.

Boyles, R. E., Pfeiffer, B. K., Cooper, E. A., Rauh, B. L., Zielinski, K. J., Myers, M. T., Brenton, Z., et al., (2017). Genetic dissection of sorghum grain quality traits using diverse and segregating populations. *Theoretical and Applied Genetics., 130*, 697–716.

Burstin, J., Gallardo, K., Mir, R. R., Varshney, R. K., & Duc, G., (2011). Improving protein content and nutrition quality. In: Pratap, A., & Kumar, J., (eds.), *Biology and Breeding of Food Legumes*. CAB International.

Chand, S., Chandra, K., & Khatik, C. L., (2020). Varietal release, notification and denotification system in India. In: *Plant Breeding-Current and Future Views* (pp. 1–12). IntechOpen.

Chen, Q., Wang, W., Wang, C., Zhang, M., Yu, J., Zhang, Y., Yuan, B., et al., (2020). Validation of QTLs for fiber quality introgressed from *Gossypium mustelinum* by selective genotyping. *Genes, Genome, Genetics, 10*(7), 2377–2384.

Chukwu, S. C., Radii, M. Y., Ramlee, S. I., Ismail, S. I., Oladosu, Y., Okporie, E., Onyishi, G., et al., (2019). Marker-assisted selection and gene pyramiding for resistance to bacterial leaf blight disease of rice (*Oryza sativa* L.). *Biotechnology and Biotechnological Equipment, 33*(1), 440–455.

Cobb, J. N., Biswas, P. S., & Damien, P. J., (2019). Back to the future, revisiting MAS as a tool for modern plant breeding. *Theoretical and Applied Genetics, 132*, 647–667.

Collard, B. C. Y., & Mackill, D. J., (2008). Marker-assisted selection, an approach for precision plant breeding in the twenty-first century. *Philosophical Transactions of the Royal Society of Britain, 363*, 557–572.

Cong, B., Liu, J., & Tanksley, S. D., (2002). Natural alleles at a tomato fruit size quantitative trait locus differ by heterochronic regulatory mutations. *Proceeding of the National Academy of Science, USA, 99*, 13606–13611.

Costantini, L., Battilana, J., Lamaj, F., Fanizza, G., & Grando, M. S., (2008). Berry and phenology-related traits in grapevine (*Vitis vinifera* L.), from quantitative trait loci to underlying genes. *BMC Plant Biology, 8*, 1–17.

Debnath, S., & Guha, S., (2015). In: Sharangi, A. B., Datta, S., (eds.), *Breeding Methods for Quality Improvement in Horticultural Crops*: *Value Addition of Horticultural Crops, Recent Trends and Future Directions* (pp. 201–211). Springer India.

Demmig-Adams, B., & Adams, W. W., (2002). Antioxidants in photosynthesis and human nutrition. *Science, 298*, 2149–2153.

Distelfeld, A., Uauy, C., Fahima, T., & Dubcovsky, J., (2006). Physical map of the wheat high-grain protein content gene *Gpc*-B1 and development of a high-throughput molecular marker. *New Phytologist, 169*, 753–763.

Dormatey, R., Sun, C., Ali, K., Coulter, J. A., Bi, Z., & Bai, J., (2020). Gene pyramiding for sustainable crop improvement against biotic and abiotic stresses. *Agronomy, 10,* 1255.

Eathington, S. R., Dudley, J. W., & Rufener, G. K., (1997). Usefulness of marker-QTL associations in early generation selection. *Crop Science, 37*, 1686–1693.

Emebiri, L., Michael, P., Moody, D. B., Ogbonnaya, F. C., & Black, C., (2009). Pyramiding QTLs to improve malting quality in barley: Gains in phenotype and genetic diversity. *Molecular Breeding, 23*, 219–228.

Fang, D. D., Zeng, L., Thyssen, G. N., Delhom, C. D., Bechere, E., Jones, D. C., & Li, P., (2021). Stability and transferability assessment of the cotton fiber strength QTL. *The Crop Journal, 9*, 380–386.

Fang, X., Liu, X., Wang, X., Wang, W., Liu, D., Zhang, J., Liu, D., et al., (2017). Fine-mapping qFS07. 1 controlling fiber strength in upland cotton (*Gossypium hirsutum* L.). *Theoretical and Applied Genetics, 130*, 795–806.

Felderhoff, T. G., Murray, S. C., Klein, P. E., Sharma, A., Hamblin, M. T., Kresovich, S., Vermerris, W., & Rooney, W. L., (2007). QTLs for energy-related traits in a sweet X grain sorghum [*Sorghum bicolor* (L.) Moench] mapping population. *Crop Science, 52*, 2040–2049.

Foolad, M. R., (2007). Genome mapping and molecular breeding of tomato. *International Journal of Plant Genomics, 6*, 43–58.

Fox, P. N., Crossa, J., & Romagosa, I., (1997). Genotype × environment and adaptation. In: Kempton, R., & Fox, P., (eds.), *Statistical Methods for Plant Variety Evaluation* (pp. 117–138) Chapman and Hall: London.

Francia, E., Tacconi, G., Crosatti, C., Barabaschi, D., Bulgarelli, Dall'Aglio, D. E., & Vale, G., (2005). Marker assisted selection in crop plants. *Plant Cell, Tissue and Organ Culture, 82*, 317–342.

Frary, A., Doganlar, S., Frampton, A., Fulton, T., Uhlig, J., Yates, H., & Tanksley, S., (2003). Fine mapping of quantitative trait loci for improved fruit characteristics from *Lycopersicon chmielewskii* chromosome 1. *Genome, 46,* 235–243.

Frary, A., Nesbitt, T. C., Frary, A., Grandillo, S., Van, D. K. E., Cong, B., Liu, J., et al., (2000). Cloning and transgenic expression of *fw2.2*, a quantitative trait locus key to the evolution of tomato fruit. *Science, 289,* 85–87.

Freyre, R., & Douches, D. S., (1994). Development of a model for marker-assisted selection of specific gravity in diploid potato across environments. *Crop Science, 34*, 1361–1368.

Fridman, E., Pleban, T., & Zamir, D., (2000). A recombination hotspot delimits a wild-species quantitative trait locus for tomato sugar content to 484 bp within an invertase gene. *Proceeding of the National Academy of Science, USA, 97*, 4718–4723.

Gali, K. K., Sackville, A., Tafesse, E. G., Reddy, L. V. B., McPhee, K., Hybl, M., Mikić, A., Smykal, P., McGee, R., et al., (2019). Genome-wide association mapping for agronomic and seed quality traits of field pea (*Pisum sativum* L.). *Frontiers in Plant Science, 10*, 1538.

Gautam, T., Dhillon, G. S., Saripalli, G., Rakhi Singh, V. P., Prasad, P., Kaur, S., Chhuneja, P., et al., (2020). Marker-assisted pyramiding of genes/QTL for grain quality and rust resistance in wheat (*Triticum aestivum* L.). *Molecular Breeding, 40*, 49.

Gelli, M., Konda, A. R., Liu, K., Zhang, C., Clemente, T. E., Holding, D. R., & Dweikat, I. M., (2017). Validation of QTL mapping and transcriptome profiling for identification of candidate genes associated with nitrogen stress tolerance in sorghum. *BMC Plant Biology, 17*, 123.

Giampieri, F., Alvarez-Suarez, J. M., Cordero, M. D., Gasparrini, M., Forbes-Hernandez, T. Y., Afrin, S., Santos-Buelga, C., et al., (2017). Strawberry consumption improves aging-associated impairments, mitochondrial biogenesis and functionality through the AMP-activated protein kinase signaling cascade. *Food Chemistry, 234,* 464–471.

Gimelfarb, A., & Lande, R., (1994). Simulation of marker-assisted selection in hybrid populations. *Genetic Researches, 63*, 39–47.

Goel, S., Singh, K., Singh, B., Grewal, S., Dwivedi, N., Alqarawi, A. A., Abd_Allah, E. F., Ahamd, P., & Singh, N. K., (2019). Analysis of genetic control and QTL mapping of essential wheat grain quality traits in a recombinant inbred population. *PLoS ONE, 14*(3), e0200669.

Gokidi, Y., Bhanu, A. N., & Singh, M. N., (2016). Marker assisted recurrent selection: An overview. *Advances in Life Science, 5*(17), 6493–4699.

Grandillo, S., Ku, H. M., & Tanksley, S. D., (1996). Characterization of *fs8.1*, a major QTL influencing fruit shape in tomato. *Molecular Breeding, 2*, 251–260.

Grandillo, S., Ku, H. M., & Tanksley, S. D., (1999). Identifying the loci responsible for natural variation in fruit size and shape in tomato. *Theoretical and Applied Genetics, 99*, 978–987.

Gu, Q., Ke, H., Liu, Z., Lv, X., Sun, Z., Zhang, M., Chen, L., et al., (2020). A high-density genetic map and multiple environmental tests reveal novel quantitative trait loci and candidate genes for fiber quality and yield in cotton. *Theoretical and Applied Genetics 133*, 3395–3408.

Guan, Y. A., Wang, H. L., Qin, L., Zhang, H. W., Yang, Y. B., Gao, F. J., Li, R. Y., & Wang, H. G., (2011). QTL mapping of bio-energy related traits in sorghum. *Euphytica, 182*, 431–440.

Guindo, D., Teme, N., Vaksmann, M., Doumbia, M., Vilmus, I., Guitton, B., Sissoko, A., et al., (2019). Quantitative trait loci for sorghum grain morphology and quality traits: Toward breeding for a traditional food preparation of West-Africa. *Journal of Cereal Science, 85*, 256–272.

Guo, W. Z., Zhang, T. Z., Ding, Y. Z., Zhu, Y. C., Shen, X. L., & Zhu, X. F., (2005). Molecular marker assisted selection and pyramiding of two QTLs for fiber strength in upland cotton. *Acta Genetica Sinica, 32*(12), 1275–1285.

Guo, Y., Zhang, G., Guo, B., Qu, C., Zhang, M., Kong, F., Zhao, Y., & Li, S., (2020). QTL mapping for quality traits using a high-density genetic map of wheat. *PLoS One, 15*, e0230601.

Habyarimana, E., & Lopez-Cruz, M., (2019). Genomic selection for antioxidant production in a panel of sorghum bicolor and *S. bicolor* × *S. halepense* Lines. *Genes, 10*, 841.

Habyarimana, E., Dall'Agata, M., De Franceschi, P., & Baloch, F. S., (2019). Genome-wide association mapping of total antioxidant capacity, phenols, tannins, and flavonoids in a panel of *Sorghum bicolor* and *S. bicolor* × *S. halepense* populations using multi-locus models. *PLoS One, 14*(12), e0225979.

Harjit-Singh, P. M., Prasad, M., Varshney, R. K., Roy, J. K., Baylan, H. S., Dhaliwa, H. S., & Gupta, P. K., (2011). STMS markers for grain protein content and their validation using near-isogenic lines in bread wheat. *Plant Breeding, 120*, 273–278.

Henning, S. M., Seeram, N. P., Zhang, Y., Li, L., Gao, K., Lee, R. P., Wang, D. C., et al., (2010). Strawberry consumption is associated with increased antioxidant capacity in serum. *Journal of Medical Foods, 13*, 116–122.

Holland, J. B., (2007). Genetic architecture of complex traits in plants. *Current Opinion in Plant Biololy, 10*, 156–161.

Hospital, F., Moreau, L., Charcosset, A., & Gallais, A., (1997). More on the efficiency of marker assisted selection. *Theoretical and Applied Genetics, 95*, 1181–1189.

Ijaz, B., Zhao, N., Kong, J., & Hua, J., (2019). Fiber quality improvement in upland cotton (*Gossypium hirsutum* L.): Quantitative trait loci mapping and marker assisted selection application. *Frontiers in Plant Science, 10*, 1585.

Ilahy, R., Hdider, C., Lenucci, M. S., Tlili, I., & Dalessandro, G., (2011). Antioxidant activity and bioactive compound changes during fruit ripening of high-lycopene tomato cultivars. *Journal of Food Composition and Analysis, 24*, 588–595.

Jamaloddin, M., Durga, R. C. V., Swathi, G., Anuradha, C., Vanisri, S., Rajan, C. P. D., et al., (2020). Marker assisted gene pyramiding (MAGP) for bacterial blight and blast resistance into mega rice variety "tellahamsa". *PLoS One, 15*(6), e0234088.

James, J., Douglas, L., & Shaw, V., (2009). Plant breeders' perspectives on improving yield and quality traits in horticultural food crops. *HortScience, 44*, 20–22.

Jena, K. K., & Mackill, D. J., (2008). Molecular markers and their use in marker-assisted selection in rice. *Crop Science, 48*, 1266–1276.

Jiang, G. L., Shi, J., & Ward, R. W., (2007). QTL analysis of resistance to fusarium head blight in the novel wheat germplasm CJ 9306. I. Resistance to fungal spread. *Theoretical and Applied Genetics, 116*, 3–13.

Joshi, R. K., & Kayak, S., (2010). Gene pyramiding-A broad spectrum technique for developing durable stress resistance in crops. *Biotechnology and Molecular Biology Review, 5*(3), 51–60.

Kassem, M. A., Meksem, K., Iqbal, M. J., Njiti, V. N., Banz, W. J., Winters, T. A., Wood, A., & Lightfoot, D. A., (2004). Definition of soybean genomic regions that control seed phytoestrogen amounts. *Journal of Biomedicine and Biotechnology*, *1*, 52–60.

Khanna, K. R., & Singh, S. P., (1991). Genetics of fatty acid composition and breeding for modified seed oils. In: Khanna, K. R., (ed.), *Biochemical Aspects of Crop Improvement* (pp. 255–281). CRC Press, Florida.

Knight, D. C., & Eden, J. A., (1996). A review of the clinical effects of phytoestrogens. *Obstetrics and Gynecololgy, 87,* 897–904.

Ku, H. M., Grandillo, S., & Tanksley, S. D., (2000). *fs8*.1, a major QTL, sets the pattern of tomato carpel shape well before anthesis. *Theoretical and Applied Genetics, 101*, 873–878.

Kumar, A., Jain, S., Elias, E. M., Ibrahim, M., & Sharma, L. K., (2018). An overview of QTL identification and marker-assisted selection for grain protein content in wheat. In: Sengar, R., & Singh, A., (eds.), *Eco-Friendly Agro-Biological Techniques for Enhancing Crop* Productivity (pp. 245–276). Springer, Singapore.

Kumar, N., Rana, M., Kumar, B., Chand, S., Shiv, A., Wani, S. H., & Kumar, S., (2021). Genomic selection for wheat improvement. *Physiological, Molecular, and Genetic Perspectives of Wheat Improvement*, 175–207.

Kumawat, G., Kumawat, C. K., Chandra, K., Pandey, S., Chand, S., Mishra, U. N., Lenka, D., & Sharma, R., (2020). Insights into marker assisted selection and its applications in plant breeding. In: *Plant Breeding-Current and Future Views* (pp. 1–21). IntechOpen.

Labadie, M., Vallin, G., Petit, A., Ring, L., Hoffmann, T., Gaston, A., Potier, A., et al., (2020). Metabolite quantitative trait loci for flavonoids provide new insights into the genetic architecture of strawberry (*Fragaria* × *ananassa*) fruit quality. *Journal of Agricultural Food Chemistry, 68*, 6927−6939.

Lacape, J. M., Llewellyn, D., Jacobs, J., Arioli, T., Becker, D., Calhoun, S., Al-Ghazi, Y., et al., (2010). Meta-analysis of cotton fiber quality QTLs across diverse environments in a *Gossypium hirsutum* × *G. barbadense* RIL population. *BMC Plant Biology, 10*, 132.

Lafiandra, D., Sanguineti, M. C., Maccaferri, M., & Deambrogio, E., (2007) Molecular markers and QTL analysis for grain quality improvement in wheat. In: Varshney, R. K., & Tuberosa, R., (eds.), *Genomics Assisted Crop Improvement* (Vol. 2, pp. 25–50). Springer.

Lande, R., & Thompson, R., (1990). Efficiency of marker-assisted selection in the improvement of quantitative traits. *Genetics, 124*, 743–756.

Lecomte, L., Duffé, P., Buret, M., Servin, B., Hospital, F., & Causse, M., (2004). Marker-assisted introgression of five QTLs controlling fruit quality traits into three tomato lines revealed interactions between QTLs and genetic backgrounds. *Theoretical and Applied Genetics, 109*, 658–668.

Lerceteau-Köhler, E., Moing, A., Guerin, G., Renaud, C., Petit, A., Rothan, C., & Denoyes, B., (2012). Genetic dissection of fruit quality traits in the octoploid cultivated strawberry highlights the role of homoeo-QTL in their control. *Theoretical and Applied Genetics, 124*, 1059–1077.

Li, H., Liu, H., Han, Y., Wu, X., Teng, W., Liu, G., & Li, W., (2010). Identification of QTL underlying vitamin E contents in soybean seed among multiple environments. *Theoretical and Applied Genetics, 120*, 1405–1413.

Li, L., Tacke, E., Hofferbert, H. R., Lubeck, J., Strahwald, J., Draffehn, A. M., Walkemeier, B., & Gebhardt, C., (2013). Validation of candidate gene markers for marker-assisted selection of potato cultivars with improved tuber quality. *Theoretical and Applied Genetics*, *126*, 1039–1052.

Lippman, Z., & Tanksley, S. D., (2001). Dissecting the genetic pathway to extreme fruit size in tomato using a cross between the small-fruited wild species *Lycopersicon pimpinellifolium* and *L. esculentum* var. *Giant Heirloom. Genetics, 158*, 413–422.

Liu, D., Liu, F., Shan, X., Zhang, J., Tang, S., Fang, X., Liu, X., Wang, W., Tan, Z., Teng, Z., et al., (2015). Construction of a high-density genetic map and lint percentage and cottonseed nutrient trait QTL identification in upland cotton (*Gossypium hirsutum* L.). molecular genetics and genomics. *Molecular Genetics and Genomics, 290*(5), 1683–1700.

Liu, J., Cong, B., & Tanksley, S. D., (2003). Generation and analysis of an artificial gene dosage series in tomato to study the mechanisms by which the cloned quantitative trait locus *fw2.2* controls fruit size. *Plant Physiology, 13*, 292–299.

Liu, X. Y., Teng, Z. H., Wang, J. X., Wu, T. T., Zhang, Z. Q., Deng, X. P., Fang, X. M., et al., (2017). Enriching an intraspecific genetic map and identifying QTL for fiber quality and yield component traits across multiple environments in upland cotton (*Gossypium hirsutum* L.). *Molecular Genetics and Genomics, 292*(6), 1281–1306.

Lu, W., Wen, Z., Li, H., Yuan, D., Li, J., Zhang, H., Huang, Z., et al., (2013). Identification of the quantitative trait loci (QTL) underlying water soluble protein content in soybean. *Theoretical and Applied Genetics, 126*, 425–433.

Luo, Y., Ma, T., Zhang, A., Ong, K. H., Li, Z., Yang, J., & Yin, Z., (2016). Marker-assisted breeding of the rice restorer line Wanhui 6725 for disease resistance, submergence tolerance and aromatic fragrance. *Rice, 9*.

Ma, L. L., Su, Y., Nie, H., Cui, Y., Cheng, C., Ijaz, B., & Hua, J., (2020). QTL and genetic analysis controlling fiber quality traits using paternal backcross population in upland cotton. *Journal of Cotton Research, 3*, 22.

Ma, W., Zhang, W., & Gale, K. R., (2003). Multiplex-PCR typing of high molecular weight glutenin alleles in wheat. *Euphytica, 134*, 51–60.

Ma, Y., Coyne, C. J., Grusak, M. A., Mazourek, M., Cheng, P., Main, D., & McGee, R. J., (2017). Genome-wide SNP identification, linkage map construction and QTL mapping for seed mineral concentrations and contents in pea (*Pisum sativum* L.). *BMC Plant Biolology, 17*, 43.

Malhotra, A., & Coupland, J. N., (2004). The effect of surfactants on the solubility, zeta potential, and viscosity of soy protein isolates. *Food Hydrocolloids, 18*, 101–108.

Massa, A. N., Manrique-Carpintero, N. C., Coombs, J. J., Zarka, D. G., Boone, A. E., Kirk, W. W., Hackett, C. A., et al., (2015). Genetic linkage mapping of economically important traits in cultivated tetraploid potato (*Solanum tuberosum* L.*). Genes, Genomes, Genetics (Bethesda), 5*, 2357–2364.

Massafaro, M., Thompson, A., Tuinstra, M., Dilkes, B., & Weil, C. F., (2016). Mapping the increased protein digestibility trait in the high-lysine sorghum mutant P721Q. *Crop Science, 56*, 2647–2651.

McCord, P. H., Sosinski, B. R., Haynes, K. G., Clough, M. E., & Yencho, G. C., (2011). QTL mapping of internal heat necrosis in tetraploid potato. *Theoretical and Applied Genetics, 122*, 129–142.

Menéndez, C. M., Ritter, E., Schäfer-Pregl, R., Walkemeier, B., Kalde, A., Salamini, F., & Gebhardt, C., (2002). Cold sweetening in diploid potato, mapping quantitative trait loci and candidate genes. *Genetics, 162*, 1423–1434.

Mezzetti, B., Giampieri, F., Zhang, Y. T., & Zhong, C. F., (2018). Status of strawberry breeding programs and cultivation systems in Europe and the rest of the world. *Journal of Berry Research, 8*, 205–221.

Mishra, G. P., & Singh, R. K., (2015). Molecular breeding approach for crop improvement of quality traits. *Journal of Biotechnology and Crop Science, 4*(5), 4–21.

Miura, H., & Tanii, S., (1994). Endosperm starch properties in several wheat cultivars preferred for Japanese noodles. *Euphytica, 72*, 171–175.

Mohammadi, M., Mehrazar, E., Izadi-Darbandi, A., & Najafian, G., (2013). Marker assisted selection for complementation of *Glu-A1b, Glu-B1i,* and *Glu-D1d* alleles in early segregating generations of wheat. *Wheat Information Services, 116*, 3–8.

Monari, A. M., Simeone, M. C., Urbano, M., Margotta, B., & Lafiandra, D., (2005). Molecular characterization of new waxy mutants identified in bread and durum wheat. *Theoretical and Applied Genetics, 110*, 1481–1489.

Munne-Bosch, S., & Alegre, L., (2002). The function of tocopherols and tocotrienols in plants. *Critical Reviews in Plant Sciences, 21*, 31–57.

Munro, I. C., Harwood, M., Hlywka, J. J., Stephen, A. M., Doull, J., Flamm, W. G., & Adlercreutz, C., (2003). Soy isoflavones, a safety review. *Nutrition Reviews, 61*, 1–33.

Nakamura, T., Vrinten, P., Saito, M., & Konda, M., (2002). Rapid classification of partial waxy wheat using PCR-based markers. *Genome, 45*, 1150–1156.

Nakamura, T., Yamamori, M., Hirano, H., & Hidaka, S., (1993). Identification of three *Wx* proteins in wheat (*Triticum aestivum* L.). *Biochemical Genetics, 31*, 75–86.

Natarajan, S., Hossain, M. R., Kim, H. T., Jesse, D. M. I., Ferdous, M. J., Jung, H. J., Park, J. I., & Nou, I. S., (2020). ddRAD-seq derived genome-wide SNPs, high density linkage map and QTLs for fruit quality traits in strawberry (*Fragaria* x *ananassa*). *3 Biotech, 10*, 353.

Nesbitt, T. C., & Tanksley, S. D., (2001). *fw2.2* directly affects the size of developing tomato fruit, with secondary effects on fruit number and photosynthate distribution. *Plant Physiology, 127*, 575–583.

Perretant, M. R., Cadalen, T., Charmet, G., Sourdille, P., Nicolas, P., Boeuf, C., Tixier, M. H., et al., (2000). QTL analysis of bread-making quality in wheat using a doubled haploid population. *Theoretical and Applied Genetics, 100*, 1167–1175.

Prasad, K. N., Kumar, A., Kochupillai, V., & Cole, W. C., (1999). High doses of multiple antioxidant vitamins, essential ingredients in improving the efficacy of standard cancer therapy. *Journal of American College of Nutrition, 18*, 13–25.

Primomo, V. S., Poysa, V., Ablett, G. R., Jackson, C. J., Gijzen, M., & Rajcan, I., (2005). Mapping QTL for individual and total isoflavone content in soybean seeds. *Crop Science, 45*, 2454–2464.

Pryor, W. A., (2000). Vitamin E and heart disease, basic science to clinical intervention trials. *Free Radical Biology and Medicine, 28*, 141–164.

Qi, M., Hettiarachchy, N. S., & Kalapathy, U., (1997). Solubility and emulsifying properties of soy protein isolates modified by pancreatin. *Journal of Food Science, 6*, 110–115.

Radovanovic, N., & Cloutier, S., (2003). Gene-assisted selection for high molecular weight glutenin subunits in wheat doubled haploid breeding programs. *Molecular Breeding, 12*, 51–59,

Ramalingam, J., Raveendra, Ch., Savitha, P., Vidya, V., Chaithra, T. L., Velprabakaran, S., Saraswathi, R., et al., (2020). Gene pyramiding for achieving enhanced resistance to bacterial blight, blast, and sheath blight diseases in rice. *Frontiers in Plant Science, 11*, 591457.

Reiprecht, Y., Poysa, V. W., Yu, K., Rajcan, I., Ablett, G. R., & Pauls, K. P., (2006). Seed and agronomic QTL in low linolenic acid, lipoxygenase-free soybean (*Glycine max* (L.) Merrill) germplasm. *Genome, 49*, 1510–1527.

Rhee, K. C., (1994). Functionality of soy proteins. In: Hettiarchchy, N. A., & Ziegler, G. R., (eds.), *Protein Functionality in Food Systems* (pp. 311–324). Marcel Dekker, New York.

Rhodes, D. H., Hoffmann, L. J., Rooney, W. L., Ramu, P., Morris, G. P., & Kresovich, S., (2014). Genome-wide association study of grain polyphenol concentrations in global sorghum [*Sorghum bicolor* (L.) Moench] germplasm. *Journal of Agricultural Food Chemistry, 62*, 10916–10927.

Rhodes, D. H., Hoffmann, L., Rooney, W. L., Herald, T. J., Bean, S., Boyles, R., Brenton, Z. W., & Kresovich, S., (2017). Genetic architecture of kernel composition in global 706 sorghum germplasm. *BMC Genomics, 18*, 15.

Ribaut, J. M., & Ragot, M., (2006). Marker-assisted selection to improve drought adaptation in maize: The backcross approach, perspectives, limitations, and alternatives. *Journal of Experimental Botany, 58*, 351–360.

Richardson, K. L., Vales, M. I., Kling, J. G., Mundt, C. C., & Hayes, P. M., (2006). Pyramiding and dissecting disease resistance QTL to barley stripe rust. *Theoretical and Applied Genetics, 113*, 485–495.

Ritter, K. B., Jordan, D. R., Chapman, S., Godwin, I. D., Mace, E. S., & McIntyre, C. L., (2008). Identification of QTL for sugar-related traits in a sweet 3 grain sorghum (*Sorghum bicolor* L. Moench) recombinant inbred population. *Molecular Breeding, 22*, 367–384.

Sacco, A., Di Matteo, A., Lombardi, N., Trotta, N., Punzo, B., Mari, A., & Barone, A., (2013). Quantitative trait loci pyramiding for fruit quality traits in tomato. *Molecular Breeding, 31*, 217–222.

Saliba-Colombani, V., Causse, M., Langlois, D., Philouze, J., & Buret, M., (2001). Genetic analysis of organoleptic quality in fresh market tomato. 1. Mapping QTLs for physical and chemical traits. *Theoretical and Applied Genetics, 102*, 259–272.

Santos-Bue, L. C., & Scalbert, A., (2000). Proanthocyanidins and tannin-like compounds–nature, occurrence, dietary intake and effects on nutrition and health. *Journal of Science, Food and Agriculture, 80*, 1094–1117.

Sarika, K., Hossain, F., Muthusamy, V., Zunjare, R. U., Baveja, A., Goswami, R., Bhat, J. S., et al., (2018). Marker-assisted pyramiding of *opaque2* and novel *opaque16* genes for further enrichment of lysine and tryptophan in sub-tropical maize. *Plant Science, 272*, 142–152.

Schäfer-Pregl, R., Ritter, E., Concilio, L., Hesselbach, J., Lovatti, L., Walkemeier, B., Thelen, H., et al., (1998). Analysis of quantitative trait loci (QTLs) and quantitative trait alleles (QTAs) for potato tuber yield and starch content. *Theoretical and Applied Genetics, 97*, 834–846.

Schönhals, E. M., Ortega, F., Barandalla, L., Aragones, A., Ruiz De Galarreta, J. I., Liao, J. C., Sanetomo, R., et al., (2016). Identification and reproducibility of diagnostic DNA markers for tuber starch and yield optimization in a novel association mapping population of potato (*Solanum tuberosum* L.). *Theoretical and Applied Genetics, 129*, 767–785.

Shang, L. G., Abduweli, A., Wang, Y. M., & Hua, J. P., (2016). Genetic analysis and QTL mapping of oil content and seed index using two recombinant inbred lines and two backcross populations in upland cotton. *Plant Breeding, 135*(2), 224–231.

Shi, Y., Li, W., Li, A., Ge, R., Zhang, B., Li, J., Liu, G., et al., (2015). Constructing a high-density linkage map for *Gossypium hirsutum* × *Gossypium barbadense* and identifying QTLs for lint percentage. *Journal of Integrated Plant Biology, 57*(5), 450–467.

Shi, Y., Liu, A., Li, J., Zhang, J., Li, S., Zhang, J., Ma, L., et al., (2020). Examining two sets of introgression lines across multiple environments reveals background-independent and stably expressed quantitative trait loci of fiber quality in cotton. *Theoretical and Applied Genetics, 133*, 2075–2093.

Shiringani, A. L., Frisco, M., & Friedt, W., (2010). Genetic mapping of QTLs for sugar-related traits in a RIL population of *Sorghum bicolor* L. moench. *Theoretical and Applied Genetics, 121*, 323–336.

Si, Z., Chen, H., Zhu, X., Cao, Z., & Zhang, T., (2017). Genetic dissection of lint yield and fiber quality traits of *G. hirsutum* in *G. barbadense* background. *Molecular Breeding, 37*, 9.

Śliwka, J., Sołtys-Kalina, D., Szajko, K., Wasilewicz-Flis, I., Strzelczyk-Żyta, D., Zimnoch-Guzowska, E., Jakuczun, H., & Marczewski, W., (2016). Mapping of quantitative trait loci for tuber starch and leaf sucrose contents in diploid potato. *Theoretical and Applied Genetics, 129*, 131–140.

Smith, C., & Simpson, S. P., (1986). The use of genetic polymorphisms in livestock improvement. *Journal of Animal Breeding and Genetics, 103* 205–217.

Snape, J., Fish, L., Leader, D., Bradburne, R., & Turner, A., (2005). The impact of genomics and genetics on wheat quality improvement. *Turkish Journal of Agriculture and Forestry, 29*, 97–103.

Soller, M., (1978). The use of loci associated with quantitative effects in dairy cattle improvement. *Animal Production, 27*, 133–139.

Sołtys-Kalina, D., Szajko, K., Sierocka, I., Sliwka, J., Strzelczyk-Zyta, D., Wasilewicz-Flis, I., Jakuczun, H., et al., (2015). Novel candidate genes *AuxRP* and *Hsp90* influence the chip color of potato tubers. *Molecular Breeding, 35*, 224–236.

Sowokinos, J., (2001). Biochemical and molecular control of cold induced sweetening in potatoes. *American Journal of Potato Researches, 78*, 221–236.

Sturer, C. W., Moll, R. H., Goodman, M. M., Schaffer, H. E., & Weir, B. S., (1980). Allozyme frequency changes associated with selection for increased grain yield in maize (*Zea mays* L.). *Genetics, 95*, 225–236.

Sun, Y. D., Liang, Y., Wu, J. M., Li, Y. Z., Cui, X., & Qin, L., (2012) Dynamic QTL analysis for fruit lycopene content and total soluble solid content in a *Solanum lycopersicum* × *S. pimpinellifolium* cross. *Genetics and Molecular Research, 11*(4), 3696–3710.

Tanksley, S. D., (2004). The genetic, developmental, and molecular bases of fruit size and shape variation in tomato steven. *The Plant Cell, 16*, 181–189.

Tanksley, S. D., Medino-Filho, H., & Rick, C. M., (1981). The effect of isozyme selection on metric characters in an interspecific backcross of tomato, Basis of an early screening procedure. *Theoretical and Applied Genetics, 60*, 291–296.

Tarán, B., Buchwaldt, L., Tullu, A., Banniza, S., Warkentin, T. D., & Vandenberg, A., (2003). Using molecular markers to pyramid genes for resistance to *Ascochyta* blight and anthracnose in lentil (*Lens culinaris* Medik.). *Euphytica, 134*(2), 223–230.

Tarpley, L., & Vietor, D., (2007). Compartmentation of sucrose during radial transfer in mature sorghum culm. *BMC Plant Biology, 7*, 33.

Terasawa, Y., Ito, M., Tabiki, T., Nagasawa, K., Hattam, K., & Nishio, Z., (2016). Mapping of a major QTL associated with protein content on chromosome 2B in hard red winter wheat (*Triticum aestivum* L.). *Breeding Science, 66*, 471–480.

Thorup, T. A., Tanyolac, B., Livingstone, K. D., Popovsky, S., Paran, I., & Jahn, M., (2000). Candidate gene analysis of organ pigmentation loci in the *Solanaceae*. *Proceeding of the National Academy of Science, USA, 97*, 11192–11197.

Tulipani, S., Mezzetti, B., Capocasa, F., Bompadre, S., Beekwilder, J., De Vos, C. H., Capanoglu, E., Bovy, A., & Battino, M., (2008). Antioxidants, phenolic compounds, and nutritional quality of different strawberry genotypes. *Journal of Agriculture and Food Chemistry, 56*, 696–704.

Turner, A. S., Bradburne, R. P., Fish, L., & Snape, J. W., (2004). New quantitative trait loci influencing grain texture and protein content in bread wheat. *Journal of Cereal Science, 40*, 51–60.

Uauy, C., Distelfeld, A., Fahima, T., Blechl, A., & Dubcovsky, J., (2006). A *NAC* gene regulating senescence improves grain protein, zinc, and iron content in wheat. *Science, 314*, 1298–1301.

Ubayasena, L., Bett, K., Taran, B., & Warkentin, T., (2011). Genetic control and identification of QTLs associated with visual quality traits of field pea (*Pisum sativum* L.). *Genome, 54*, 261–272.

Vallarino, J. G., Pott, D. M., Cruz-Rus, E., Miranda, L., Medina-Minguez, J. J., Valpuesta, V., Fernie, A. R., et al., (2019). Identification of quantitative trait loci and candidate genes for primary metabolite content in strawberry fruit. *Horticulture Research, 6*, 4.

Van De, W. M. H., Jacobsen, E., & Visser, R. G., (2001). Multiple allelism as a control mechanism in metabolic pathway, GBSSI allelic composition affect s the activity of granule-bound starch synthase I and starch composition in potato. *Molecular Genetics and Genomics, 265*(6), 1011–1021.

Van, D. K. E., & Tanksley, S. D., (2003). The making of a bell pepper-shaped tomato fruit, identification of loci controlling fruit morphology in yellow stuffer tomato. *Theoretical and Applied Genetics, 107*, 139–147.

Van, D. K. E., Lippman, Z. B., & Tanksley, S. D., (2002). Extremely elongated tomato fruit controlled by four quantitative trait loci with epistatic interactions. *Theoretical and Applied Genetics, 104*, 241–247.

Verma, S., Zurn, J. D., Salinas, N., Mathey, M. M., Denoyes, B., Hancock, J. F., Finn, C. E., et al., (2017). Clarifying sub-genomic positions of QTLs for flowering habit and fruit quality in U.S. strawberry (*Fragaria×ananassa*) breeding populations using pedigree-based QTL analysis. *Horticulture Research, 4*, 17062.

Walsh, D. J., Cleary, D., McCarthy, E., Murphy, S., & FitzGerald, R. J., (2003). Modification of the nitrogen solubility properties of soy protein isolate following proteolysis and transglutaminase cross-linking. *Food Research International, 36*, 677–683.

Wang, H., Huang, C., Guo, H., Li, X., Zhao, W., Dai, B., Yan, Z., & Lin, Z., (2015). QTL mapping for fiber and yield traits in upland cotton under multiple environments. *PLoS One, 10*, e0130742.

Wang, H., Huang, C., Zhao, W., Dai, B., Shen, C., Zhang, B., Li, B., & Lin, Z., (2016). Identification of QTL for fiber quality and yield traits using two immortalized backcross populations in upland cotton. *PLoS One, 11*(12), e0166970.

Wang, P., Xing, Y., Li, Z., & Yu, S., (2012). Improving rice yield and quality by QTL pyramiding. *Molecular Breeding, 29*, 903–913.

Weaver, C. A., Hamaker, B. R., & Axtell, J. D., (1998). Discovery of grain sorghum germ plasm with high uncooked and cooked *in vitro* protein digestibility's. *Cereal Chemistry, 75*(5), 665–670.

Werij, J., Furrer, H., Van, E. H., Visser, R., & Bachem, C., (2012). A limited set of starch related genes explain several interrelated traits in potato. *Euphytica*, 1–16.

Werij, J., Kloosterman, B., Celis-Gamboa, C., De Vos, C., America, T., Visser, R., & Bachem, C., (2007). Unravelling enzymatic discoloration in potato through a combined approach of candidate genes, QTL, and expression analysis. *Theoretical and Applied Genetics, 115*, 245–252.

Wickneswari, R., Bhuiyan, M., Kalluvettankuzhy, S., Lim, L., Thomson, M., Narimah, K., & Abdullah, Z., (2012). Identification and validation of quantitative trait loci for agronomic traits in advanced backcross breeding lines derived from *Oryza rufipogon* × *Oryza sativa* cultivar MR219. *Plant Molecular Biology Report, 30*, 929–939.

Winn, J. A., Esten, M. R., Robbins, A. L., Rooney, W. L., & Hays, D. B., (2009). QTL mapping of a high protein digestibility trait in *Sorghum bicolor*. *International Journal of Plant Genomics, 2009*, 1–6.

Wittaker, J. C., Curnow, R. N., Haley, C. S., & Thompson, R., (1995). Using marker-maps in marker-assisted selection. *Genetic Research, 66*, 255–265.

Wrolstad, R. E., & Shallenberger, R. S., (1981). Free sugars and sorbitol in fruits – a compilation from the literature. *Journal of Association of Official Analytical Chemists, 64*, 91–103.

Wrolstad, R. E., Putnam, T. P., & Varseveld, G. W., (1970). Color quality of frozen strawberries, effect of anthocyanin, pH, total acidity and ascorbic acid variability. *Journal of Food Science, 35*, 448–452.

Wu, M., & Kubota, C., (2008). Effects of high electrical conductivity of nutrient solution and its application timing on lycopene, chlorophyll and sugar concentrations of hydroponic tomatoes during ripening. *Scientia Horticulturae, 116*, 122–129.

Xie, D., Han, Y., Zeng, Y., Chang, W., Teng, W., & Li, W., (2012). SSR- and SNP-related QTL underlying linolenic acid and other fatty acid contents in soybean seeds across multiple environments. *Molecular Breeding, 30*, 169–179.

Xie, W., Feng, Q., Yu, H., Huang, X., Zhao, Q., Xing, Y., Yu, S., Han, B., & Zhang, Q., (2010). Parent-independent genotyping for constructing an ultrahigh-density linkage map based on population sequencing. *Proceeding of National Academy of Science, USA., 107*, 10578–10583.

Xu, Y., & Crouch, J. H., (2008). Marker-assisted selection in plant breeding: From publications to practice. *Crop Science, 48*, 391–407.

Yamamori, M., & Quynh, N. T., (2000). Differential effects of Wx-A1, -B1 and -D1 protein deficiencies on apparent amylose content and starch pasting properties in common wheat. *Theoretical and Applied Genetics, 100*, 32–38.

Yamamori, M., Nakamura, T., Endo, T., & Nagamine, T., (1994). Waxy protein deficiency and chromosomal location of coding genes in common wheat. *Theoretical and Applied Genetics, 89*, 179–184.

Yang, X., Wang, Y., Zhang, G., Wang, X., Wu, L., Ke, H., Liu, H., & Ma, Z., (2016). Detection and validation of one stable fiber strength QTL on C9 in tetraploid cotton. *Molecular Genetics and Genomics, 291*, 1625–1638.

Yates, H. E., Frary, A., Doganlar, S., Frampton, A., Eannetta, N. T., Uhlig, J., & Tanksley, S. D., (2004). Comparative fine mapping of fruit quality QTLs on chromosome 4 introgressions derived from two wild tomato species. *Euphytica, 135*, 283–296.

Ye, Z. H., Lu, Z. Z., & Zhu, J., (2003). Genetic analysis for developmental behavior of some seed quality traits in upland cotton (*Gossypum hirsutum* L.). *Euphytica, 129*, 183–191.

Yoo, S. H., & Jane, J. L., (2002). Structural and physical characteristics of waxy and other wheat starches. *Carbohydrate Polymers, 49*, 297–305.

Yu, J. W., Yu, S. X., Fan, S. L., Song, M. Z., Zhai, H. H., Li, X. L., & Zhang, J. F., (2012). Mapping quantitative trait loci for cottonseed oil, protein and gossypol content in a

Gossypium hirsutum x Gossypium barbadense backcross inbred line population. *Euphytica, 187*(2), 191–201.

Zeng, M., Morris, C. F., Batty, I. L., & Wrigley, C. W., (1997). Sources of variation for starch gelatinization, pasting and gelation properties in wheat. *Cereal Chemistry, 74*, 63–71.

Zhang, Z., Li, J., Muhammad, J., Cai, J., Jia, F., & Shi, Y., (2015). High resolution consensus mapping of quantitative trait loci for fiber strength, length and micronaire on chromosome 25 of the upland cotton (*Gossypium hirsutum* L.). *PloS One, 10*, e0135430.

CHAPTER 13

Recent Advances in Plant Breeding for Pulse Improvement

ANKITA MISHRA, D. B. K. V. MANI, and MANASI DASH

Department of Plant Breeding and Genetics, College of Agriculture, OUAT, Bhubaneswar, Odisha, India

ABSTRACT

Plant breeding has always played a vital role in crop improvement and maintaining food security all over the world. However, in the current scenario, where there is less control over the increasing population, it is becoming very difficult to manage the food quality and quantity. In addition, crop growth is affected due to drought, salinity, heat, cold, and heavy metal stresses, thereby resulting in significant yield loss. Pulses have the capability to alleviate global malnutrition in developing and underdeveloped countries due to the presence of high amount of proteins, vitamins, and minerals. Considering the importance of pulses, conventional breeding methods have produced many improved varieties conferring to better yield, fine quality and resistance to biotic and abiotic stresses. Even though conventional breeding approaches has tremendous potential, the use of such methods is limited due to extended time period, lengthier selection time, laborious, linkage drag, and uncontrolled genetic recombination. In this context, molecular and genomics technologies can be a great supplement to the traditional plant breeding methods. In the last 5–10 years, pulse molecular breeding and genomics has shown dramatic progress. These approaches include marker-assisted selection (MAS), gene pyramiding, genomic wide association studies (GWAS), genome editing (GE), 'omics' techniques and speed breeding. MAS has assisted in the selection of plants at the juvenile stage and led to advanced

Smart Breeding: Molecular Interventions and Advancements for Crop Improvement. Kailash Chandra, Subhash Chand, Ravi Prakash Saini, and Rohit Sharma (Eds.)

marker-assisted backcrossing (MABC) and marker-assisted recurrent selection (MARS). Furthermore, gene pyramiding has led to the development of varieties with introgression of more than two genes governing one or more traits. Gene pyramiding has the potential of developing a 'super pulse' crop having all characteristics of an ideal ideotype. Recent advances in CRISPR/Cas GE tool have enabled precise editing in crop genome. Advances in 'omics' approaches like genomics, transcriptomics, phenomics, proteomics, metabolomics, and ionomics has enriched the understanding about the complex traits in pulses. These novel approaches help to find marker-trait-association and improve selection efficiency for economically important traits. In this chapter, we will discuss about the details of advanced breeding methods and genomics assisted breeding in major pulses and their future prospects for accelerated crop improvement.

13.1 INTRODUCTION

According to FAO, there are a total of 11 pulses cultivated in different parts of the world, covering an area of 59 Mha and account to 45 million tons (MT) (FAOSTAT, 2019). In India, the most important pulse crops are chickpea (38%), pigeon pea (16%), mung bean (12%), urd bean (10%), lentil (7%) and field pea (5%) (Solanki et al., 2011). The major production share of nearly 75% comes from the states of Madhya Pradesh, Maharashtra, Rajasthan, Uttar Pradesh and Andhra Pradesh. In the present scenario of climate change and increased population rate, it has become challenging to meet the national pulse requirement.

Plant breeding has played a significant role since its inception to meet the food and feed requirement of burgeoning human and livestock population, respectively, using extensive farm trials for varietal stability, uniformity, and adaptation (Chand et al., 2020). Conventional breeding has played a great part in pulse crop improvement. Backcrossing in pulse crop has been used to transfer many oligogenic traits. Besides, hybridization has led to the development of new varieties in different pulses. Despite the increased yield and quality using traditional tools, it has not been able to meet the demand of the ever-increasing global population. The breeding arena has substantially changed after the discovery of molecular markers and in turn became the greatest tool for selection at the genotypic level. The marker-assisted selection (MAS) approaches such as marker-assisted backcrossing (MABC), marker-assisted recurrent selection (MARS) and gene pyramiding are advance and important methodologies to assist plant breeding activities (Kumawat et al., 2020). Moreover, with the discovery of advance technologies in genotyping and phenotyping, there has been a substantial shift in pulse genomics and 'omics' assisted breeding. Furthermore, the study of genome, transcriptome, proteome, metabolome, and ionome of a crop has led to a better understanding of all the factors responsible for the expression of quantitative traits. The discovery of genome editing (GE) tools like TAL effector proteins (TALEN), zinc finger nucleases (ZNFs) or clustered regularly interspaced short palindromic repeats (CRISPR) has enabled to edit or modify plant genome (Kumar et al., 2021). Even with the futuristic tools, progress is slow due to the involvement of crossing between parents and then selfing to obtain homozygous lines. Therefore, scientists have tried to reduce the crop generation time using new breeding methods, i.e., speed breeding. In this method, the environmental conditions around the crop are artificially regulated by adjusting the light intensity, photoperiod duration, temperature, and humidity, and resulting into early flowering by hormonal spray and shortening of maturity period. In this chapter, therefore, we have made an attempt to bring together and organize the updated literature for pulse crop improvement from traditional breeding methods to the modern gene editing techniques.

13.2 CLASSICAL BREEDING

Classical methods of breeding still play a crucial role in pulse crop improvement programs. Presence of optimum genetic diversity and selection process are the two essential criteria in conventional plant breeding methods (Khatik et al., 2021). Mostly pulse breeders use mass selection, pedigree selection, bulk selection, single seed decent and backcross (BC) methods to improve the designated traits.

Mass selection is the most commonly used breeding method for both self and cross-pollinated crops. In this method, plants having similar phenotypes are selected and their seeds are bulked. Selection is done for easily observable phenotypic characters and the variety developed by mass selection is a mixture of different pure lines which possess considerable genetic variation. For example, 'Kalu' variety of black gram (1989), 'Dhanush' and 'Trishul' of chickpea (1979) and 'Sindur' of lentil (1979) were developed by using mass selection (Joshi et al., 2017). In addition, pure line selection is applicable for only self-pollinated crops. For example, 'Akash' (1990) of cowpea and

'Bageswori' and 'Rampur Arahar-1' (1991) of pigeon pea were the varieties developed by using pureline selection (Joshi et al., 2017). IPA 206 variety of pigeon pea was developed by crossing between Bahar and AC 314-314 using the pedigree method for high yield (Singh et al., 2019).

In chickpea, efforts are being made for pyramiding of race-specific resistant genes to provide broad-spectrum resistance against fusarium wilt (FW). Studies on inheritance indicated that three loci are controlling resistance to race 1 (Singh et al., 1987) as well as to race 2 (Kumar, 1998). In another study, two recessive genes, *'er1'* and *'er2'* and two independent dominant genes, *'Pm1'* and *'Pm2'* were identified for imparting resistance against powdery mildew (PM) in pea and mung bean, respectively. The genetic studies on PM helped in developing stable resistant cultivars in mung bean, urd bean and pea. Prominent cultivars of urd bean have also been developed combining resistance with high yield (LBG 17, LBG 402, CO 5, WBU 108, and WBU 26). Among them, LBG 17 has revolutionized rice fallow – urd bean cultivation coastal districts of Andhra Pradesh (Krishnaiah et al., 1978).

BC method is predominantly used to transfer mono/oligogenic characters governed by one or few genes like disease resistance, male sterility, etc., from donor to recipient parent. The F_1 and the subsequent selected progenies are repeatedly back crossed to the recipient parent to recover its whole genome with trait of interest (Kumawat et al., 2020). The resultant lines are called near isogenic lines (NILs) and could be generated by 6–8 generations of repeated back crossing. In lentil, crossing between the wild relative (*Lens. ervoides*-a donor) and cultivar LR-59-81 (recipient-recurrent parent) resulted into advanced BC population or near isogenic lines, that are resistant to anthracnose (*Colletotrichum lentis*) and Stemphylium blight (*Stemphylium botryosum*) disease (Gela et al., 2021a).

Natural variability in crops has been exhausted since the green revolution due to the development of pureline cultivars and improved inbred lines, therefore, new genetic variability has to be created for further improvement. To achieve desired genetic variability, hybridization or crossing between genetically dissimilar or distantly related individuals is the best method. Hybridization can be easily done between two different varieties of the same crop (inter-varietal hybridization), with some difficulties between two species of a crop (inter-specific hybridization) or with extreme difficulties between two different genus (inter-generic hybridization). In pigeon pea, a hybrid 'GTH 1,' having 27% yield superiority over the traditional cultivars has been developed by introgression of a CMS trait (Nadarajan et al., 2010). On the other hand, a cross between *Vigna radiata* and *Vigna trilobata* resulted into decreased hybrid germination, hybrid pollen fertility, hybrid lethality (up to 53.84%), and hybrid breakdown (46.15%) (Pandiyan et al., 2012). A cross combinations between the three lines (adzuki bean – *V. angularis*, black gram – *V. mungo*, ricebean-*V. umbellata*) with mung bean (*V. radiata*) was carried out to determine the cross barriers to hybridization and asses the possibility to overcome those barriers. Out of all cross combinations, the reciprocal cross between adzuki bean and ricebean yielded complete fertile off-spring while in the remaining crosses, methods were used to overcome hybridization barriers (Chang et al., 1976). Many pulse varieties have been developed using traditional breeding methods (Table 13.1).

13.3 MUTATION BREEDING

Mutations have played a significant role in crop evolution since long back and are considered to be the paramount source of new species origin. It is now well established that this technique coupled with hybridization is an important source to create new variation for the improvement of agricultural crops. Mutation breeding dates back when X-rays was discovered by Rontgen in 1895 which led to generate genetic changes in fruit fly (*Drosophila melanogaster*) by Muller (1927) and in barley by Stadler (1928). In pulses, both physical (x-ray, gamma ray, etc.) and chemical (ethyl methane sulfonate (EMS), methyl methane sulfonate (MMS), hydrazine hydrate, etc.) mutagens are used to generate variability. The past research findings indicate that induced mutations in small seeded crops is effective by EMS, whereas in bold seeded types gamma rays are effective (Pathania et al., 2006). The frequency and spectrum of receiving morphological mutants such as dwarf plant, compact branching, prostrate, gigas, white flower, non-flowering, and sterile is high with EMS than hydrazine hydrate (HZ) (Solanki et al., 2011). The EMS and gamma rays induce high chlorophyll mutations (Barshile et al., 2006) as well as glabrous mutants (Pundir et al., 1989). Moreover, both EMS and gamma rays can be used to induce mutants with increased seed size and shape (Wani et al., 2008). The HZ is found to be highly effective followed by sodium azide (SA) and EMS with respect to seedling injury, pollen sterility and meiotic abnormalities (Khan et al., 2005). Mutation induction in three varieties (Nirmal, Biol-212 and Berhampur Local) of grass pea (*Lathyrus sativus* L.) using gamma rays suggested that optimum dose of mutagen can be effective for induction of large-scale mutations in grass pea (Singh et al., 2019).

TABLE 13.1 Recently Developed Pulse Varieties Through Classical Breeding Approaches

Crop	Variety	Released Year	Parentage	Characteristics	References
Urd bean	**VBN 9** and **VBN 10**	2019	Mash 114/Vamban 3; Vamban 1/UH 04-04	Resistant to mung bean yellow mosaic virus (MYMV), urd bean leaf crinkle virus, leaf curl virus and powdery mildew.	NPRC, Vamban (2019)
Mung bean	**VBN 4**	2019	PDM 139/BB 2664	Moderately resistance to MYMV.	NPRC, Vamban (2019)
Cowpea	**VBN 3**	2018	TLS 38/VCP 16-1	Multiple resistance to bean common mosaic virus, rust, and anthracnose.	NPRC, Vamban (2019)
Field pea	**IPFD 18-11** and **IPFD 18-14**	–	DDR 30/VRP 22	Flowering less than 36 days.	Parihar et al. (2021)
Pigeon pea	**IPH 15-03**	–	PA 163A/AK 250189R	Higher grain yield, early maturation and high seed weight.	Bohra et al. (2019)

The first Indian mutant chickpea variety 'Kiran' (RSG-2) was released (1984), followed by 'Ajay,' 'Atul' and 'Girnar' (1985). After these, gamma rays have been used in the generation of disease resistant variety, i.e., 'CM 98' against Ascochyta blight and wilt in chickpea (Haq et al., 1999). In Pigeon pea, streptomycin sulfate and SA have been used effectively to develop male sterile plants (Pandey et al., 1996). The mutagens like gamma rays, EMS and epichlorohydrin caused genetic variation with significant increase in pod number per plant, seeds per pod, 100-seed weight and seed yield (SY) in green gram (Singh et al., 2001). Also, gamma irradiation was reported to increase the protein content with SY in mung bean (Chakraborty et al., 1998). The EMS with increasing concentration showed a consistent reduction in plant height, days to 50% flowering and days to maturity, whereas it showed the opposite trend on fertile branches, pods per plant and height of plant (Wani et al., 2004). Mutagen such as gamma rays and EMS used in black gram had induced early mutants with increased number of pods, seeds per pod, 100-seed weight and protein content (Sharma et al., 2007). Gamma rays at the dose of 15 kR can induce xantha, albo-xantha, and trigina types of chlorophyll mutants in lentil (Paul et al., 2002). Similarly, in field pea, gamma rays induce different types of chlorophyll mutants like Albina, xantha, and chlorine type (Paul & Singh, 2005). In India, a large number of pulse varieties possessing higher yield and other desirable characteristics have been developed using different mutagens (Table 13.2).

A significant development has been made to improve different traits in many pulse crops through conventional approaches; however, it has numerous limitations. The major constraints are – labor intensive and time consuming. The hybridization process becomes complicated due to the innate fertilization barrier in crops, and the gene introgression takes more than six generation to achieve the desired genotype. The major setback of wild gene introgression through backcrossing is linkage drag where the undesirable characters of donor parents also introgressed into the genetic makeup of the recipient parent. Moreover, mutagenesis can be hazardous and oncogenic to the handler of mutagens. With these major limitations, traditional plant breeding methods need assistance from innovative methods to accelerate the breeding cycle with assured accuracy to meet the global pulse demand.

13.4 MOLECULAR BREEDING

In MAS, desired plants are selected based on molecular markers closely linked to the gene of interest (Kumawat et al., 2020). In classical breeding,

TABLE 13.2 Details of Pulse Varieties Developed Through Mutation Breeding (IAEA, Mutant Variety Database)

Name of Variety	Year of Release	Center of Research	Parent Variety	Mutagenic Agent	Area Recommended	Varietal Characteristics	References
Chickpea							
Kiran (RSG-2)	1984	ARS Durgapura, Jaipur	RS-10	Fast neutron	–	High number of pods, early maturity, high yield with salinity tolerance	Solanki et al. (2011)
Ajay (Pusa-408)	1985	IARI, New Delhi	G-130	60 kR, γ-rays	–	High yielding with 140–150 days maturity	Solanki et al. (2011)
Atul (Pusa-413)					Northeast	High yielding, wilt resistant, 130–140 days maturity	
Girnar (Pusa-417)	1985	IARI, New Delhi	BG-203	60 kR, γ-rays	–	High yielding, wilt resistant, semi-erect, dwarf, 110–130 days maturity	Solanki et al. (2011)
BGM-547	2005	IARI, New Delhi	BG-256	300 Gy	Northwest Plain zone	High yielder, mature within 135 days, tolerant to wilt, root rot and stunt diseases	Solanki et al. (2011)
Pusa-547	2006	I.A.R.I., New Delhi	BG-256	–	–	Bold seeds with thin testa as well as good cooking quality	Kharkwal et al. (2008)
BGM-547		Variety registration and certification centers	BG-256	–	–	Tolerant to wilt, root rot and stunt diseases	
TAEK-SAGEL			AK 71114	–	–	Ascochyta blight resistance and better quality	Sagel et al. (2009)
CM-2008	2008	NIAB, Pakistan	Punjab-1	–	–	Improved seed size, resistance to wilt	Shah et al. (2010)
Pigeon Pea							
Trombay Vishakha-1	1976	BARC, Bombay	T-21	Fast neutron	–	Increased seed size by 35% compared to parental line	BARC-NABD, Bioscience (2019)

TABLE 13.2 *(Continued)*

Name of Variety	Year of Release	Center of Research	Parent Variety	Mutagenic Agent	Area Recommended	Varietal Characteristics	References
CO-3	1977	TNAU, Coimbatore	CO-1	CO-1 0.6% EMS	Tamil Nadu	High yielding, bold seeded	Solanki et al. (2011)
CO-5	1984	TNAU, Coimbatore	CO-1	16 kR gamma rays	Tamil Nadu	Early type with photo-insensitive and drought tolerant	Solanki et al. (2011)
TAT-5	1984	PKV Akola and BARC, Mumbai	T-21	1.5 Krad Fast neutrons	–	Early type (140 days) with increased seed size	BARC-NABD, Bioscience (2019)
TAT-10	1985	PVK Akola and BARC, Mumbai	Cross of T-21 Mutants (TT-2 × TT-8)	2.5 Krad Fast neutron	Maharashtra	Extra early type (115–120 days) with medium large grains	BARC-NABD, Bioscience (2019)
TT-401	2007	BARC, Mumbai	ICPL84008 × TT-6	–	Central zone	Resistance to pod borer and pod fly, tolerant to fusarium	Danish et al. (2019)
TJT-501	2009	BARC and JNKV, Khargone	ICPL84008 × TT-6	–	Central zone	Tolerance to phytophthora blight, high yield, early maturity	Danish et al. (2019)
PKV-TARA	2013	PVK Akola and BARC, Mumbai	ICPL84008 × TT-6	–	Maharashtra	High yielding tolerant to fusarium wilt and sterility mosaic diseases	BARC-NABD, Bioscience (2019)
Mung Bean							
CO-4	1982	TNAU, Coimbatore	CO-1	20 kR Gamma rays	–	High yielder, 85 days maturity and drought tolerant	Solanki et al. (2011)

TABLE 13.2 *(Continued)*

Name of Variety	Year of Release	Center of Research	Parent Variety	Mutagenic Agent	Area Recommended	Varietal Characteristics	References
Pant M 2	1982	GBPAUA and Pantnagar	ML-26	10 kR Gamma rays	–	High yielder, resistant to YMV	Solanki et al. (2011)
TAP-7	1983	BARC, Mumbai, and PVK Akola	S-8	Gamma rays	Maharashtra, Karnataka	Early by 5–7 days and 23% high yield than Kopergaon	BARC-NABD, Bioscience (2019)
MUM 2	1992	–	K 851	–	Northeast Plain zone	Early, dwarf type, resistant to MYMV	Solanki et al. (2011)
BM 4	1992	–	T 44	–	Central zone	Bushy erect type, tolerant to MYMV and powdery mildew	Solanki et al. (2011)
TARM-2	1994	BARC, Mumbai, and PKV Akola	RUM-5 × TPM-1	–	Maharashtra	High yielding, resistant to powdery mildew	BARC-NABD, Bioscience (2019)
LGG 407	1995	RARS, Guntur, A.P.	Pant M 2	–	Andhra Pradesh and Gujarat	Tolerant to MYMV	Solanki et al. (2011)
TARM-18	1997	BARC, Mumbai, and PKV Akola	PDM54 × TARM-2	–	Maharashtra	Resistant to powdery mildew	BARC-NABD, Bioscience (2019)
OUM 11-5	2002	–	Mutant of Dhauli	–	South zone	Resistant to PM, MYMV, and CLS	Solanki et al. (2011)
TMB-37	2005	BARC, Mumbai, and ANGRAU	Kopergaon × TARM-2B	–	Northeast	Early maturing and tolerant to YMV.	BARC-NABD, Bioscience (2019)
TM-96-2 (Trombay Pesara)	2007				Plain zone	Resistant to powdery mildew, large seed size, suitable for rice fallows	
TJM-3	2007	BARC, Mumbai	Kopergoan × TARM-1B	–	Madhya Pradesh	Early maturity, large seeds, Resistance to YMV, Rhizoctonia root rot and powdery mildew.	Danish et al. (2019)

TABLE 13.2 *(Continued)*

Name of Variety	Year of Release	Center of Research	Parent Variety	Mutagenic Agent	Area Recommended	Varietal Characteristics	References
TM 2000-2 (Pairy mung)	2010	BARC, Mumbai	TARM-1 × JL-781	–	Chhattisgarh	Resistance to powdery mildew	Danish et al. (2019)
Urd Bean							
CO-4	1978	TNAU, Coimbatore	CO-1	0.02% MMS	–	Early type erects and determinate type	Solanki et al. (2011)
TAU-1	1985	PKV Akola and BARC, Mumbai	U-196 (mutant) × T-9	Gamma rays	–	High yield, large-seeded resistant to powdery mildew	BARC-NABD, Bioscience (2019)
Sarla	1985	–	Mutant of T 9	–	–	MYMV resistant	Solanki et al. (2011)
TAU-2	1992	BARC, Mumbai	–	–	Maharashtra	High yielding	BARC-NABD, Bioscience (2019)
TPU-4	1992	BARC, Mumbai, and MPKV, Rahuri	U-201 (mutant) × T-9	–	Central zone	Large seeded and high yielding	BARC-NABD, Bioscience (2019)
Vamban 2	1997	TNAU, Vamban	Mutant of T-9	–	Tamil Nadu	Pods glabrous, tolerant to drought	Solanki et al. (2011)
TU-94-2	1998	BARC, Mumbai	TPU3xTAU3 (mutations)		South zone	High yielding and resistant to YMV	BARC-NABD, Bioscience (2019)
DU-1	2007	–	–	–	–	High yield, resistance to insects	Danish et al. (2019)
TU-40	2013	BARC, Mumbai	TU-94-2 × *V. mungo* var *silvestris*	–	Tamil Nadu, Andhra Pradesh, Karnataka, and Odisha	Resistant to YMV and powdery mildew, high yield	BARC-NABD, Bioscience (2019)

TABLE 13.2 *(Continued)*

Name of Variety	Year of Release	Center of Research	Parent Variety	Mutagenic Agent	Area Recommended	Varietal Characteristics	References
Lentil							
Ranjan (S-256)	1981	Berhampur, Orissa	B-77	Radiation	West Bengal	High yielding, spreading type 110 days of maturity	Solanki et al. (2011)
PL 77-2	1984	Pantnagar	BR-25	–	–	Spreading type, high yielding tolerant to wilt and Ascochyta blight	
Rajendra Masoor 1	1996	–	–	Gamma rays	–	Cold tolerant	
Cowpea							
COCP 702	2002	–	–	–	–	Good quality with high yield	Danish et al. (2019)
TRC77-4 (Kalleshwari)	2007	BARC, Mumbai, and IGKVV, Raipur	Mutant of V-130	–	Chhattisgarh	Dual purpose, high yielding	Danish et al. (2019)
TC-901	2018	BARC, Mumbai	Mutant of EC394763	–	Gujarat, Rajasthan, and central zone	Early and synchronous maturity, tolerant to YMV and root-rot diseases, high yielding	BARC-NABD, Bioscience (2019)

selection is mainly based on morphological features of plants; however, these characters are known to be affected by the environment and hence do not provide a concrete basis for selection of superior plant types. With the availability of molecular markers and genetic maps, the scenario of plant breeding has completely changed over time. Molecular markers provide an advantage of being stable over environments and easy detectability. Molecular markers such as restriction fragment length polymorphisms (RFLP); amplified fragment length polymorphisms (AFLP); microsatellites or simple sequence repeats (SSR); random amplified polymorphic DNA (RAPD); cleaved amplified polymorphic sequences (CAPS); single strand conformation polymorphisms (SSCP); single nucleotide polymorphisms (SNPs), etc., help in selection of superior plants in a particular population. MAS can be used for traits that are governed not only by oligogenes but also polygenes. In breeding programs, several factors are responsible for their effective application (Francia et al., 2005) like availability of genetic map showing the *QTLs* for trait of interest, tight linkage between the marker and *QTL*, easy detection, effective cost and time. The steps in MAS are illustrated in Figure 13.1.

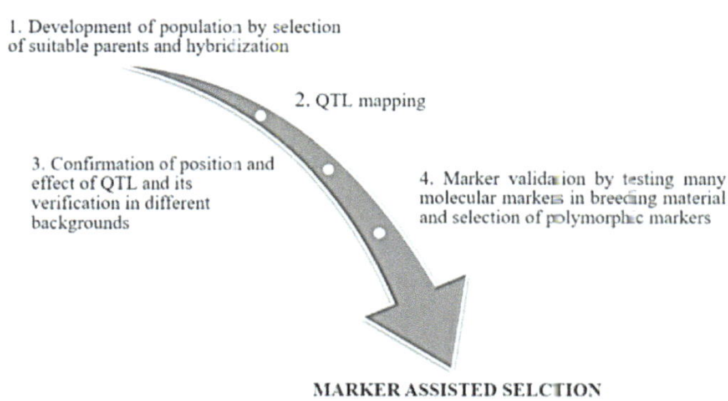

FIGURE 13.1 Steps in marker-assisted selection.

Source: Modified from: Collard & Mackill (2008).

13.4.1 IDENTIFICATION AND GENETIC DIVERSITY STUDY OF CULTIVAR WITH MARKERS

Molecular markers are being used to identify cultivars, *QTL* linked to traits of interest, genetic diversity study, purity assessment and many more. Level of pre-existing crop genetic diversity helps to decide the effective breeding programs to increase crop productivity over the existing cultivars. Hybridization requires parents with broad genetic base and thereby, there is a need to study the genetic diversity in the breeding materials to be used (Reif et al., 2005). There are numerous reports on diversity studies among breeding materials with the help of DNA markers.

In pigeon pea, Khoiriyah & co-workers (2016) divided the 30 gamma-irradiated seeds into two main clusters with a similarity coefficient ranging from 0.43 to 1.00. The RAPD markers were previously used in molecular studies and replaced by microsatellite and SNP markers recently. Later, diversity study using SSR was widely used in molecular breeding. Afzal & associates (2018) characterized chickpea genotypes using SRAP (Sequence-related amplified polymorphism) and SSR markers and showed eight genotypes having the highest level of genetic diversity. Similarly, 167 chickpea genotypes were categorized into two clusters using SSR and EST-SSR markers and grouped into four clusters on the basis of morphological attributes, hence, showed weak correlation between molecular and morphological findings (Seyedimoradi et al., 2019). However, Sachdeva & co-workers (2019) observed a strong correlation between the 2D plot of PCoA analysis and molecular dendrogram by using 125 SSR markers in 40 chickpea varieties. They found that 25 markers generated 90 alleles having average genetic diversity of 0.485 and PIC of 0.437. After SAHN (sequential agglomerative hierarchical non-overlapping) grouping, 29 varieties were grouped in cluster I and remaining in cluster II. Ahmad & associates (2017) used SCoT (start codon targeted) markers to classify 35 chickpea varieties into three major clusters and observed a weak relation between morphological and molecular findings. Kanavi & co-workers (2019) recorded genetic similarity range of 0.35–1.00 with maximum divergence between KKM-3 and TM-962 of green gram. The genotypes (55) were grouped into eight clusters and showed moderate level of diversity as green gram is self-pollinated in nature. Further, 30 cowpea genotypes were grouped into six clusters using 13 polymorphic microsatellite markers and genotypes CP 2-1, KB CP 39 and K 13 CP 25 showed highest dissimilarity (Devi et al., 2020).

13.4.2 LINKAGE MAPPING USING DNA MARKERS

The success of marker-assisted breeding programs is dependent on the identification of molecular markers linked to the QTL(s) of interest. To identify the tightly linked markers, it is utmost important to construct a linkage map of molecular markers. Hence, plants with contrasting traits are used as parents to develop a biparental mapping population like RILs, NILs, F_2, BC, and DH population. This population is genotyped with number of polymorphic markers and the marker-trait association (MTA) is calculated on basis of chi-square test and the genetic distance between trait-marker is determined (Kumar et al., 2009).

Ates & associates (2018a) developed a consensus map of lentil using RIL population and DArT (Diversity Arrays Technology) markers which covered 977.47 cM with an average distance of 0.10 cM between markers. In addition, it showed 7 LGs in 7 chromosomes in lentil. In green gram, linkage map comprising of 11 LGs was made for mung bean yellow mosaic Indian virus (MYMIV) resistance by using 224 molecular markers (Singh et al., 2018). Verma & co-workers (2015) constructed intra-specific linkage map in lentil from a cross between *L. culinaris* cv. Precoz and *L. culinaris* cv. L830 using SSRs. Among 647 markers, 216 were present on seven LGs covering a distance of 1183.7 cM with an average marker density of 5.48 cM. The intraspecific genetic map constructed in chickpea from four crosses viz., ICCV 96029/CDC Frontier, BGD 132/CDC Frontier, ICC 16541/CDC Frontier and ICC 5810/CDC Frontier had average distance of 248.8 cM, 331.4 cM, 311.1 cM and 385.1 cM, respectively. A consensus map traversing 363.8 cM with 109 loci was constructed by assimilating four genetic maps (Mallikarjuna et al., 2017). Similarly, interspecific linkage map in pigeon pea derived from interspecific F_2 progenies from a cross of *Cajanus cajan* cv. ICP-26 and *C. scarabaeoides* acc. ICPW-94 was constructed with 31 SCoT, 148 RAPD and 6 ISSR markers to study six trait loci. These 191 markers were assigned into 11 LGs, ranging from 320.63 and 43.38 cM covering 1624.71 cM of the genome with a mean marker interval of 8.51 cM (Sahu et al., 2015). To increase the accuracy of marker-gene association and to develop fine mapping, single nucleotide polymeric markers can be used. Recently a high-density genetic map of Chick pea was constructed using 3,818 SNPs, and mapped across eight different LGs with a large variation in marker density (Barmukh et al., 2021a).

13.4.3 QTL MAPPING

Identification of *QTL* region responsible for expression of a particular trait using linkage map is a major breakthrough in molecular breeding. The basic principle of this technique is the simultaneous segregation of marker locus and *QTL* based on the assumption that there is the linkage between the marker and *QTL* (Chandra et al., 2017). Statistical methods like single marker analysis (SMA), simple interval mapping (SIM) and composite interval mapping (CIM) are used to determine relation between markers and traits in linkage map and discover the QTLs/genes responsible for that particular trait (Kumawat et al., 2020). The genotypic and phenotypic data are fed into software like QTL cartographer, MapMaker, QGene, MapQTL, etc., to establish marker-trait relation (Kumar et al., 2011). Numerous research has been done to date for various traits in different pulse crop. In chickpea, a study was done to identify heat tolerance *QTL* using RIL lines developed from cross ICC 4567 (heat sensitive) and ICC 15614 (heat tolerant) using SNPs which resulted in discovery of four *QTLs* located in *CaLG05* (Paul et al., 2018). In another study six *QTLs* were identified for manganese concentration in seeds of 120 RILs of lentil with phenotypic variation of 15.3–24.1% and LOD score 3.00–4.42 (Ates et al., 2018b). These QTLs can be transferred to elite cultivars to develop micronutrient rich bio-fortified lentil genotypes. Two novel *QTLs* for seed hardness in *Phaseolus vulgaris* RIL population was detected in chromosome 1 and 2 (Sandhu et al., 2018). Two major *QTLs* (qLeaf_or-1, qLeaf_or-2) for pea leaf shape were traced on linkage group 7 (Zheng et al., 2018). Similarly, three *QTLs* for bean weevil resistance in common bean were detected using SNPs in chromosome Pv04 and Pv06 (Kamfwa et al., 2018). In chickpea, a high-density genetic map was developed for ascophyta blight (AB) resistance with the discovery of genome-wide SNPs. QTL mapping was done using RILs which led to the identification of eight *QTLs* for AB resistance on chromosomes 2, 3, 4, 5, and 6 (Deokar et al., 2019). Recently two *QTLs* for resistance to another biotic stress in lentil, i.e., lentil anthracnose (*Colletotrichum lentis*) were identified through *QTL* mapping of RIL population. These *QTLs* are present in chromosome 3 and 7 with phenotypic variation of 20.1–31.2% and 8.3–18.4%, respectively (Gela et al., 2021b). Nine *QTLs* with main-effect and 955 epistatic QTLs (E-QTLs) with 42.49% and 38.05% phenotypic variance, respectively, were identified for pod borer (*Helicoverpa armigera*) in chickpea with interspecific RIL population (Barmurk et al., 2021b). In pea, a region on chromosome 4 was identified for partial resistance to Aphanomyces root rot isolate *Ae-MDCR1* (Wu et al., 2021).

13.4.4 MARKER-ASSISTED BACKCROSSING (MABC)

The important *QTLs*/genes identified through *QTL* mapping can be transferred to desired genotypes that lacks the gene of interest. With the discovery of markers, MABC has become easier and more efficient to transfer a particular

QTL. The three general levels of marker-aided backcrossing are foreground, recombinant, and background selection. Foreground selection involves the selection of only the target gene with the help of flanking markers, background selection helps selecting the backcross (BC) progeny having highest percentage of recurrent parent genome (RPG) while recombinant selection involves both selection of the plants carrying the gene of interest and recombination events between target gene and its flanking markers (Collard et al., 2008; Hasan et al., 2015;).

MABC is an efficient method to transfer traits related to abiotic stress tolerance. For drought stress tolerance, the important trait is the root architecture. A chickpea variety ICC 4985 was used as a donor for a successful transfer of *QTL* linked to extensive rooting system, into varieties ICCV 92944 and ICCV 00108 (Kosgei et al., 2017). Varshney & associates (2013) led an accelerated introgression of '*QTL-hotspot*' for root traits conferring drought tolerance (DT) in chickpea from donor ICC 4985 into elite cultivar JG 11. Root trait phenotyping of these ILs showed higher rooting depth, better rooting density and higher root dry weight. To develop drought tolerant cowpea genotypes for intercropping, Batieno & co-workers (2016) introgressed *QTL* for DT, striga, and root-knot nematode resistance into Burkina Faso cultivar. In BC_3F_1, six families were selected after screening for desired traits, to produce seeds for the next generation. Recently, Bharadwaj & co-workers (2021) had successfully introgressed of '*QTL-hotspot*' of chickpea ICC 4958 harboring several *QTLs* for DT into three chickpea varieties viz., Pusa 372, Pusa 362 and DCP 92-3 and developed 61 introgression lines (20 BC_3F_3 in Pusa 372; 20 BC_2F_3 in Pusa 362, and 21 BC_3F_3 in DCP 92-3, with >90% RPG recovery.

MABC is also a very efficient method to transfer biotic stress related genes to produce resistant varieties, which will boost the crop yield under unfavorable conditions. FW and AB are major diseases leading to crop loss in chickpea. Varshney & co-workers (2014a) introgressed resistance to FW and AB into a high yielding cultivar of chickpea, i.e., 'C 214.' Two parallel backcrossing was done to transfer *Foc1* locus and two *QTLs*, *AB-QTL-I*, and *AB-QTL-II* into C 214 for FW and AB resistance, respectively. They recovered three resistant lines (with 92.7–95.2% RPG) to race 1 of FW, and seven resistant lines (81.7–85.40% RPG) to AB. In a similar way, two fusaria wilt resistant introgression lines were developed using MABC, i.e., 'Super annigeri 1' and 'JG 74315-14.' These are the superior lines of Annigeri 1 and JG 74, respectively, by transfer of genomic region of *WR 315* conferring resistance to FW (Mannur et al., 2019). Pratap & co-workers (2017) transferred gene *Foc2* (resistance gene of FW) into chickpea (donor variety

'Vijay' and recipient variety Pusa 256) and identified five highly resistant lines with *Foc2* gene.

Viruses are the most destructive pathogen to crop plants. In pea, pea seed-borne mosaic virus (PSbMV) resistant lines were recovered by transfer of recessive resistant gene *Sbm-1* from donor 'Lifter' (Scegura, 2017). After an attempt to identify QTLs for bacterial leaf blight (BLB) resistance in cowpea, Dinesh & associates (2016) located three major *QTLs*, one on LG 8 and other two on LG 11. The *QTL* with high phenotypic variation (30.58%) was on LG 8 and this was transferred by MABC from V-16 to C-152. This was the very first use of MAS in cowpea breeding. After being able to successfully transfer BLB resistance, Dinesh & co-workers (2018) transferred mosaic virus resistance in the same cowpea cultivar (C-152). They used V-57817 as donor parent and screened using SSR marker MA 15 linked to the resistance. They reported six progenies with the resistant gene with RPG recovery of 84.09 to 93.19%, i.e., almost same phenotype with C-152.

13.4.5 GENE PYRAMIDING

Marker-assisted gene pyramiding is the introgression of several genes altogether into a single dominant variety using molecular markers to sustain yield productivity and other relevant traits. Mostly, pyramiding is used to combine multiple disease resistance genes in one cultivar to create a broad base and durable resistance for many pathotypes. Two genes, *ra1* and *AbR1*, conferring resistance to AB and a *QTL* for anthracnose resistance was pyramided into a lentil line. Among the RILs developed, only 11 had all three resistant genes (Taran et al., 2003). The broad resistance against lepidopteron pests were observed in chickpea when the modified insecticidal genes, *cry1Ab* and *cry1Ac* of *Bacillus thuringiensis* were pyramided into chickpea and showed higher resistance against *Helicoverpa armigera* larvae as compared to level expression of a single gene (Mehrotra et al., 2011). Successful pyramiding of resistance genes to anthracnose (*Co-2* ^{C}, *Co-2* ^{A252} and *Co-3/9*) and Bean Common Mosaic (*I* and *bc-3*) was achieved in bean through six parallel backcrossing programs (Ferreira et al., 2012). Lines that are resistant to both angular leaf spot (ALS) and common bacterial blight (CBB) were developed in common bean by pyramiding of three genes (*Phg-2* gene for ALS resistance and two for CBB resistance) into cultivar 'Redwolaita' (Rezene et al., 2019). This method can be used within a short time due to the exclusive presence of high RGP. Improved mung bean breeding lines were developed with resistance to cercospora leaf spot (CLS) and PM having a high RGP recovery of 85 to 100% in the BC_2F_1 plants (Witsarut et al., 2019). Climbing bean populations targeting six major resistance genes for anthracnose (*Co-42*, *Co-5*), ALS (*Phg-2*), bean common mosaic virus (*I* and *bc3*) and *Pythium ultimum* root rots (P.ult) were developed by MABC. Even after third BC, phenotypic similarity between pyramided lines and recurrent parent was still low for the number of pods per plant (12.5%) and weight of 100 seeds (10.7%) (Okii, 2021). Whereas the seed color was fully recovered on some of the BC_3 progenies with combined resistance genes for BCMV and BCMN after pyramiding *I* and *bc-3* gene into Rwandan variety (Ruhimbana & Mutlu, 2019).

Multiple pathogen attack on common bean affects the yield of crop and sometimes leads to complete crop loss. Besides, marker-assisted gene pyramiding contributes to resistance against many pathogens at once. To access the efficiency of genetic pyramids produced from four-way cross in *Phaseolus vulgaris*, Okii & associates (2017) studied the agronomic characteristics and witnessed varying levels of resistance to foliar diseases (Anthracnose and BCMNV). Pyramided disease resistance genes did not affect yield related characters, excluding days to flowering and number of flower buds. Effectiveness of pyramided lines conferring resistance to anthracnose disease in common bean was tested against lines having a single gene. Gene *Co-42* and *Co-5* showed resistance to all races (352, 713, 767, and 2047), whereas *Co-43* and *Co-9* conferred resistance only to race 352. The pyramids *Co-42* + *Co-5* + *Co-9* and *Co-42* + *Co-5* had the lowest disease severity. However, *Co-9* wherever pyramided has shown an antagonistic effect. Therefore, pyramids other than those with *Co-9* showed resistance to the disease (Kiryowa et al., 2021). This also indicates that some genes have different effects in different genetic combinations.

13.5 'OMICS' TECHNOLOGY

Omics technology is a broad term that involves genomics, transcriptomics, proteomics, metabolomics, ionomics, phenomics, etc., which is an integrated multi-disciplinary technology. It is useful to understand the complex quantitative traits in a better way.

13.5.1 GENOMICS

Genomics-assisted breeding (GAB) is the use of genomic tools in crop improvement. The main objective is to determine the relationship between genotype and phenotype in breeding programs. It includes techniques like MABC, MARS, high density sequence-based mapping, association mapping (AM) and genomic selection (GS). For major effect *QTL*, the MABC method is efficient in transferring to elite cultivar whereas for small effect *QTL*, MARS is the best approach (Ribaut & Ragot, 2006). Rapid progress has been witnessed in pulse genomics in the last few years for pulse crop improvement with the advancement in high-throughput (HT) DNA sequencing and phenotyping techniques.

13.5.1.1 ASSOCIATION MAPPING (AM)

In recent years, the first ever AM was done for loci linked with FW resistance (*Foc2*) in chickpea using SSR markers. Three significant MTA for FW resistance was identified. There were three SSR markers, i.e., *CESSR433*, *NCPGR21* and *ICCM0284*, that could be used for developing FW-resistant chickpea varieties (Jha et al., 2021a). Similarly, the first report on identification of favorable alleles for *Fusarium udum* variant 2 resistances in pigeon pea was done using AM. They identified six SSR markers (*HASSRs8*, *18*, *30*, *121*, *128*, *174*) showing 6% variation, among which *HASSR18* was most stable and two of these were further validated through analysis of F_2 individuals (Patil et al., 2017). Association studies for heat tolerance resulted in identification of SSR markers, i.e., *ICCM0297*, *NCPGR150*, *TAA160*, and *NCPGR156* that showed significant MTA under both normal and late sowing conditions (Jha et al., 2021b). Likewise, a total of 32 MTAs was detected: six for chlorophyll content, two each for photochemical reflectance and canopy temperature, seven for stem length and nine for pod number (Tafesse et al., 2020). Around 60 accessions of cowpea were used to conduct AM for DT traits. Genome wide association studies (GWAS) allows high resolution mapping for identification of candidate gene/s for phenotypic traits in a large natural population, however, for identification of rare alleles in a population, it requires quite large population size with adequate variation. One *QTL* (*HB5.1*) was detected after GWAS for halo-blight resistance to race 6 in common bean which had significant influence on SY (Tock et al., 2017). Similarly, two significant MTA on chromosome Pv-02 was identified for anthracnose (25% variation) and one MTA for ALS in chromosome Pv-10 (19% phenotypic variation) in common bean (Fritsche-Neto et al., 2019). In another GWAS for seed quality traits in pea, Gali et al. (2019) found MTAs for plant height (*Chr5LG3_566189651* and *Chr5LG3_572899434*), lodging resistance (*Chr2LG1_409403647*), grain yield (*Chr1LG6_57305683* and *Chr1LG6_366513463*), seed starch content (*Chr1LG6_176606388*, *Chr2LG1_457185*, *Chr3LG5_234519042*, and *Chr7LG7_8229439*) and seed protein concentration (*Chr3LG5_194530376*). The MLM approach identified association of SSR markers, *PBALC 224* with grain diameter along with *GLLC 614* and *PBALC 29* for grain weight (GW) in lentil (Singh et al., 2019). A SNP locus (*Ca1_12310101*) on Ca1 associated with three traits, i.e., Phosphorus-use efficiency, dry weight of shoot, and shoot P content was identified (Thudi et al., 2021). Likewise, in another GWAS for root traits exposed many candidate loci, which could play an essential role in determining root hair length in low P condition for chickpea. Association peaks for root hair length in +P condition were seen mainly from chromosomes 5, 4, and 6 (Kohli et al., 2020). AM identified eight SNPs for Cu, P, and K concentration in chickpea seed (Ozkuru et al., 2019). Likewise, AM analysis in lentil identified three and four SSRs associated with grain Fe and Zn concentration, respectively (Singh et al., 2017). These identified *QTLs* for micronutrient concentration can be used in developing bio-fortified crops.

13.5.1.2 SEQUENCE-BASED HIGH DENSITY LINKAGE MAPPING

The rapid advancement in next-generation sequencing (NGS) has led to the development of high-density genetic maps with SNP markers by genotype-by-sequencing (GBS) technology. Even crops with no reference genome, GBS can be used for genotyping. Therefore, this is one of the best methods for identification of candidate gene/s. In the modern era, the GBS approach has been used to distinguish genome-wide SNPs in chickpea, measure genetic diversity, determination of population structure, and develop high-density linkage maps, *QTL* analysis, GWAS, and GS. The high-density linkage map developed by RAD-Seq has provided an idea about recombination frequency and hotspots availability across the genome of chickpea (Deokar et al., 2014). The first *QTL*-seq study in chickpea was conducted to determine gene expressing 100-seed weight by the use of extreme phenotype from an intraspecific mapping population of cross ICC 7184 and ICC 15061 (Das et al., 2015a). An intraspecific high density linkage map consisting of 839 SNPs spanning over a distance of 1088.04 cM with an average marker density of 1.30 cM was constructed to study genomic regions controlling seed Fe and Zn concentration in chickpea (Sab et al., 2020). GBS approach was used to identify and genotype SNPs in three mapping population of pigeon pea, i.e., two RILs and F_2. The study produced three high density maps having 964, 1101, and 557 SNPs with marker distance 1.16, 0.84, and 2.60 cM, respectively (Saxena et al., 2017). Moreover, this was the first report of *QTL* mapping for FW resistance in pigeon pea. In pigeon pea, Yadav et al. (2019) constructed a high density genetic and *QTL* map for high-selfing flower and seed quality traits with 6818 SNPs in 974 cM distance. They identified a consistent *QTL qCl3.2* for cleistogamous flower on CcLG03. A high-resolution map with 4867 SNPs was developed and a total of four *QTLs* for fertility restoration (Rf) in pigeon pea were identified (Saxena et al., 2020). A high-density genetic map based on 9,493 SNPs was constructed through SLAF sequencing in cowpea and discovered 59 SNPs that may be associated with fresh pod color trait located on LG6 (Li et al., 2021).

13.5.1.3 GENOMIC SELECTION (GS)

Many QTLs of small additive effects are not detected by MAS; hence, GS can be used for these QTLs. The GS was popularly used in animal breeding, but now it is being used for many crops breeding programs also. Because of the advancement in genotyping and phenotyping, GS is being implemented to improve genetic gain in crop plants. GS helps in calculation of genomic estimated breeding values (GEBVs) after phenotyping and genotyping the 'training population' (Meuwissen et al., 2001). GS uses a large number of markers information data across the genome to understand the genetic value of an individual (GEBV) by using appropriate statistical methods. For prediction of precise GEBV, Bayesian or BLUP methods are thought to be the best models (Zhang et al., 2010a, b). The advantage of this method is that it helps in effective selection, reduces breeding cycle along with high resource use efficiency that leads to high genetic gain (Bhat et al., 2018). Moreover, a combination of MAS and GS is recommended for crop improvement in pigeon pea (Varshney et al., 2014b). A collection of 320 chickpea breeding lines were phenotyped and genotyped using DArT markers and the data was used to estimate prediction estimate using six statistical GS models. GS model tested four characters, i.e., SY, 100-seeds weight, days to 50% flowering and days to maturity. Prediction accuracies varied from 0.138 (seed yield) to 0.912 (100-seed weight). Even kinship matrix results showed the existence of two different groups within selected lines (Roorkiwal et al., 2016). Inclusion of G×E interaction in GS models can improve the prediction accuracy which will help in the selection of lines in a particular environment. Morphological data of 320 chickpea lines were also recorded in two different environments for eight traits and genotyped using DArT and GBS and it was proved that prediction accuracies using DArT Seq was higher when compared with GBS (Roorkiwal et al., 2018). In a recent study, it was suggested that prediction accuracy can be increased by incorporation of GWAS results in GS models for complex traits (Li et al., 2018).

13.5.2 TRANSCRIPTOMICS

Genes express differentially in response to stress across organs, tissues or cells to get adapted. Hence, the study of transcriptome profile is an important part of understanding the function of genes in biological processes. Till date many techniques have been developed to study the transcriptome with two basic approaches, i.e., microarray technology (hybridization based) and sequencing based methods like RNA sequencing (RNA-Seq), serial analysis of gene expression (SAGE), massively parallel signature sequencing (MPSS). Sequence based profiling provides complete characterization of the transcriptome while the hybridization-based profiling discovers the information on how gene expression changes in different developmental stages and how it varies within or between species. As the transcripts are produced after splicing of intron sequence from genomic DNA, the cost of sequencing transcriptome is cheaper than genome. Furthermore, the candidate genes identified by genomic tools will be subjected to expression analysis by RT-PCR for confirmation.

In chickpea, high throughput sequencing (454/Roche) identified 5,615 differentially expressed genes (DEGs) of which 1696 genes were expressed during early infection while 1,338 and 2,581 genes were expressed in middle stage and late stage of infection during nodule development, respectively (Kaur et al., 2017). The transcriptome data generated using RNA-seq for DT is the first in faba bean (Khan et al., 2019). Transcriptome analysis of two drought tolerant chickpea varieties generated 152 million high reads, moreover, 97 genes of 1,674 total drought-related genes expressed in both genotypes (Kumar et al., 2019). Bhaskarla et al. (2020) studied root transcriptomics in chickpea and identified specific DEGs in response to drought stress. RNA-sequencing analysis identified a total of 2,161 and 1,873 DEGs (at the vegetative stage), and 4,109 and 3,772 DEGs (at reproductive stage), in the tolerant and sensitive genotypes, respectively. Likewise, in another transcriptome profiling in chickpea for cold stress, 526 genes showed significant expressions in tolerant genotype (Sel96Th11439), of which 261 and 265 genes were up and down regulated, respectively. Based on identified transcript patterns, those genes involved in the biosynthetic pathway of some metabolites were significantly increased in tolerant genotype whereas decreased in sensitive genotype (Khaledian et al., 2020). Around 34 pigeon pea homologous genes were identified, few encoding *MYB* and *bHLH* transcription factors which were expressed low in sterile bud implying role in pollen sterility (Saxena et al., 2020). Similarly, comparative transcriptome analysis during a flower development stage of A4-CMS line ICPA 2043 and its maintainer of pigeon pea identified 10 DEGs (Bohra et al., 2021). Illumina sequencing detected 0.11 million transcripts in two species of pigeon pea, i.e., *Cajanus cajan* and *C. platycarpus* with an annotation of 0.09 million (82%) transcripts using BlastX whereas, gene ontology and Mapman analysis showed higher number of transcripts in the wild relatives pertaining to signaling, transcription factors and stress responsive genes (Rathinam et al., 2019; Ramakrishna et al., 2021). Transcriptome profiling of Ascophyta infected lentil revealed that there are higher expression levels of structural defense related genes in resistant genotypes (Khorramdelazad et al., 2018). Similar response to Ascophyta infection were observed in lentil by Garcia et al. (2019). They witnessed the up-regulation of the jasmonic acid pathway, lignin biosynthesis pathway and salicylic acid mechanism in resistant genotypes, while they were down-regulated in the susceptible ones. In common bean, transcriptome study revealed many defenses associated genes against spider mite (Hoseinzadeh et al., 2020).

13.5.3 PROTEOMICS AND METABOLOMICS

Proteomics and metabolomics are emerging 'omics' technology that permits large scale precise information on proteins and metabolites in pulses, respectively. Therefore, it is important to study the proteome and metabolome along with transcriptome and genome to pave an improvement in nutritive value of pulses.

The stem proteome of *Phaseolus vulgaris* were analyzed after separation of the proteins by SDS-PAGE. The identified protein groups were then categorized into different groups according to their presumed functions mainly into energy uptake, photosynthetic process, protein synthesis, proteolysis, and stress related proteins (Zadražnik et al., 2017). In chickpea, Parankusam et al. (2017) identified a set of 482 heat-responsive proteins in the heat tolerant genotypes. Proteomic analysis using 2D high resolution electrophoresis followed by MALDI-TOF-MS analysis (matrix assisted laser desorption ionization-time of flight mass spectrometry (MS)) helped in identification of 373 pigeon pea seed proteins and also a large number of stress-related proteins (Krishnan et al., 2017). In chickpea, Barua et al. (2019) identified 4,832 dehydration responsive nuclear proteome and 478 phosphosites. Among the identified proteins, there were groups of kinases, phosphatases, and transcription factors. Additionally, 660 proteins were identified which are not being characterized, of which splicing related proteins were dominant. After proteomic comparison of two contrasting cowpea varieties, 108 differentially expressed proteins related to DT were identified. Drought stress-response peptides, including glutamine synthetase, chaperonin, malate dehydrogenase, heat-shock proteins (hsp), and RUBISCO were expressed in both genotypes (Lima et al., 2019). Proteomics study for resistance to halo blight in common bean identified 160 proteins which were similar to salicylic acid responders EDS1 and NDR1, ethylene and jasmonic acid biosynthesis enzymes. In addition, they also identified 29 proteins specific to hypersensitive immunity (Cooper et al., 2020). Comparative proteomics approach was conducted to study protein abundance in chickpea roots under drought stress. Stress related protein contributed 16% of total protein, 13% involved in protein metabolism, 13% in reactive oxygen metabolism (ROS) and 9% in signal transduction (Gupta et al., 2020). In a recent comparative TMT (Tandem Mass Tag) proteomic analysis for *H. armigera* resistance in *C. scarabaeoides.* Ngugi-Dawit et al. (2021) observed that *H. armigera* resistance appeared to be related to increase of secondary metabolites precursors, antioxidants, and phenylpropanoid pathway.

Two main approaches used for the generating the metabolic spectral data are nuclear magnetic resonance (NMR) and mass spectrometry (MS). Simultaneous measurement of a large number of metabolites without having prior knowledge on the set can be achieved by the process of untargeted metabolomic analysis. However, this requires extensive use of bioinformatics tools (Bueno et al., 2020). However, targeted metabolomic analysis is categorized by the measurement of a predefined group of metabolites with a higher level of precision and accuracy (Brunetti et al., 2018). Acetylsalicylic acid was found to be increased in all the parts of pigeon pea when stress is applied but, when the seeds were subjected to halo priming then up-regulated metabolites were down regulated in the leaves. However, beta-cyanoalanine showed a positive trend with increase in salt concentrations (Biswas et al., 2018). Metabolic profiling to study variation in metabolite levels in chickpea varieties under drought stress led to the identification of 20 known metabolites. The drought stress led to increase of allantoin, proline, arginine, histidine, isoleucine, and tryptophan whereas, decrease in levels choline, phenylalanine, gamma-aminobutyric acid, alanine, phenylalanine, tyrosine, glucosamine, guanine, and aspartic acid (Khan et al., 2019). LC-MS based untargeted metabolomics approach was conducted to study the composition of bioactive compounds in three pulses, i.e., chickpea, lentil, and bean. The most discriminative compound in the crops were phenolic acids, flavanols, and prenol lipids and eucomic acids (Llorach et al., 2019). Accumulation of osmoprotectant like L-proline, sugars, and sugar alcohols was higher in desi chickpea and metabolic pathway analysis showed involvement of inositol phosphate in plant defense response (Nisa et al., 2020). Tolerant A116 cowpea genotype showed accumulation in 13 out of 14 responsive metabolites of DT especially increase in sucrose, fructose, urea, alanine, and putrescine (Gomes et al., 2020). The drought-adaptive nature of two contrasting lentil varieties were observed by differential accumulation of certain metabolites like D-fructose, α-trehalose, myo-inositol, and L-tryptophan (Foti et al., 2021).

13.5.4 PHENOMICS

Phenotyping is a crucial chunk of plant breeding experiments, and by manual ways, it takes a lot of time and manpower. Moreover, precise measurement of many traits like morphological (above- and below-ground traits such as shoot and root architecture, respectively), physiological, and biochemical are important in crop improvement. Therefore, in recent years accurate and rapid phenotyping methods have been developed. These methods include use of un-intrusive imaging techniques including color imaging for plant biomass and plant morphology; near infrared imaging (NIR) for measuring tissue and soil water contents; far infrared imaging for plant canopy temperature; fluorescence imaging for physiological study of photosynthesis, automated weighing and watering for enforcing drought/salinity stress (Kumar et al., 2015).

In several pea and chickpea varieties, plant responses were evaluated using UAV (unmanned aerial vehicle) integrated with multispectral imaging sensors. Assessment showed that the vegetation indices are strongly correlated with SY of pulse crops (Sankaran et al., 2016). Similarly, a quad copter UAV and a five-band multispectral camera was used to obtain image data of chickpea and dry pea for crop performance evaluation. The images were first processed in Pix4D mapper to generate orthomosaic images and image-based features from the UAV data were analyzed using Pearson's correlation (Zhang et al., 2021). It is observed that phenomics technologies can be used to save time and labor. The improved efficacy of high-throughput phenotyping (HTP) was demonstrated for salinity screening in lentil using red-green-blue (RGB) images on a LemnaTec Scanalyzer 3D phenomics platform (Dissanayake et al., 2020). Phenotyping of roots can be a tedious and difficult job. Many new technologies are being developed to characterize crop root system architecture (CRSA). An open source phenomics software 'DIRT' (digital imaging of root traits) is an online platform to store images of plant roots, measure dicot and monocot root characteristics and share the stored records. All stored and computed data is easily available to the public and scientific community (Das et al., 2015b). This technology was used for storing root images data to study the roots of bean and cowpea (Burridge et al., 2016). Phenomics technologies can support plant breeders in assessing the morphological and few physiological performances of breeding materials and help to accelerate the development of better performing varieties of pulse and other crops.

13.6 SPEED BREEDING

The main constraint of most of the breeding programs is the duration to achieve the breeding goal. To overcome the time constraint, a system called speed breeding is trending to reduce the breeding cycles in different crops. The main aim of speed breeding is to shorten the breeding cycle, rapid development of homozygous lines, accelerated cultivar development, and faster seed harvest (Watson et al., 2018). This technique uses optimal light intensity, quality of light, day-length time and controlled temperature to increase the rate of photosynthesis, which in turn fastens the flowering, plant growth and maturity (Cazzola et al., 2021). Furthermore, the integration of speed breeding with conventional methods, genetic engineering, and MAS increases the effective selection of elite cultivar with good characteristics (Wanga et al., 2021). Speed breeding techniques have also been standardized in some pulses. For example, rapid generation experiment was conducted on an F_2 population derived from a *L. culinaris*/*L. ervoides* cross and screening for aphanomyces root rot was done. Plants scoring less than 4.0 were advanced to the next generations through a modified single-seed decent (SSD) method in a growth chamber with controlled conditions of 20 hours of photoperiod. Seeds were germinated in a 100 μM gibberellin solution and thereby, five generations were possible within a year (Lulsdort & Banniza, 2018). Samineni et al. (2020) developed a protocol for increasing number of generations from one to 6–7 per year in chickpea. The experiment was conducted in the glasshouse with the photoperiod control system. It was observed that with enhanced photoperiod, the reduction in flowering was 8–19, 7–16, and 11–27 days in early, medium, and late varieties, respectively. Cazzola et al. (2020) developed a simple, cost-effective, and hydroponics system with 22-hours photoperiod, temperature of 20±2°C, flurprimidol anti-gibberellin to obtain early grain harvest in pea. This method showed higher efficiencies than the traditional SSD in the field by growing five generations of crop per year. Mobini et al. (2020) also have reported a cost and space-efficient *in vivo* protocol that increases pod setting. They examined the time of 6-benzylaminopurine (BAP), a cold treatment and their combination in two faba bean genotypes; and observed improved pollen germination on application of BAP. Cold treatment was amenable to reduce the time required for first seed formation and increased pollen viability. Thus, this accelerated the breeding generations in faba bean.

13.7 GENOME EDITING (GE)

Genome editing (GE) is a modern breeding tool that helps in precise modification in target genome and therefore it has immense potential in crop improvement. Several tools like ZFNs, TALENs, RNA-guided endonucleases (RGENs), and CRISPR (clustered regularly interspaced short palindromic repeats)/Cas9 (CRISPR associated protein 9) have been used for modification in different crops. Among all these, CRISPR is becoming more popular because of its simplicity and effectiveness.

The CRISPR method requires two main components: (i) sgRNA – a short synthetic guide RNA of 20 nucleotides that has to bind to target DNA; and (ii) Cas9 enzyme that cleaves the targeted DNA near to PAM region (protospacer adjacent motifs). The steps for CRISPR are very simple which include: (a) identifying a PAM sequence in target genome; (b) synthesizing a guide RNA; (c) cloning of sgRNA into vector; (d) transfer to host cell and generation of transformants; (e) screening of transformants; and finally (f) validation for the edited gene (Jaganathan et al., 2018). Tao et al. (2018) cloned a full-length DNA sequence of CASase gene of lathyrus which was 3,134 bp in length and encoded for β-cyanoalanine synthase (CASase). All the five selected sgRNA worked well with the target sequence. The CRISPR system was able to construct four knock-out vectors and was confirmed by colony PCR and sequencing. The first successful CRISPR editing of cowpea was done by Ji et al. (2019). They used the CRISPR/Cas9-mediated GE technology to competently disturb the symbiotic nitrogen fixation (SNF) gene in cowpea. Guide RNAs (gRNAs) targeting symbiosis receptor-like kinase (SYMRK) achieved complete blockage of nodule formation when both alleles are interrupted. In order to achieve higher gene editing efficiency, three gRNAs targeting both DNA strands at different sites of VuSYMRK exon 2 were used which led to the formation of InDels in the gene (Ji et al., 2019). The first Report of CRISPR/Cas9 mediated DNA-free editing of 4CL (4-coumarate ligase) and RVE7 (Reveille 7) genes in chickpea protoplasts was conducted by Badhan et al. (2021). The 4CL encodes for enzyme involved in phenylpropanoid metabolism in the lignin biosynthesis pathway whereas, RVE7 encodes for enzymes responsible for regulating circadian rhythm in plants. Sanger sequencing was performed to detect mutations in the transfected protoplast. It was hence observed that RVE7 sgRNA 1 showed 76%, 77%, and 79% indel percentage for the treated protoplast samples compared to the 99% for the control sample (Indels represent the editing efficiency). However, all other 4CL test samples showed no indel percentage. It was concluded that RVE7 showed high efficiency editing as compared to 4CL. Juranić et al. (2020) developed a transient method of testing CRISPR/Cas9 constructs in a short time of 48 hours via Agrobacterium infiltration of detached leaflets. This method of CRISPR/Cas-mediated gene editing was successfully demonstrated by Che et al. (2021) in cowpea. Transgenic cowpea plantlets were developed from the nodes of cotyledon at frequencies of 4% to 37% across a wide range of genotypes. The combination of CRISPR tool with other advance breeding approaches like speed breeding will be the most efficient method for developing elite genotypes in a safer and quicker way.

13.8 CONCLUSION

Pulses are the major source of protein in the human diet. But with the ever-increasing population, the pulse production is not able to meet the global demand. Globally, breeding and biotechnological approaches are being used to boost the pulse production. Crop improvement depends mainly in availability of genetic variability in a population and selection of best genotype is the crucial step. Nevertheless, traditional approaches of breeding take care of both the aspect, however, the approaches have their own limitations like time-consuming, accumulation of undesirable effects, fertilization barrier and many more. To overcome these constraints, molecular breeding and other advance breeding strategies like omics technology, speed breeding and GE plays a complementary role. With the discovery of molecular markers, there has been many developments in molecular breeding platform, moreover, progression in HT genotyping and phenotyping, now it is the era of genomics.

KEYWORDS

- **CRISPR/Cas**
- **genomic wide association studies**
- **marker-assisted backcrossing**
- **marker-assisted selection**
- **omics technology**
- **pulse breeding**
- **speed breeding**

REFERENCES

Afzal, M., Alghamdi, S. S., Migdadi, H. M., Khan, M. A., & Farooq, M., (2018). Morphological and molecular genetic diversity analysis of chickpea genotypes. *Int. J. Agric. Biol, 20*, 1062–1070.

Ahmad, K. M. S., & Talebi, R., (2017). Genetic diversity assessment in chickpea using morphological characters and start codon targeted (SCoT) molecular markers. *Annual Research & Review in Biology*, 1–8.

Ates, D., Aldemir, S., Alsaleh, A., Erdogmus, S., Nemli, S., Kahriman, A., & Tanyolac, B., (2018a). A consensus linkage map of lentil based on DArT markers from three RIL mapping populations. *PloS One, 13*(1), e0191375.

Ates, D., Aldemir, S., Yagmur, B., Kahraman, A., Ozkan, H., Vandenberg, A., & Tanyolac, M. B., (2018b). QTL mapping of genome regions controlling manganese uptake in lentil seed. *G3: Genes, Genomes, Genetics, 8*(5), 1409–1416.

Badhan, S., Ball, A. S., & Mantri, N., (2021). First report of CRISPR/Cas9 mediated DNA-free editing of 4CL and RVE7 genes in chickpea protoplasts. *International Journal of Molecular Sciences, 22*(1), 396.

BARC-NABD, Bioscience, (2019). http://www.barc.gov.in/group/66_s246.pdf (accessed on 04 January 2022).

Barmukh, R., Roorkiwal, M., Jaba, J., Chitikineni, A., Mishra, S. P., Sagurthi, S. R., & Varshney, R. K., (2021b). Development of a dense genetic map and QTL analysis for pod borer *Helicoverpa armigera* (Hübner) resistance component traits in chickpea (*Cicer arietinum* L.). *The Plant Genome, 14*(1), Art-e20071.

Barmukh, R., Soren, K. R., Madugula, P., Gangwar, P., Shanmugavadivel, P. S., Bharadwaj, C., & Varshney, R. K., (2021a). Construction of a high-density genetic map and QTL analysis for yield, yield components and agronomic traits in chickpea (*Cicer arietinum* L.). *Plos One, 16*(5), e0251669.

Barshile, J. D., Auti, S. G., Dalve, S. C., & Apparao, B. J., (2006). Mutagenic sensitivity studies in chickpea employing SA, EMS and gamma rays. *Indian Journal of Pulses Research, 19*(1), 43.

Barua, P., Lande, N. V., Subba, P., Gayen, D., Pinto, S., Keshava, P. T. S., & Chakraborty, N., (2019). Dehydration-responsive nuclear proteome landscape of chickpea (*Cicer arietinum* L.) reveals phosphorylation-mediated regulation of stress response. *Plant, Cell & Environment, 42*(1), 230–244.

Batieno, B. J., Danquah, E., Tignegre, J. B., Huynh, B. L., Drabo, I., Close, T. J., & Ouedraogo, T. J., (2016). Application of marker-assisted backcrossing to improve cowpea (*Vigna unguiculata* L. Walp) for drought tolerance. *Journal of Plant Breeding and Crop Science, 8*(12), 273–286.

Bharadwaj, C., Tripathi, S., Soren, K. R., Thudi, M., Singh, R. K., Sheoran, S., & Varshney, R. K., (2021). Introgression of "QTL-hotspot" region enhances drought tolerance and grain yield in three elite chickpea cultivars. *The Plant Genome, 14*( ), e20076.

Bhaskarla, V., Zinta, G., Ford, R., Jain, M., Varshney, R. K., & Mantri, N., (2020). Comparative root transcriptomics provide insights into drought adaptation strategies in chickpea (*Cicer arietinum* L.). *International Journal of Molecular Sciences, 21*(5), 1781.

Bhat, J. A., Shivaraj, S. M., Ali, S., Mir, Z. A., Islam, A., & Deshmukh, R., (2018). Genomic resources and omics-assisted breeding approaches for pulse crop improvement. In: *Pulse Improvement* (pp. 13–55). Springer, Cham.

Biswas, S., Biswas, A. K., & De, B., (2018). Metabolomics analysis of *Cajanus cajan* L. seedlings unraveled amelioration of stress induced responses to salinity after halopriming of seeds. *Plant Signaling & Behavior, 13*(7), e1489670.

Bohra, A., Rathore, A., Gandham, P., Saxena, R. K., Naik, S. S., Dutta, D., & Singh, N. P., (2021). Genome-wide comparative transcriptome analysis of the A4-CMS line ICPA 2043 and its maintainer ICPB 2043 during the floral bud development of pigeonpea. *Functional & Integrative Genomics, 21*(2), 251–263.

Bohra, A., Singh, I. P., Satheesh, N. S. J., Singh, F., Datta, D., Kumar, A., & Singh, N. P., (2019). IPH 15–03: Pigeonpea Hybrid for Cultivation in North-Western Plain Zone. *IIPR*, Newsletter. *30*(2).

Brunetti, A. E., Neto, F. C., Vera, M. C., Taboada, C., Pavarini, D. P., Bauermeister, A., & Lopes, N. P., (2018). An integrative omics perspective for the analysis of chemical signals in ecological interactions. *Chemical Society Reviews, 47*(5), 1574–1591.

Bueno, P. C., & Lopes, N. P., (2020). Metabolomics to characterize adaptive and signaling responses in legume crops under abiotic stresses. *ACS Omega, 5*(4), 1752–1763.

Burridge, J., Jochua, C. N., Bucksch, A., & Lynch, J. P., (2016). Legume shovelomics: High—throughput phenotyping of common bean (*Phaseolus vulgaris* L.) and cowpea

(*Vigna unguiculata* subsp, unguiculata) root architecture in the field. *Field Crops Research, 192*, 21–32.
Cazzola, F., Bermejo, C. J., Gatti, I., & Cointry, E., (2021). Speed breeding in pulses: An opportunity to improve the efficiency of breeding programs. *Crop and Pasture Science, 72*(3), 165–172.
Cazzola, F., Bermejo, C. J., Guindon, M. F., & Cointry, E., (2020). Speed breeding in pea (*Pisum sativum* L.), an efficient and simple system to accelerate breeding programs. *Euphytica, 216*(11), 1–11.
Chakraborty, R. K., Bhowmik, A., Hossain, T., & Mian, M. A. K., (1998). Induction of mutation in mungbean (Vigna radiata) through gamma irradiation. *Annals of Bangladesh Agriculture, 8*(2), 129–136.
Chand, S., Chandra, K., & Khatik, C. L., (2020). Varietal release, notification and denotification system in India. In: *Plant Breeding-Current and Future Views*. IntechOpen.
Chandra, K., & Pandey, A., (2017). QTL mapping in crop improvement: A basic concept. *International Journal of Current Microbiology and Applied Sciences, 6*(12), 835–842.
Chang-soon, A. H. N., (1976). *Interspecific Hybridization Among Four Species of the Genus Vigna Savi* (p. 7714591). University of Hawai'i at Manoa. ProQuest Dissertations Publishing.
Che, P., Chang, S., Simon, M. K., Zhang, Z., Shaharyar, A., Ourada, J., & Jones, T. J., (2021). Developing a rapid and highly efficient cowpea regeneration, transformation and genome editing system using embryonic axis explants. *The Plant Journal, 106*(3), 817–830.
Collard, B. C., & Mackill, D. J., (2008). Marker-assisted selection: An approach for precision plant breeding in the twenty-first century. *Philosophical Transactions of the Royal Society B: Biological Sciences, 363*(1491), 557–572.
Cooper, B., Campbell, K. B., Beard, H. S., Garrett, W. M., & Ferreira, M. E., (2020). The proteomics of resistance to halo blight in common bean. *Molecular Plant-Microbe Interactions, 33*(9), 1161–1175.
Danish, M., Sirohi, S. P. S., & Mohit, (2019). Mutation breeding in pulses to curb malnutrition. *Int. J. Curr. Microbiol. App. Sci., 8*(5), 1068–1096.
Das, A., Schneider, H., Burridge, J., Ascanio, A. K. M., Wojciechowski, T., Topp, C. N., & Bucksch, A., (2015b). Digital imaging of root traits (DIRT): A high-throughput computing and collaboration platform for field-based root phenomics. *Plant Methods, 11*(1), 1–12.
Das, S., Upadhyaya, H. D., Bajaj, D., Kujur, A., Badoni, S., Kumar, V., & Tyagi, A. K., (2015a). Deploying QTL-seq for rapid delineation of a potential candidate gene underlying major trait-associated QTL in chickpea. *DNA Research, 22*(3), 193–203.
Deokar, A. A., Ramsay, L., Sharpe, A. G., Diapari, M., Sindhu, A., Bett, K., & Tar'an, B., (2014). Genome wide SNP identification in chickpea for use in development of a high-density genetic map and improvement of chickpea reference genome assembly. *BMC Genomics, 15*(1), 1–19.
Deokar, A., Sagi, M., & Tar'an, B., (2019). Genome-wide SNP discovery for development of high-density genetic map and QTL mapping of *Ascochyta* blight resistance in chickpea (*Cicer arietinum* L.). *Theoretical and Applied Genetics, 132*(6), 1861–1872.
Devi, S. M., & Jayamani, P., (2020). Molecular genetic diversity in cowpea {*Vigna unguiculata* (L.) Walp.} genotypes. *Journal of Food Legumes, 33*(1), 6–9.
Dinesh, H. B., Lohithaswa, H. C., Viswanatha, K. P., Singh, P., & Rao, A. M., (2016). Identification and marker-assisted introgression of QTL conferring resistance to bacterial leaf blight in cowpea (*Vigna unguiculata* (L.) Walp.). *Plant Breeding, 135*(4), 506–512.

Dinesh, H. B., Lohithaswa, H. C., Viswanatha, K. P., Singh, P., Manjunatha, L., Ambika, D. S., & Kumar, M. K., (2018). Genetic analysis and marker assisted backcrossing for transfer of mosaic virus resistance in cowpea [*Vigna unguiculata* (L.) Walp.]. *Legume Research-An International Journal, 41*(5), 663–668.

Dissanayake, R., Kahrood, H. V., Dimech, A. M., Noy, D. M., Rosewarne, G. M., Smith, K. F., & Kaur, S., (2020). Development and application of image-based high-throughput phenotyping methodology for salt tolerance in lentils. *Agronomy, 10*(12), 1992.

FAOSTAT (2012). *FAO Statistics Division.* http://faostat.fao.org/beta/en/#home (accessed on 04 January 2022).

Ferreira, J. J., Campa, A., Pérez-Vega, E., Rodríguez-Suárez, C., & Giraldez, R., (2012). Introgression and pyramiding into common bean market class fabada of genes conferring resistance to anthracnose and potyvirus. *Theoretical and Applied Genetics, 124*(4), 777–788.

Foti, C., Kalampokis, I. F., Aliferis, K. A., & Pavli, O. I., (2021). Metabolic responses of two contrasting lentil genotypes to PEG-induced drought stress. *Agronomy, 11*(6), 1190.

Francia, E., Tacconi, G., Crosatti, D., Barabaschi, D., Bulgarelli, E., Dall'Aglio, & Giampiero, V., (2005). Marker assisted selection in crop plants. *Plant Cell, Tissue and Organ Culture, 82*(3), 317–342.

Fritsche-Neto, R., Souza, T. L. P. O. D., Pereira, H. S., Faria, L. C. D., Melo, L. C., Novaes, E., & Jannink, J. L., (2019). Association mapping in common bean revealed regions associated with anthracnose and angular leaf spot resistance. *Scientia Agricola, 76*, 321–327.

Gali, K. K., Sackville, A., Tafesse, E. G., Lachagari, V. B., McPhee, K., Hybl, M., & Warkentin, T. D., (2019). Genome-wide association mapping for agronomic and seed quality traits of field pea (*Pisum sativum* L.). *Frontiers in Plant Science, 10*, 1538.

García, P., Francisca, V. R., Francisco, V. B., Luis E. S. De. M., Carlos, G. P., & Ana, I. G. R. H., (2019). Transcriptome profiling of lentil in response to *Ascochyta lentis* infection. *Spanish Journal of Agricultural Research, 17*(4), 7.

Gela, T. S., Adobor, S., Khazaei, H., & Vandenberg, A., (2021a). *An Advanced Lentil Backcross Population Developed from a Cross Between Lens Culinaris × L. Ervoides for Future Disease Resistance and Genomic Studies.* bioRxiv preprint. doi: https://doi.org/10.1101/2021.01.13.426580.

Gela, T. S., Koh, C. S., Caron, C. T., Chen, L. A., Vandenberg, A., & Bett, K. E., (2021b). QTL mapping of lentil anthracnose (*Colletotrichum lentis*) resistance from *Lens ervoides* accession IG 72815 in an interspecific RIL population. *Euphytica, 217*(4), 1–11.

Gomes, A. M., Rodrigues, A. P., António, C., Rodrigues, A. M., Leitão, A. E., Batista-Santos, P., & Ramalho, J. C., (2020). Drought response of cowpea (*Vigna unguiculata* (L.) Walp.) landraces at leaf physiological and metabolite profile levels. *Environmental and Experimental Botany, 175*, 104060.

Gupta, S., Mishra, S. K., Misra, S., Pandey, V., Agrawal, L., Nautiyal, C. S., & Chauhan, P. S., (2020). Revealing the complexity of protein abundance in chickpea root under drought-stress using a comparative proteomics approach. *Plant Physiology and Biochemistry, 151*, 88–102.

Haq, M. A., Sadiq, M., & Hassan, M., (1999). CM 98 (CM 31-1/85): A Very High-Yielding, Disease-Resistant Mutant Variety of Chickpea. *International Chickpea and Pigeonpea Newsletter. 6,* 9–10.

Hasan, M. M., Rafii, M. Y., Ismail, M. R., Mahmood, M., Rahim, H. A., Alam, M. A., & Latif, M. A., (2015). Marker-assisted backcrossing: A useful method for rice improvement. *Biotechnology & Biotechnological Equipment, 29*(2), 237–254.

Hoseinzadeh, A. H., Soorni, A., Shoorooei, M., Torkzadeh, M. M., Maali, A. R., Allahyari, H., & Mohammadi, R., (2020). Comparative transcriptome provides molecular insight into defense-associated mechanisms against spider mite in resistant and susceptible common bean cultivars. *PloS One, 15*(2), e0228680.

Jaganathan, D., Ramasamy, K., Sellamuthu, G., Jayabalan, S., & Venkataraman, G., (2018). CRISPR for crop improvement: An update review. *Frontiers in Plant Science, 9*, 985.

Jha, U. C., Jha, R., Bohra, A., Manjunatha, L., Saabale, P. R., Parida, S. K., & Singh, N. P., (2021a). Association mapping of genomic loci linked with *Fusarium* wilt resistance (Foc2) in chickpea. *Plant Genetic Resources*, 1–8.

Jha, U. C., Jha, R., Thakro, V., Kumar, A., Gupta, S., Nayyar, H., & Singh, N. P., (2021b). Discerning molecular diversity and association mapping for phenological, physiological and yield traits under high temperature stress in chickpea (*Cicer arietinum* L.). *Journal of Genetics, 100*(1), 1–15.

Ji, J., Zhang, C., Sun, Z., Wang, L., Duanmu, D., & Fan, Q., (2019). Genome editing in cowpea *Vigna unguiculata* using CRISPR-Cas9. *International Journal of Molecular Sciences, 20*(10), 2471.

Joshi, B. K., (2017). Plant breeding in Nepal: Past, present and future. *Journal of Agriculture and Forestry University, 1*, 1–33.

Juranić, M., Nagahatenna, D. S., Salinas-Gamboa, R., Hand, M. L., Sánchez-León, N., Leong, W. H., & Koltunow, A. M., (2020). A detached leaf assay for testing transient gene expression and gene editing in cowpea (*Vigna unguiculata* [L.] Walp.). *Plant Methods, 16*(1), 1–17.

Kamfwa, K., Beaver, J. S., Cichy, K. A., & Kelly, J. D., (2018). QTL mapping of resistance to bean weevil in common bean. *Crop Science, 58*(6), 2370–2378.

Kanavi, M. S. P., Rangaiah, S., & Krishnaprasad, B. T., (2019). Molecular characterization and genetic diversity analysis through SSR markers in germplasm lines of green gram [*Vigna radiata* (L.) Wilczek]. *International Journal of Bio-resource and Stress Management, 10*(3), 298–305.

Kaur, S., Singh, S., & Tandon, G., (2017). Effect of mesorhizobium in chickpea root nodulation using transcriptome profiling approach. *Indian Journal of Agricultural Sciences, 87*(11), 1507–1514.

Khaledian, Y., Maali-Amiri, R., Mohammadi, R., & Mantri, N., (2020). Transcriptome profiling of the early-response in chickpea to cold stress. *Iranian Journal of Field Crop Science, 51*(3), 1–16.

Khan, M. A., Alghamdi, S. S., Ammar, M. H., Sun, Q., Teng, F., Migdadi, H. M., & Al-Faifi, S. A., (2019). Transcriptome profiling of faba bean (*Vicia faba* L.) drought-tolerant variety hassawi-2 under drought stress using RNA sequencing. *Electronic Journal of Biotechnology, 39*, 15–29.

Khan, N., Bano, A., Rahman, M. A., Rathinasabapathi, B., & Babar, M. A., (2019). UPLC-HRMS-based untargeted metabolic profiling reveals changes in chickpea (*Cicer arietinum*) metabolome following long-term drought stress. *Plant, Cell & Environment, 42*(1), 115–132.

Khan, S., Wani, M. R., & Parveen, K., (2005). Chlorophyll mutations induced in mungbean by chemical mutagens. *Adv. Pl. Sci., 18*(1), 343–347.

Kharkwal, M. C., Gopalakrishna, T., & Pawar, S. E., (2008). *Mutation Breeding for Improvement of Food Legumes* (pp. 221). Indian Society of Genetics and Plant Breeding, New Delhi – 1994.

Khatik, C. L., Chandra, K., Khan, M., Yadav, V., Jatav, H. S., Khan, M. A., Chand, S., & Mishra, C. D., (2021). Evaluation of mung bean germplasm [*Vigna radiata* (L.) Wilczek] for yield in transitional plain of inland drainage zone of Rajasthan India. *Legume Research*. doi: 10.18805/LR-4543.

Khoiriyah, N., Yuniastuti, E., & Purnomo, D., (2018). Genetic diversity of pigeon pea (*Cajanus cajan* (l.) Millsp.) based on molecular characterization using randomly amplified polymorphic DNA (RAPD) markers. In: *IOP Conference Series: Earth and Environmental Science* (Vol. 129, No. 1, p. 012016). IOP Publishing.

Khorramdelazad, M., Bar, I., Whatmore, P., Smetham, G., Bhaaskaria, V., Yang, Y., & Ford, R., (2018). Transcriptome profiling of lentil (*Lens culinaris*) through the first 24 hours of *Ascochyta lentis* infection reveals key defense response genes. *BMC Genomics, 19*(1), 1–21.

Kiryowa, M., Nkalubo, S. T., Mukankusi, C., Male, A., Gibson, P., Tukamuhabwa, P., & Rubaihayo, P., (2021). Effectiveness of pyramided genes in conferring resistance to anthracnose disease in common bean populations. *Journal of Plant Breeding and Crop Science, 13*(1), 1–13.

Kohli, P. S., Verma, P. K., Verma, R., Parida, S. K., Thakur, J. K. & Giri, J., (2020). Genome-wide association study for phosphate deficiency responsive root hair elongation in chickpea. *Functional & Integrative Genomics, 20*(6), 775–786.

Kosgei, A. J., Kimurto, P. K., Gaur, P. M., Yeboah, M. A., Offei, S. K., Danquah, E. Y., & Varshney, R. K., (2017). *Introgression of Drought Tolerance Traits into Adapted Kenyan Chickpea Varieties Using Marker Assisted Backcrossing (MABC). 32*.

Krishnaiah, V. V., Venkateswarlu, P., Prasad, M. M. K. D., & Rao, Y. H., (1978). *A New Strain of Black Gram (Vigna Mungo (L.) Hepper) Resistant to Powdery mildew (Erysiphe Polygoni)* (No. RESEARCH).

Krishnan, H. B., Natarajan, S. S., Oehrle, N. W., Garrett, W. M., & Darwish, O., (2017). Proteomic analysis of pigeonpea (*Cajanus cajan*) seeds reveals the accumulation of numerous stress-related proteins. *Journal of Agricultural and Food Chemistry, 65*(23), 4572–4581.

Kumar, J., Choudhary, A. K., Solanki, R. K., & Pratap, A., (2011). Towards marker-assisted selection in pulses: A review. *Plant Breeding, 130*(3), 297–313.

Kumar, J., Pratap, A., & Kumar, S., (2015). *Phenomics in Crop Plants: Trends, Options and Limitations* (Vol. 8, p. 296). New Delhi: Springer India.

Kumar, J., Pratap, A., & Solanki, R. K., (2010). Basics of molecular genetic mapping and QTL analysis in plants. *Adv. Biotechnol.*, 35–52.

Kumar, M., Chauhan, A. S., Yusuf, M. A., Sanyal, I., & Chauhan, P. S., (2019). Transcriptome sequencing of chickpea (*Cicer arietinum* L.) genotypes for identification of drought-responsive genes under drought stress condition. *Plant Molecular Biology Reporter, 37*(3), 186–203.

Kumar, N., Rana, M., Kumar, B., Chand, S., Shiv, A., Wani, S. H., & Kumar, S., (2021). Genomic selection for wheat improvement. *Physiological, Molecular, and Genetic Perspectives of Wheat Improvement*, 175–207.

Kumar, S., (1998). Inheritance of resistance to *Fusarium oxysporum* f. sp. *ciceri* (race2) in chickpea. *Plant Breeding, 117*, 139–142.

Kumawat, G., Kumawat, C. K., Chandra, K., Pandey, S., Chand, S., Mishra, U. N., & Sharma, R., (2020). Insights into marker assisted selection and its applications in plant breeding. In: *Plant Breeding-Current and Future Views*. IntechOpen.

Li, G., Khan, T. A., Guo, J., Wenlong, L., Fu, M., Li, T., & Akram, W., (2021). Development of SNP-based high-density genetic map and gene mapping of pod color trait in Cowpea (*Vigna unguiculata* L. Walp.). *The Journal of Horticultural Science and Biotechnology, 96*(1), 87–94.

Li, Y., Ruperao, P., Batley, J., Edwards, D., Khan, T., Colmer, T. D., & Sutton, T., (2018). Investigating drought tolerance in chickpea using genome-wide association mapping and genomic selection based on whole-genome resequencing data. *Frontiers in Plant Science, 9*, 190.

Lima, E. N., Silva, M. D. S., De Abreu, C. E. B., Mesquita, R. O., Lobo, M. D. P., Monteiro-Moreira, A. D. O., & Bertini, C. D. M., (2019). Differential proteomics in contrasting cowpea genotypes submitted to different water regimes. *Genetics and Molecular Research, 18*(4).

Llorach, R., Favari, C., Alonso, D., Garcia-Aloy, M., Andres-Lacueva, C., & Urpi-Sarda, M., (2019). Comparative metabolite fingerprinting of legumes using LC-MS-based untargeted metabolomics. *Food Research International, 126*, 108666.

Lulsdorf, M. M., & Banniza, S., (2018). Rapid generation cycling of an F2 population derived from a cross between *Lens culinaris* Medik. and Lens ervoides (Brign.) Grande after *Aphanomyces* root rot selection. *Plant Breeding, 137*(4), 486–491.

Mallikarjuna, B. P., Samineni, S., Thudi, M., Sajja, S. B., Khan, A. W., Patil, A., Kannalli, P. V., et al., (2017). Molecular mapping of flowering time major genes and QTLs in chickpea (*Cicer arietinum* L.). *Frontiers in Plant Science, 8*, 1140.

Mannur, D. M., Babbar, A., Thudi, M., Sabbavarapu, M. M., Roorkiwal, M., Sharanabasappa, B. Y., & Varshney, R. K., (2019). Super Annigeri 1 and improved JG 74: Two *Fusarium* wilt-resistant introgression lines developed using marker-assisted backcrossing approach in chickpea (*Cicer arietinum* L.). *Molecular Breeding, 39*(1), 1–13.

Mehrotra, M., Singh, A. K., Sanyal, I., Altosaar, I., & Amla, D. V., (2011). Pyramiding of modified cry1Ab and cry1Ac genes of *Bacillus thuringiensis* in transgenic chickpea (*Cicer arietinum* L.) for improved resistance to pod borer insect *Helicoverpa armigera*. *Euphytica, 182*(1), 87–102.

Meuwissen, T. H., Hayes, B. J., & Goddard, M. E., (2001). Prediction of total genetic value using genome-wide dense marker maps. *Genetics, 157*(4), 1819–1829.

Mobini, S., Khazaei, H., Warkentin, T. D., & Vandenberg, A., (2020). Shortening the generation cycle in faba bean (*Vicia faba*) by application of cytokinin and cold stress to assist speed breeding. *Plant Breeding, 139*(6), 1181–1189.

Nadarajan, N., & Gupta, S., (2010). Role of classical breeding in improvement of pulse crops. *Electronic Journal of Plant Breeding, 1*(4), 1099–1106.

Ngugi-Dawit, A., Njaci, I., Higgins, T. J., Williams, B., Ghimire, S. R., Mundree, S. G., & Hoang, L. T. M., (2021). Comparative tmt proteomic analysis unveils unique insights into *Helicoverpa armigera* (Hübner) resistance in *Cajanus scarabaeoides* (L.) Thouars. *International journal of Molecular Sciences, 22*(11), 5941.

Nisa, Z. U., Arif, A., Waheed, M. Q., Shah, T. M., Iqbal, A., Siddiqui, A. J., & Musharraf, S. G., (2020). A comparative metabolomic study on desi and Kabuli chickpea (*Cicer arietinum* L.) genotypes under rainfed and irrigated field conditions. *Scientific Reports, 10*(1), 1–11.

NPRC, Vamban: https://tnau.ac.in/prs-vamban/varieties-released/ (accessed on 04 January 2022).

Okii, D., (2021). *Towards Effective Gene Pyramiding for Disease Resistance and Trait Selection in Common Bean (Phaseolus vulgaris L.).* Doctoral dissertation, Makerere University.

Okii, D., Tukamuhabwa, P., Tusiime, G., Talwana, H., Odong, T., Mukankusi, C., & Gepts, P., (2017). Agronomic qualities of genetic pyramids of common bean developed for multiple-disease-resistance. *African Crop Science Journal, 25*(4), 457–472.

Ozkuru, E., Ates, D., Nemli, S., Erdogmus, S., Karaca, N., Yilmaz, H., & Tanyolac, B., (2019). Association mapping of loci linked to copper, phosphorus, and potassium concentrations in the seeds of *C. arietinum* and *C. reticulatum*. *Genomics, 111*(6), 1873–1881.

Pandey, N., Ojha, C. B., Jha, V. B., & Singh, N. B., (1996). Effect of chemical mutagens on the rate of germination, seedling mortality, and induced sterility in pigeonpea. *International Chickpea and Pigeonpea Newsletter, 3*, 65–67.

Pandiyan, M., Senthil, N., Suersh, R., Chakravarthy, N., Packiaraj, D., & Jagadeesh, S., (2012). Interspecific hybridization of *Vigna radiata* x *Vigna trilobata*. *Euphytica, 35*, 1017–1022.

Parankusam, S., Bhatnagar-Mathur, P., & Sharma, K. K., (2017). Heat responsive proteome changes reveal molecular mechanisms underlying heat tolerance in chickpea. *Environmental and Experimental Botany, 141*, 132–144.

Parihar, A. K., Dixit, G. P., Singh, A. K., Sewak, S., & Singh, N. P., (2021). Extra Early Flowering Genotypes Developed in Fieldpea (*Pisum sativum* L.), *IIPR*, Newsletter. *32*(1).

Pathania, A., & Sood, B. C., (2006). Spectrum and frequency of mutations induced by gamma-rays and EMS in chickpea (*Cicer arietinum* L.). *Research on Crops, 7*(2), 453.

Patil, P. G., Dubey, J., Bohra, A., Mishra, R. K., Saabale, P. R., Das, A., & Singh, N. P., (2017). Association mapping to discover significant marker-trait associations for resistance against *Fusarium* wilt variant 2 in pigeonpea [*Cajanus cajan* (L.) Millspaugh] using SSR markers. *Journal of Applied Genetics, 58*(3), 307–319.

Paul, A., & Singh, D. P., (2002). Induced chlorophyll mutations in lentil (*Lens culinaris* Medik). *Indian Journal of Genetics and Plant Breeding, 62*(3), 263–264.

Paul, A., & Singh, D. P., (2005). Frequency, spectrum and segregation pattern of chlorophyll and macromutations in field pea (*Pisum satvum*). *Legume Research-An International Journal, 28*(2), 94–98.

Paul, P. J., Samineni, S., Thudi, M., Sajja, S. B., Rathore, A., Das, R. R., & Gaur, P. M., (2018). Molecular mapping of QTLs for heat tolerance in chickpea. *International journal of Molecular Sciences, 19*(8), 2166

Pratap, A., Chaturvedi, S. K., Tomar, R., Rajan, N., Malviya, N., Thudi, M., & Singh, N. P., (2017). Marker-assisted introgression of resistance to fusarium wilt race 2 in Pusa 256, an elite cultivar of desi chickpea. *Molecular Genetics and Genomics, 292*(6), 1237–1245.

Pundir, R. P. S., & Reddy, K. N., (1989). Induction, genetics and possible use of glabrousness in chickpea. *Euphytica, 42*(1), 141–144.

Ramakrishna, G., Kaur, P., Singh, A., Yadav, S. S., Sharma, S., Singh, N. K., & Gaikwad, K., (2021). Comparative transcriptome analyses revealed different heat stress responses in pigeonpea (*Cajanus cajan*) and its crop wild relatives. *Plant Cell Reports, 40*(5), 881–898.

Rathinam, M., Mishra, P., Vasudevan, M., Budhwar, R., Mahato, A., Prabha, A. L., & Sreevathsa, R., (2019). Comparative transcriptome analysis of pigeonpea, *Cajanus cajan* (L.) and one of its wild relatives *Cajanus platycarpus* (Benth.) Maesen. *PLoS One, 14*(7), e0218731.

Reif, J. C., Hamrit, S., Heckenberger, M., Schipprack, W., Maurer, H. P., Bohn, M., & Melchinger, A. E., (2005). Trends in genetic diversity among European maize cultivars and their parental components during the past 50 years. *Theoretical and Applied Genetics, 111*(5), 838–845.

Rezene, Y., Tesfaye, K., Mukankusi, C., Ratz, B., & Gepts, P., (2019). Marker-assisted pyramiding resistance genes against angular leaf spot and common bacterial blight disease into preferred common bean cultivar. *Molecular Plant Breeding, 10.*

Ribaut, J. M., & Ragot, M., (2007). Marker-assisted selection to improve drought adaptation in maize: The backcross approach, perspectives, limitations, and alternatives. *Journal of Experimental Botany, 58*(2), 351–360.

Roorkiwal, M., Jarquin, D., Singh, M. K., Gaur, P. M., Bharadwaj, C., Rathore, A., & Varshney, R. K., (2018). Genomic-enabled prediction models using multi-environment trials to estimate the effect of genotype× environment interaction on prediction accuracy in chickpea. *Scientific Reports, 8*(1), 1–11.

Roorkiwal, M., Rathore, A., Das, R. R., Singh, M. K., Jain, A., Srinivasan, S., & Varshney, R. K., (2016). Genome-enabled prediction models for yield related traits in chickpea. *Frontiers in Plant Science, 7*, 1666.

Ruhimbana, C., & Mutlu, N., (2019). Marker-assisted pyramiding potyvirus resistance genes into Rwandan common bean (*Phaseolus vulgaris* L.) genotypes. *Mediterranean Agricultural Sciences, 32*(3), 381–385.

Sab, S., Lokesha, R., & Mannur, D. M., (2020). Genome-wide SNP discovery and mapping QTLs for seed iron and zinc concentrations in chickpea (*Cicer arietinum* L.). *Frontiers in Nutrition, 7.*

Sachdeva, S., Bharadwaj, C., Singh, S., Roorkiwal, M., Sharma, V. I. N. A. Y., Singh, A. N. U. P. A. M., & Varshney, R. K., (2019). Yield plasticity and molecular diversity analysis in chickpea (*Cicer arietinum*). *Indian Journal of Agricultural Sciences (TSI), 89*(5), 834–841.

Sahu, A. R., Mishra, R. R., Rath, S. C., & Panigrahi, J., (2015). Construction of interspecific genetic linkage map of pigeonpea using SCoT, RAPD, ISSR markers and simple inherited trait loci. *The Nucleus, 58*(1), 23–31.

Samineni, S., Sen, M., Sajja, S. B., & Gaur, P. M., (2020). Rapid generation advance (RGA) in chickpea to produce up to seven generations per year and enable speed breeding. *The Crop Journal, 8*(1), 164–169.

Sandhu, K. S., You, F. M., Conner, R. L., Balasubramanian, P. M., & Hou, A., (2018). Genetic analysis and QTL mapping of the seed hardness trait in a black common bean (*Phaseolus vulgaris*) recombinant inbred line (RIL) population. *Molecular Breeding, 38*(3), 1–13.

Sangle, S. M., (2018). Induction of some promising mutants in pigeon pea. *Int. J. Res. Anal. Rev., 5*, 585–589.

Sankaran, S., Khot, L. R., Quirós, J., Vandemark, G. J., & McGee, R. J., (2016). UAV-based high-throughput phenotyping in legume crops. In: *Autonomous Air and Ground Sensing Systems for Agricultural Optimization and Phenotyping* (Vol. 9866, p. 98660G). International Society for Optics and Photonics.

Saxena, R. K., Molla, J., Yadav, P., & Varshney, R. K., (2020). High resolution mapping of restoration of fertility (Rf) by combining large population and high-density genetic map in pigeonpea [*Cajanus cajan* (L.) Millsp]. *BMC Genomics, 21*(1), 1–8.

Saxena, R. K., Singh, V. K., Kale, S. M., Tathineni, R., Parupalli, S., Kumar, V., & Varshney, R. K., (2017). Construction of genotyping-by-sequencing based high-density genetic maps and QTL mapping for fusarium wilt resistance in pigeonpea. *Scientific Reports, 7*(1), 1–11.

Saxena, S., Sahu, S., Kaila, T., Nigam, D., Chaduvla, P. K., Rao, A. R., & Gaikwad, K., (2020). Transcriptome profiling of differentially expressed genes in cytoplasmic male-sterile line and its fertility restorer line in pigeon pea (*Cajanus cajan* L.). *BMC Plant Biology, 20*(1), 1–24.

Scegura, A. L., (2017). *Marker Assisted Backcross Selection for Virus Resistance in Pea (Pisum sativum L.).* Doctoral dissertation, North Dakota State University.

Seyedimoradi, H., Talebi, R., Kanouni, H., Naji, A. M., & Karami, E., (2019). Agro-morphological description, genetic diversity and population structure of chickpea using genomic-SSR and ESR-SSR molecular markers. *Journal of Plant Biochemistry and Biotechnology, 28*(4), 483–495.

Shah, T. M., Atta, B. M., Alam, S. S., Ali, H., Haq, M. A., & Hassan, M., (2010). High yielding Kabuli mutant chickpea (*Cicer arietinum* L.) variety" cm 2008". *Pakistan Journal of Botany, 42*(5), 3533–3545.

Sharma, A. K., Singh, V. P., & Sarma, M. K., (2007). Induced seed and pod color mutations in urdbean (*Vigna mungo* (L.) Hepper). *Indian Journal of Genetics and Plant Breeding, 67*(3), 270.

Singh, A., Dikshit, H. K., Mishra, G. P., Aski, M., & Kumar, S., (2019). Association mapping for grain diameter and weight in lentil using SSR markers. *Plant Gene, 20*, 100204.

Singh, A., Sharma, V., Dikshit, H. K., Aski, M., Kumar, H., Thirunavukkarasu, N., & Sarker, A., (2017). Association mapping unveils favorable alleles for grain iron and zinc concentrations in lentil (*Lens culinaris* subsp. culinaris). *Plos One, 12*(11), e0188296.

Singh, G., Sareen, P. K., Saharan, R. P., Singh, A., Singh, G., & Singh, A., (2001). Induced variability in mungbean (*Vigna radiata* (L.) Wilczek). *Indian Journal of Genetics and Plant Breeding, 61*(3), 281–282.

Singh, N., Mallick, J., Sagolsem, D., Mandal, N., & Bhattacharyya, S., (2018). Mapping of molecular markers linked with MYMIV and yield attributing traits in mungbean. *Indian J. Genet, 78*(1), 118–126.

Singh, P. K., & Sadhukhan, R., (2019). Ems and gamma radiation induced mutation in grasspea (*Lathyrus sativus* L.). *Legume Research-An International Journal, 42*(3), 300–307.

Singh, S., Bisht, I. S., & Dabas, B. S., (1987). Screening of urdbean germplasm for field resistance to yellow mosaic virus. *Indian Journal of Mycology and Pathology, 17*, 216–217.

Solanki, R. K., Gill, R. K., Verma, P., & Singh, S., (2011). Mutation breeding in pulses: An overview. *Breeding of Pulse Crops* (pp. 85–103). Kalyani Publishers, Ludhiana.

Tafesse, E. G., Gali, K. K., Lachagari, V. B., Bueckert, R., & Warkentin, T. D., (2020). Genome-wide association mapping for heat stress responsive traits in field pea. *International Journal of Molecular Sciences, 21*(6), 2043.

Tao, Y., Liu, F., Bi, C., Ren, X., Qu, R., Li, K., & Xu, Q., (2018). Amplification of CASase gene from *Lathyrus sativus* and construction of its knock-out vector via CRISPR/Cas9 system. *Journal of Northwest A & F University-Natural Science Edition, 46*(8), 23–38.

Tar'an, B., Buchwaldt, L., Tullu, A., Banniza, S., Warkentin, T. D., & Vandenberg, A., (2003). Using molecular markers to pyramid genes for resistance to *Ascochyta* blight and anthracnose in lentil (*Lens culinaris* Medik). *Euphytica, 134*(2), 223–230.

Thudi, M., Chen, Y., Pang, J., Kalavikatte, D., Bajaj, P., Roorkiwal, M., & Varshney, R. K., (2021). Novel genes and genetic loci associated with root morphological traits, phosphorus-acquisition efficiency and phosphorus-use efficiency in chickpea. *Frontiers in Plant Science, 12*.

Tock, A. J., Fourie, D., Walley, P. G., Holub, E. B., Soler, A., Cichy, K. A., & Miklas, P. N., (2017). Genome-wide linkage and association mapping of halo blight resistance in common bean to race 6 of the globally important bacterial pathogen. *Frontiers in Plant Science, 8*, 1170.

Varshney, R. K., Gaur, P. M., Chamarthi, S. K., Krishnamurthy, L., Tripathi, S., Kashiwagi, J., & Jaganathan, D., (2013). Fast-track introgression of "QTL-hotspot" for root traits and other drought tolerance traits in JG 11, an elite and leading variety of chickpea. *The Plant Genome, 6*(3).

Varshney, R. K., Mohan, S. M., Gaur, P. M., Chamarthi, S. K., Singh, V. K., Srinivasan, S., & Kaur, L., (2014a). Marker-assisted backcrossing to introgress resistance to *Fusarium* wilt race 1 and *Ascochyta* blight in C 214, an elite cultivar of chickpea. *The Plant Genome, 7*(1).

Varshney, R. K., Terauchi, R., & McCouch, S. R., (2014b). Harvesting the promising fruits of genomics: Applying genome sequencing technologies to crop breeding. *PLoS Biology, 12*(6), e1001883.

Verma, P., Goyal, R., Chahota, R. K., Sharma, T. R., Abdin, M. Z., & Bhatia, S., (2015). Construction of a genetic linkage map and identification of QTLs for seed weight and seed size traits in lentil (*Lens culinaris* Medik.). *PloS One, 10*(10), e0139666.

Wanga, M. A., Shimelis, H., Mashilo, J., & Laing, M. D., (2021). Opportunities and challenges of speed breeding: A review. *Plant Breeding, 140*(2), 185–194.

Wani, A. A., & Anis, M., (2008). Gamma ray-and EMS-induced bold-seeded high-yielding mutants in chickpea (*Cicer arietinum* L.). *Turkish Journal of Biology, 32*(3), 161–166.

Wani, M. R., & Khan, S., (2004). Ethylmethane sulphonate induced quantitative variability in mung bean. *Bionotes*, 111.

Watson, A., Ghosh, S., Williams, M. J., Cuddy, W. S., Simmonds, J., Rey, M. D., & Hickey, L. T., (2018). Speed breeding is a powerful tool to accelerate crop research and breeding. *Nature Plants, 4*(1), 23–29.

Witsarut, C., (2019). *Development of Mungbean Breeding Lines with Improved Resistance to Cercospora Leaf Spot and Powdery Mildew by molecular Marker Assisted Gene Pyramiding.* Doctoral dissertation, School of Crop Production Technology Institute of Agricultural Technology Suranaree University of Technology.

Wu, L., Fredua-Agyeman, R., Hwang, S. F., Chang, K. F., Conner, R. L., McLaren, D. L., & Strelkov, S. E., (2021). Mapping QTL associated with partial resistance to *Aphanomyces* root rot in pea (*Pisum sativum* L.) using a 13.2 K SNP array and SSR markers. *Theoretical and Applied Genetics*, 1–26.

Yadav, P., Saxena, K. B., Hingane, A., Kumar, C. S., Kandalkar, V. S., Varshney, R. K., & Saxena, R. K., (2019). An "Axiom Cajanus SNP array" based high density genetic map and QTL mapping for high-selfing flower and seed quality traits in pigeonpea. *BMC genomics, 20*(1), 1–10.

Zadražnik, T., Egge-Jacobsen, W., Meglič, V., & Šuštar-Vozlič, J., (2017). Proteomic analysis of common bean stem under drought stress using in-gel stable isotope labeling. *Journal of Plant Physiology, 209*, 42–50.

Zhang, C., McGee, R. J., Vandemark, G., & Sankaran, S., (2021). Crop performance evaluation of chickpea and dry pea breeding lines across seasons and locations using phenomics data. *Frontiers in Plant Science, 12*, 61.

Zhang, X., Scheuring, C. F., Zhang, M., Dong, J. J., Zhang, Y., Huang, J. J., & Zhang, H. B., (2010a). A BAC/BIBAC-based physical map of chickpea, *Cicer arietinum* L. *BMC Genomics, 11*(1), 1–10.

Zhang, Z., Ersoz, E., Lai, C. Q., Todhunter, R. J., Tiwari, H. K., Gore, M. A., & Buckler, E. S., (2010b). Mixed linear model approach adapted for genome-wide association studies. *Nature Genetics, 42*(4), 355–360.

Zheng, Y., Xu, F., Li, Q., Wang, G., Liu, N., Gong, Y., & Xu, S., (2018). QTL mapping combined with bulked segregant analysis identify SNP markers linked to leaf shape traits in *Pisum sativum* using SLAF sequencing. *Frontiers in Genetics, 9*, 615.

CHAPTER 14

Role of Next-Generation Sequencing Technology in Crop Improvement

DIVYA CHAUHAN,[1] GEETA PRASAD,[2] DEVENDRA SINGH,[3] and HIMANSHU PANDEY[4]

[1]*Banasthali Vidyapeeth, Jaipur, Rajasthan, India*

[2]*ICAR–National Bureau of Plant Genetic Resources, New Delhi, India*

[3]*Institute of Biosciences and Technology, Shri Ramswaroop Memorial University Barabanki, Uttar Pradesh, India*

[4]*P.G. Department of Agriculture Khalsa College Amritsar, Punjab, India*

ABSTRACT

Several species, including crops with significant economic importance, have been utilized in whole-genome sequencing technology. The evolution of next-generation sequencing (NGS) technology has reached the sequencing of single-nucleotide using massively parallel sequencing platforms. This development has provided us with a better understanding of studying genomics on different levels, including transcriptomics, metabolomics, and proteomics. Moreover, the identification of molecular markers generated by NGS technology helps in finding genetic diversity among different crop species. The growing demand and easy accessibility of high throughput platforms with reduced costs have enabled sequencing technology significant for food security. Here we discuss the current status of whole-genome sequencing, crop resequencing, and applications of NGS in crop improvement.

Smart Breeding: Molecular Interventions and Advancements for Crop Improvement. Kailash Chandra, Subhash Chand, Ravi Prakash Saini, and Rohit Sharma (Eds.)

14.1 INTRODUCTION

DNA (Deoxyribonucleic acid) was first identified as the genetic material in 1944 by O. T. Avery. In 1953, Watson & Crick determined its double-helical structure (Church et al., 1984). DNA sequencing technology has become a revolution in genetics and genomics. In 1977, Frederick Sanger developed DNA sequencing technology, identifying the nucleotide sequences (linear) by electrophoretic parting of randomly terminated extension products. This sequencing is considered the "first-generation" of DNA sequencing technologies (Sanger et al., 1977). Still, this traditional sequencing is used for several uses; the majority of expressed gene and DNA sequence data produced currently are generated via a variety of NGS (next-generation sequencing) technology. NGS technology includes second and third-generation sequencing methods (Meyerson et al., 2010; Schadt et al., 2010). The difference between both is that the second-generation sequencing methods require the libraries preparation of amplified sequencing before the amplified DNA clones sequencing. In contrast, 3rd generation sequencing includes single molecular sequencing without generating amplification libraries (Thompson et al., 2011).

NGS, also called a high-throughput (HT) sequencing technique, is currently in high demand due to its potential of producing large sequence data in a short duration of time. Therefore, due to its powerful and flexible nature, NGS is considered an indispensable tool for a broad spectrum of scientific research in biological science in the present era. It gives the chance to explore genetic diversity in plants and their wild relatives on a much bigger scale that was impossible with the earlier technologies (Varshney et al., 2009). Rapid progress in NGS technology and bioinformatics tools led to generating reference and *de novo*-based genome sequences for plants and animals. NGS is also employed for whole genome sequencing (WGS) since NGS supports WGS analysis utilizing substantial computational and biological resources to produce large sequence data (Hugo et al., 2012). Besides using NGS for whole-genome sequencing, it can also be used for RNA sequencing (RNA-Seq) (Marco-Puche et al., 2019), whole-exome sequencing (WES) (Bahareh et al., 2014), targeted or candidate gene sequencing (CGS) (Zhong et al., 2018) and methylation sequencing (Michael et al., 2012). NGS facilitates understanding of plant genetic architecture and comportment of plant genomes. NGS-based exome sequencing, WGS, and transcriptome sequencing will help in discovering agronomically important molecular markers and genes associated with various important agronomic traits that regulate the tolerance to abiotic and biotic stresses and also the yield. Thus, next-generation sequencing technology provides better opportunities for studying crop genomics and other post-genomics like metabolomics, proteomics, transcriptomics, leading to a gene revolution of crop breeding. The chapter gives an overview of the advances and applications of NGS, and platforms related to this technology and storage for plants and crop improvements in the world of "omic" expansion.

14.2 TYPES OF NGS PLATFORMS

Since the last few years, DNA sequencing technology and its application have undergone remarkable development (Goodwin et al., 2016). Today it is considered the engine of the genome era that represents an enormous amount of genome data. Few firms have developed different high-performing sequencing instruments capable of producing sequencing reactions simultaneously. Description of the NGS platforms that are being presently used for sequencing is provided below:

- Roche sequencing;
- Illumina sequencing;
- SOLID sequencing;
- Ion torrent: Proton sequencing;
- Oxford nanopore sequencing.

14.2.1 ROCHE SEQUENCING

The initial NGS technologies were introduced in 2005 by Rothberg and co-workers, and further, it was named as 454 Genome Sequencer. It incorporates sequencing-by-synthesis methodology and employs pyrosequencing with PPi (Margulies et al., 2005). It is based on the association between emulsion PCR and pyrosequencing. Briefly, pyrosequencing involves the recognition of pyrophosphate that can be observed during the integration of a nucleotide base at the time of the newly synthesized DNA strand. At the time of sequencing, primer gets attached with a biotin-labeled single-stranded DNA all together combined with the four enzymes: luciferase, DNA polymerase, ATP sulfurylase, and apyrase, and the substrates (Pal, 2001; Gharizadeh et al., 2007). At the beginning of sequencing, the first step includes random fragmentation of DNA samples or production of amplicons of suitable size followed by library preparation where DNA fragments or amplicons are ligated to 454 sequencing adaptors. These adapters facilitate the anchorage of DNA sequences to the streptavidin beads present on primers. Each bead is generally linked with a single segment and is trapped into individual emulsion droplets. Subsequently, dNTPs are added to the PCR master mixture; later, these droplets are exposed to emPCR, which generates a large number of copies of each DNA segment on the surface of the individual bead. It is considered that a single bead relates to a single read.

Further, template beads are placed on a pico-titer plate. The sequencing cascade proceeds via the integration of nucleotides, leading to the synthesis of DNA segments using a pyrosequencing reaction. Thus, conversion of PPi to ATP is perceived by quantifying the light emitted after integrating a single nucleotide base by DNA polymerase (Mardis et al., 2008). At every step of the cycle, CCD cameras are involved in signals acquisition from spots on PTP, then captured photos are processed for base calling. Roche Sequencing can produce reads with length ranges from 450 to 700 Mb.

14.2.2 ILLUMINA (SOLEXA) SEQUENCING

Illumina technology was first introduced in 2006 to generate better-quality reads of enormous information at a decreased cost (Balasubramanian et al., 2015). It is a sequencing technology based on sequencing by synthesis (SBS). Initially, the sequencing steps involve random fragments of DNA samples into sequences followed by 5′ and 3′ adapter ligation. Later, adapter-ligated fragments are loaded on a slide with complementary adapters located on a slide followed by bridge PCR amplification that produces numerous similar replicas of sequence. Further, illumine technology works on SBS methodology that depends on the reversible terminator. This process includes four modified nucleotides, i.e., nucleotides are labeled with a specific fluorescent dye.

Additionally, nucleotides have an inactive 3′-hydroxyl group to ensure incorporation of a single nucleotide at a time. Along with dNTP sequencing primers and DNA, polymerases are incorporated as a mix, and these primers get hybridized with the sequences. Together this complex is considered as a cluster. Due to the laser effect, clusters are excited and emit a light signal specific to each nucleotide during the sequencing process. It is detected via a CCD camera that later translates signals into nucleotides sequence with the help of computer programs. The following reaction continues with removing the terminator with the fluorescent label and the starting of another cycle with the effective addition of a base nucleotide. It has been seen that illumine sequencers have given output data of higher than 600 Gbps, and lengths of short reads are about 125 bp.

14.2.3 ABI/SOLID SEQUENCING

Supported Oligonucleotide Ligation and Detection is an additional NGS technology acquired by applied biosystems (ABI) in 2007 and named ABI/SOLID sequencing technology. It is based on sequencing by ligation methodology (Bentley et al., 2008). It involves library construction of DNA fragments, attachment of adapters to the DNA sequences followed by fixing on beads and later cloned by emPCR. The following beads are immobilized onto a polyacrylamide gel-based microscope glass with a fluorescent-tagged end, are successively ligated to DNA segments, and the color emitted by the labeled nucleotide sequences is documented. This sequencing system utilizes four fluorescent dyes to characterize 16 possible combinations of two bases, and unbound probes get degenerated.

Furthermore, the sequencer keeps repeating the ligation cycle following removal of the complementary strand and begins a current sequencing cycle at the point n–1 of the template. The cycle continues unless a single base is sequenced twice. The emitted colors are captured as signals that are converted and represented as DNA bases. Thus, the process helps in determining the sequence of the DNA segment (Shendure et al., 2008). It has shown that ABI/SOLiD is the primary sequencer that can generate shorter reads with a length of 35 bp and an output of 3 Gb per run. However, recent advancement has increased the paired read length up to 2×50 with an output up to 320 Gb per run and 99.99% accuracy (Mardis et al., 2008). ABI/SOLiD is highly flexible and performs real-time run tracking without replicating the slide. Two different experimentation works can be carried out using individual SOLiD analyzers simultaneously.

14.2.4 ION TORRENT: PROTON/PGM SEQUENCING

Thermo-Fisher Scientific has developed another NGS platform known as Ion Torrent. It is a semiconductor-based technique where slight pH changes result in the discharge of H^+ ions as a by-product of nucleotide integration that is detected by the machine. The sequencing activity takes place on a CMOS chip. PGM sequencing technique recognizes incorporated bases avoiding the application of fluorescence. It directly counts the emission level of H^+ ions while dNTPs incorporation during DNA strand synthesis leads to the release of H^+ ions that consequently alters the pH in the micro-environment by creating a higher positive voltage (Rothberg et al., 2011). The CMOS chip has microwells with a bead carrying many similar DNA fragments. ISFET sensor located to the bottom of a microwell acts as a pHFET. Sequencing begins with DNA or RNA fragmentation of approximately 200–400 bp. that gets attached to an individual bead in a different well of CMOS chip, acting as a template for the sequencing reaction. Since each bead consists of large numbers of clonally amplified template DNA produced through the emPCR process, each bead is retained in its specific well and allows the limited release of H^+ ions during the incorporation of individual nucleotides.

Further, pH changes occur due to H^+ ions are being identified by the ISFET that is transformed into a voltage signal directly proportional to the number of nucleotides added. The PGM sequencers have many advantages in comparison to other NGS platforms. For instance, it generates longer read lengths up to 200 bp, 400 bp, and 600 bp sizes compared to different NGS platforms. In addition, it incorporates unchanged bases as an alternative to fluorescently labeled dNTPs, and signals are produced as hydrogen ions in place of specific fluorescence. Ion-Torrent provides enhanced enzymatic action, longer reads, and inexpensive consumables.

14.2.5 OXFORD NANOPORE SEQUENCING

This technology works on single nucleotide sequencing that does not use DNA amplification through a protein nanopore. The process involves moving DNA sequences through the pore, such that the sensor recognizes changes in the ionic current to the corresponding passing nucleotide (Deamer et al., 2016). It holds advantages over the current technologies as it sequences the strand of nucleotides regardless of the size. Therefore, the library preparation step is the real challenge, which requires loading high molecular weight DNA on flow cells. Secondly, it does not utilize an imaging device to detect the nucleotide as it works on the MinION device, which is connected to computers for visualization.

Therefore, the technology has been used from bacterial genomes to humans and plants, ranging from *A. thaliana* with a genome size of 119.5 Mbp (Michael et al., 2018) to Chrysanthemum with a 2.53 Gbp genome (Song et al., 2018). Recently, GPU computing has made new improvements in processing raw data generated by this technology (Noble et al., 2017), thereby increasing computing speed compared to CPU-based computers.

14.3 GENOMES DIRECTLY IMPACTING CROP RESEARCH

Genomics plays a pivotal role in improving crop breeding. Recent progress in genomics provides modern tools and technologies that offer precise and efficient breeding programs. Genomics tools such as functional molecular markers aid the identification of relative alleles at a locus in a single population. Besides, it also facilitates information related to statistical and inheritance phenomena in crop species, the gene-editing and identification methods for exposing the candidate genes, and genetic interactions in plants in addition to genomics breeding by a marker-assisted selection of variants. Further, advances in genomics and molecular technologies lead to the characterization and isolation of various agronomically essential genes, i.e., dormancy, increasing seed number, controlling seed shattering and size for their efficiency, and improving inbreeding (Doebley et al., 2006). The crop improvement has gained new momentum with the availability of NGS technologies that enable WGS, resequencing, and transcriptome sequencing at a faster rate at low prices.

Further, reference genomes of model plant species and key food crops impart a solid platform for improving the crop. NGS technology has commenced the availability of reference genome sequences of more than 20 crop and plant species (Table 14.1) (Van et al., 2013).

Furthermore, functional genomics approaches like GWAS (genome-wide association studies) and gene expression profiling based on microarrays assist in the investigation of novel genes linked with abiotic and biotic stresses. In addition, advancement in sequencing technologies can also imply de-novo sequencing. It is done for genome sequences of crop species; those reference genome sequences are not available on the database.

Moreover, transcriptome sequencing for gene expression analysis, targeted sequencing of specific chromosomal regions, resequencing approaches, mutation mapping, and the study of protein-DNA interaction and small RNAs on various crop species in the area of plant breeding. Moreover, these technologies are also involved in efficiently discovering and developing SSR or microsatellite loci. These markers are commonly used to construct linkage maps, study gene flow, cultivar fingerprinting MAS, and QTL mapping. Further, detection of single nucleotide polymorphism (SNP) based markers in plant species irrespective of reference genome availability or not (Kilian & Graner, 2012). With the aid of NGS, the determination of novel alleles/genes for different agronomic traits employing the genotyping-by-sequencing (GBS) approach is possible (Kole et al., 2015).

The development of technologies like NGS has contributed to a better understanding of the plant genome and has become a new promising technique for improving crops according to future needs. Despite the specific advantages, all these technologies are challenged with a few technical flaws. For example,

TABLE 14.1 A List of Different Platforms Used for Reference Genome Sequenced

Species	Genome Size	Platform Used	References
Simmondsia chinensis (jojoba)	887 Mbp	SMRT PacBio genome sequencing	Sturtevant et al. (2020)
Fontinalis antipyretica (greater water-moss)	385.2 Mbp	BGISEQ-500 and 10X	Yu et al. (2020)
Gnetum monatum (Gam nui)	4.07 Gb	Illumina sequencing	Wan et al. (2018)
Abies alba (European silver fir)	18.16 Gb	Roche-KAPA biosystem	Mosca et al. (2019)
Annona muricata (soursop)	799.11 Mb	Illumina and PacBio short reads, in combination with 10× Genomics and bio-nanodata	Strijk et al. (2021)
Macadamia integrifolia (Macadamia nut)	745 Mbp	PacBio RSII system	Nock et al. (2020)
Macadamia jansenii (Bulburin Nut Tree.)	750 Mbp	Compared nanopore, Illumina, and BGI stLRF data	Murigneux et al. (2020)
Suaeda aralocaspica (Seablite)	467 Mb	Illumina HiSeq 2000	Wang et al. (2019)
Corylus heterophylla Fisch (Asian hazel)	370.75 Mbp	Nanopore/Hi-C chromosome scale.	Zhoa et al. (2021)
Corylus mandshurica (Asian Hazel)	367.67 Mbp	Illumina HiSeq 2000, PromethION sequencer	Li et al. (2021)
Eruca sativa (salad rocket)	851 Mbp	Illumina MiSeq and HiSeq 2500.	Bell et al. (2020)
Erysimum cheiranthoides (wormseed wallflower)	175 Mbp	PacBio and Illumina transcriptome sequencing	Osuna et al. (2020)
Cucurbita argyrosperma (Cushaw Pumpkin)	255.2 Mbp	PacBio and Illumina HiSeq 4000	Barrera-Redondo et al. (2021)
Ricinus communis L. (Wild Castor)	318.13 Mbp	PacBio sequel sequencing platform	Xu et al. (2021)
Dorifera (fragrant rosewood)	653 Mbp	Illumina HiSeq 4000	Hung et al. (2020)
Eucalyptus pauciflora (Snow gum)	594.87 Mbp	Nanopore and Illumina sequencing	Wang et al. (2020)
Euca alternifolia (tea tree)	362 Mbp	PacBio sequel single-molecule real-time (SMRT) cells	Voelker et al. (2021)
Riobotrya japonica (Loquat)	760.1 Mbp	Illumina + Nanopore + Hi-C	Jiang et al. (2020)
Gillenia trifoliata (bowman's root)	320.17±4.22 Mbp	Illumina HiSeq 2000 system	Su et al. (2021)
Prunus salicina (Japanese plum)	284.2 Mbp	PacBio/Hi-C	Liu et al. (2020)
Acer truncatum (purple blow maple)	633.28 Mbp	Illumina and Pacbio	Ma et al. (2020)

it requires optimization and improvements in data analysis and management, loading of sequence, and NGS accuracy. Efforts are underway employing NGS in metagenomic studies, epigenetic modification (GM), and gene mining.

14.4 EVOLUTION OF NGS

Genome sequencing technology has paralleled the methodology used to sequence the human genome. The sequencing of the first plant genome was simultaneously completed with the drafting of the human genome (Pollack et al., 2000). The methodology followed by the Arabidopsis project developed a clone-by-clone finished genome. This strategy covers a physical map of BAC (bacterial artificial chromosome) or cosmid clones to form contigs to cover the MGS (mapped genome space) (Mozo et al., 1999). According to MSU6 (http://ww.gramene.org), the latest release of the rice genome has 373.2 Mbp of the cloneable portion of the genome. The difficulty with clone-based BAC projects has been coordinating mapping and sequencing, which has halted projects for a lifetime.

With the advancement in technology, Myers (Weber & Myers, 1997) proposed strategies and application of whole-genome sequencing (WGS) of vertebrates which were adopted by different genomic centers. This technology aims to remove the steps required for massive library construction, which are utilized by the clone-by-clone process, followed by physical mapping to reduce repetitive portions of the genome and therefore, have added computational steps for reconstructing the genome sequence. By following this strategy, different-sized DNA inserts are produced, later sequenced from both ends (Batzoglou et al., 2002). Also, a computer algorithm is used to compare these DNA sequences to rebuild a single linear DNA sequence. This technology holds an advantage as it takes less time than BAC clone library-based sequencing. However, this strategy carries drawbacks in producing a complete genome sequence.

Initial attempts on WGS sequencing showed promising results in the grapevine (Jaillon et al., 2007), papaya (Ming et al., 2008), and *Physcomiterlla* (Rensing et al., 2008) and therefore has laid the path for Sanger based reference plant genome sequencing in past three years. In recent years, the completion of draft WGS and annotation of different plants have been achieved, including *S. bicolor* and *G. max* (Paterson et al., 2009; Schmutz et al., 2010). These genomes can be used as a reference to bring advancement in genomic studies. Yet, the largest maize reference genome was produced by the clone-by-clone approach in 2009 (Schnable et al., 2009); it accounts for a total of 3 Gbp in size. Even after producing a large majority of reference sequencing in previous years, a shift has been observed to short-read pyrosequencing, which includes *de novo* references for plant genome and SBS (Fuller et al., 2009).

The second-generation technology provides an advantage over Sanger-based sequencing as Illumina produces sequence data that is 100 times cheaper than the Sanger method and is widely used to sequenced plant genomes. Presently, new projects use BAC sequences and short contigs generated from pyrosequencing, sanger-based technology, and sequencing by ligation. However, there are a few limitations, too, as the reference genome can be less accurate than sanger-based assemblies. These problems arise due to short-length sequences and combined methodologies from NGS platforms. To date, only two published genomes combine sanger-based and pyrosequencing on apple (Velasco et al., 2010) and cucumber (Huang et al., 2009). Since sequencing by ligation produces cheaper results, the only vertebrate sequenced through this technology is the panda genome (Li et al., 2012). However, the WGS strategies that will be applicable to crop-based scientific applications are yet to be explored.

14.5 NGS: A TOOL FOR GENE DISCOVERY AND HIGH-RESOLUTION MAPPING

Using NGS technologies, various approaches utilizing high-resolution mapping have resulted in rapid gene discovery. Few such important platforms will be discussed in this section. On a large scale, high-density genotyping is used to develop marker profiles for different individuals. Nowadays, SNP array-based genotyping platforms are chosen over high-density genotyping because the output generated is less demanding for analysis. Cai et al. (2017) conducted a study that revealed eight SNPs associated with (abiotic) salt stress using CottonSNP80K SNP chip. Another study in wheat resulted in the development of a TaBW280K chip containing 280,226 SNPs, and germplasm diversity was assessed by Rimbert et al. (2018).

With the advent of bulked segregant RNA seq technology, crops with complex genomes such as wheat, where resequencing remains difficult and cost-ineffective, are used for gene discovery. This technology is useful in crops lacking reference genomes and is similar to QTL-Seq. It sequences the RNA for a trait for interest from extreme bulks and identifies the target region to develop markers from the region where the gene of interest lies. A study by Liu et al. (2012) assigned a gene in maize related to MYB transcription factor lying in the 2 Mb region and identified markers associated with it. The gene in maize codes for long-chain fatty acid synthesis. Another study in wheat reported fine mapping of GPC (grain protein content) present in the 30 cM regions. The study revealed around 15 candidate genes for GPC (Ramirez-Gonzalez et al., 2015). The above-mentioned fine-mapping can help analyze germplasms to be incorporated in marker-assisted selection. Klein et al. (2018) studied mutant genes in maize that associate mapping intervals with candidate SNPs produced through whole-genome sequencing. Hence, this approach is preferred for maker-linked genes and gene discovery. As bulk segregant analysis (BSA) combines with sequencing methods, it allows fine mapping and QTL detection at a stretch in a smaller genomic segment. A 7 Mb QTL region was identified in chickpea to reveal 35 kb of the genomic region associated with 100-seed weight on CaLG01 (Das et al., 2015). Likewise, the inbreeding program QTL-Seq is used for fine mapping resistance from bacterial wilt and utilized for developing diagnostic markers in groundnut (Luo et al., 2019).

The latest breeding programs use several approaches for effective and efficient technologies combined with a whole breeding period. The advancement in NGS technologies has left a significant impact in the arena of plant genomics for non-model and model crops (Varshney et al., 2013). Moreover, the gap between phenotype and genotype can be bridged using GWAS (Genome-wide association studies) and RNA-Seq.

14.6 APPLICATIONS

14.6.1 GENETIC DIVERSITY

Changes in the genes due to different environments in crops display the ability of crops to acclimatize to changing environments. With latest practices and smaller germplasm of high-yielding crops have led to the decline in crop genetic diversity (Hyten et al., 2006). Moreover, new insect pests, diseases, and climate change can severely affect the crop population. Few studies reveal the failure of crops to a rapidly changing environment that has threatened long-term food security (Turner et al., 2009; Brown & Funk, 2008). Therefore, genetic diversity can be enhanced by using elite germplasm, landraces, and mutants. In earlier times, plant accessions were represented by their phenotypic traits, morphology, geographical distribution, and pedigree (Hammer, 2003). The development of molecular markers has significantly contributed to genetic diversity analysis. These markers have been successfully used in studies determining population structure and phylogenetic relationships; marker-assisted selection, map-based cloning, and QTL mapping (Gupta et al., 2010; Moose & Mumm, 2008). However, very few genome-based databases are available to provide information such as mutants, QTL's expression data, physical maps, and genetic markers. Therefore, crop improvement should be facilitated for developing an integrated database that curates the genome sequences.

Yet some obstructions do exist, like contributions of managers/developers behind the platform, shortage of financial support for the development of the database, and also compromises the quality of the genome sequences. Moreover, the fragmentation of the genome poses a difficulty in creating well-documented information. NGS plays a vital role in resequencing the genome and exhibits as a vital tool for crop improvement. As there are two kinds of assembly, *viz. de novo*, and reference-based, the former one is not compatible with plant genomes having short read lengths as it contains repeated sequences (Varshney et al., 2009). As a result, resequencing of those species is preferred, which have a reference genome such that identification of SNPs becomes easy and useful for marker development or exploring functional alleles (Yan et al., 2010; You et al., 2011).

Evolutionary studies and association mapping (AM) have become easy due to genome-wide SNP genotyping (Akhunov et al., 2009). As a result, many polymorphic markers are available from a diverse set of germplasm derived from SNP panels (Huang et al., 2010). Moreover, from genetic diversity analysis to the building of high-density maps from restriction site-associated DNA-based markers has gained success in NGS technologies. So far, many different crops, such as maize, soybean, and rice have given opportunities to exploit DNA-level diversity and explore relatedness among different crop species (Paterson et al., 2010). In order to understand the association of phenotype-genotype on a molecular basis, the prerequisite is the collection of large phenotypic data to effectively utilize genomic studies further. Another shortfall is a smaller number of trained breeders who can understand and apply genomic and phenomic technologies up to their potential. Hence, there is a need to have an advanced computational system that can make predictions and crosses based on the predictions associated with genomic and phenomic interaction. Scientists and engineers should collaborate to create an effective platform to fasten the collection time of phenotypes and to improve the mineral content and vigor of the plant.

14.6.2 TRANSCRIPTOME

The transcriptome is the collection of all RNA transcribed by a group of cells at a particular biological state. It includes the study of mRNAs and non-coding RNAs (nc-RNAs) responsible for encoding various proteins and acting as cellular regulators, respectively. Transcriptomic studies decipher better understanding of the transcripts and also offer essential insights on gene structure, expression, and regulation, which subsequently unveil molecular mechanisms of the specific biological processes and disease occurrence (Jain et al., 2012; Lowe et al., 2017). Recently, transcriptomics studies have advanced considerably due to the development of sequencing technology.

RNA-sequencing (RNA-seq) is an innovative approach to quantifying and mapping transcriptomes using HT sequencing technologies. It is used for finding the expressed sequence information in a particular cell type or tissue within a definite time. In this process, RNA is converted into a library of cDNA fragments with adaptors followed by sequencing of those fragments. Later, reads are obtained and aligned either to an assembled *de novo* or reference genome (Wang et al., 2019). Illumina technology is mainly useful for better analyzing plant transcriptomes at a worldwide level. Plant transcriptomics is widely applied to plants and crops and is likely to flourish in the coming years with the development of new technologies in systems biology. These technologies are helpful in gaining information on different cell types and their interaction with each other and can also reveal the selective effect of the different pathogen on specific cell types than the other. Therefore, in future research on plant sciences, transcriptomics will play a vital role.

14.7 ROLE OF MOLECULAR MARKERS IN CROP IMPROVEMENT

Molecular markers help in improving cereal breeding by analyzing germplasm and trait mapping. These markers can be deployed for parental genotype selection, removing linkage drag during the backcrossing, and performing phenotypic assays for trait selection. These markers were used for variety distinctiveness, agronomic traits, inferences of population history, and stability assessment. High-density linkage maps are a potential source for developing haplotypes for genes utilized in genome mapping (Duran et al., 2010). More significant insights into the cereal genome were obtained by comparing genetic linkage. Among different species, markers can relate co-segregation in diverse populations based on their linkage groups (LGs). Therefore, they act as a complementary device in traditional selection. It helps understand the phenotypic characteristics and genetic interactions, which guides them for an effective breeding strategy. These DNA markers hold an advantage over phenotypic markers as they are unaffected by the environment.

Over time, individual plants are observed within species as they were tagged to molecular makers. SNPs are the decisive form of molecular markers, the smallest element of inheritance, and represent a location where two individuals can be differentiated. Therefore, three different types of SNPs; the first is transversion, followed by transitions and insertions or deletions (InDels) (Edwards et al., 2007). Markers like SNPs are useful in the genetic analysis as they allow the exploitation of genetic composition to select the breeding process. With the development of technologies, different crop species with a variety of traits have the ability to expand marker-assisted selection at a reduced cost. At present, the advanced application in the breeding system uses AM and genomic selection to broaden the screening of a diverse and extensive panel of markers at little cost (Duran et al., 2010).

For instance, wheat SNPs are divided into two categories like varietal or homoeologous. The former differs on a single wheat subgenome among wheat varieties. These markers hold significance as they can identify the difference between population and individuals. Hence, it can also detect varietal SNPs. In contrast, homolog SNPs only differ between three genomes of wheat, i.e., AA, BB, and DD. These markers contain the sequence information that depicts the nature of variants. The sequence holds the information on the development of an organism and its adaptation to the environment. It finds the accurate position of markers where other markers are unable to locate (Batley & Edwards, 2007).

Moreover, there are a number of software for SNP discovery like autoSNP that provide evidence on polymorphism, gene annotation, and cultivars that help in examining the data. SNPs were discovered in various crops like barley, *Brassica* spp., and rice using this software (Duran et al., 2009).

Barbazuk et al. (2007) reported approx. 7,000 SNPs in maize line B73 and Mo17 with more than 85% success rate. Large data from the Illumina platform delivers SNPs from a wider genome. A study conducted by Lai et al. (2010) revealed 1 million SNPs in maize inbred lines associated with heterosis. A pipeline for second-generation sequencing called SGSautoSNP can predict SNPs using Illumina shotgun from the whole genome and has a success rate of more than 95%. Thus, SNP-based studies help in understanding the mechanism underlining the changes in the genome at the nucleotide level and can provide information that will be useful for marker-assisted breeding and genomic selection.

14.8 CONCLUSION

The revolution bought by NGS technologies in genome biology displays a significant role in crop improvement. This technology is much cheaper and provides faster sequencing data; however, wheat has a large genome size, and its complexity poses a limitation to NGS utility. The development of new bioinformatics software can enable NGS applications in crops like wheat to be accessible. It will also help in annotating and finding variation, and differential expression across the genome. Therefore, studies on the evolution of polyploidy and structure must be clearly known in order to do studies on the genome. The findings for genome-wide SNP markers in complex crops have already begun and are expected to reach immense heights on crop improvement. The real challenge is understanding the association of variations in sequence to the heritable agronomic traits. In the coming future, oxford nanopore technology will provide insights into the complex genome of plants. Hence developments are made in producing better base calling, which will help in transcriptome and epigenomic analysis.

KEYWORDS

- **crop improvement**
- **genome sequencing**
- **marker-assisted selection**
- **metabolomics**
- **next-generation sequencing**
- **proteomics**
- **sequencing**
- **transcriptomics**

REFERENCES

Akhunov, E., Nicolet, C., & Dvorak, J., (2009). Single nucleotide polymorphism genotyping in polyploid wheat with the Illumina golden gate assay. *TAG. Theoretical and Applied Genetics. Theoretische und Angewandte Genetik, 119*(3), 507–517.

Balasubramanian, S., (2015). Solexa sequencing: Decoding genomes on a population scale. *Clinical Chemistry, 61*(1), 21–24.

Banks, J. A., (2018). Fern genomes finally here. *Nature plants, 4*(7), 404, 405.

Barrera-Redondo, J., Sanchez-De La, V. G., Aguirre-Liguori, J. A., Castellanos-Morales, G., Gutiérrez-Guerrero, Y. T., Aguirre-Dugua, X., Aguirre-Planter, E., et al., (2021). The domestication of *Cucurbita argyrosperma* as revealed by the genome of its wild relative. *Horticulture Research, 8*(1), 1–14.

Batzoglou, S., Jaffe, D. B., Stanley, K., Butler, J., Gnerre, S., Mauceli, E., Berger, B., et al., (2002). ARACHNE: A whole-genome shotgun assembler. *Genome Research, 12*(1), 177–189.

Bell, L., Chadwick, M., Puranik, M., Tudor, R., Methven, L., Kennedy, S., & Wagstaff, C., (2020). The *Eruca sativa* genome & transcriptome: A targeted analysis of sulfur metabolism & glucosinolate biosynthesis pre & postharvest. *Frontiers in Plant Science, 11*, 1624.

Bentley, D. R., Balasubramanian, S., Swerdlow, H. P., Smith, G. P., Milton, J., Brown, C. G., Hall, K. P., et al., (2008). Accurate whole human genome sequencing using reversible terminator chemistry. *Nature, 456*(7218), 53–59.

Brown, M. E., & Funk, C. C., (2008). Food security under climate change. *Science, 319*(5863), 580, 581.

Cai, C., Zhu, G., Zhang, T., & Guo, W., (2017). High-density 80 K SNP array is a powerful tool for genotyping G. hirsutum accessions and genome analysis. *BMC Genomics, 18*(1), 654.

Church, G. M., & Gilbert, W., (1984). Genomic sequencing. *Proceedings of the National Academy of Sciences of the United States of America, 81*(7), 1991–1995.

Das, S., Upadhyaya, H. D., Srivastava, R., Bajaj, D., Gowda, C. L. L., Sharma, S., Singh, S., et al., (2015). Genome-wide insertion–deletion (InDel) marker discovery and genotyping for genomics-assisted breeding applications in chickpea. *DNA Research, 22*(5), 377–386.

Deamer, D., Akeson, M., & Branton, D., (2016). Three decades of nanopore sequencing. *Nature Biotechnology, 34*(5), 518–524.

Doebley, J. F., Gaut, B. S., & Smith, B. D., (2006). The molecular genetics of crop domestication. *Cell, 127*(7), 1309–1321.

Edwards, D., & Batley, J., (2010). Plant genome sequencing: Applications for crop improvement. *Plant Biotechnology Journal, 8*(1), 2–9.

Fuller, C. W., Middendorf, L. R., Benner, S. A., Church, G. M., Harris, T., Huang, X., Jovanovich, S. B., et al., (2009). The challenges of sequencing by synthesis. *Nature Biotechnology, 27*(11), 1013–1023.

Ganal, M. W., Altmann, T., & Röder, M. S., (2009). SNP identification in crop plants. *Current Opinion in Plant Biology, 12*(2), 211–217.

Gharizadeh, B., Ghaderi, M., & Nyrén, P., (2007). Pyrosequencing technology for short DNA sequencing and whole genome sequencing. *Seibutsu Butsuri, 47*(2), 129–132.

Goodwin, S., McPherson, J. D., & McCombie, W. R., (2016). Coming of age: Ten years of next-generation sequencing technologies. *Nature Reviews. Genetics, 17*(6), 333–351.

Gupta, P. K., Kumar, J., Mir, R. R., & Kumar, A., (2010). Marker-assisted selection as a component of conventional plant breeding. *Plant Breeding Reviews, 33*, 145.

Hammer, K., Arrowsmith, N., & Gladis, T., (2003). Agrobiodiversity with emphasis on plant genetic resources. *Naturwissenschaften, 90*(6), 241–250.

Huang, S., Li, R., Zhang, Z., Li, L. I., Gu, X., Fan, W., Lucas, W. J., et al., (2009). The genome of the cucumber, Cucumis sativus L. *Nature Genetics, 41*(12), 1275–1281.

Huang, X., Wei, X., Sang, T., Zhao, Q., Feng, Q., Zhao, Y., Li, C., et al., (2010). Genome-wide association studies of 14 agronomic traits in rice landraces. *Nature Genetics, 42*(11), 961–967.

Hung, T. H., So, T., Sreng, S., Thammavong, B., Boounithiphonh, C., Boshier, D. H., & MacKay, J. J., (2020). Reference transcriptomes & comparative analyses of six species in the threatened rosewood genus Dalbergia. *Scientific Reports, 10*(1), 1–14.

Hyten, D. L., Song, Q., Zhu, Y., Choi, I. Y., Nelson, R. L., Costa, J. M., Specht, J. E., et al., (2006). Impacts of genetic bottlenecks on soybean genome diversity. *Proceedings of the National Academy of Sciences of the United States of America, 103*(45), 16666–16671.

Jaillon, O., Aury, J. M., Noel, B., Policriti, A., Clepet, C., Casagrande, A., Choisne, N., Aubourg, S., Vitulo, N., Jubin, C., Vezzi, A., Legeai, F., Hugueney, P., Dasilva, C., Horner, D., Mica, E., Jublot, D., Poulain, J., Bruyère, C., & French-Italian Public Consortium for Grapevine Genome Characterization, (2007). The grapevine genome sequence suggests ancestral hexaploidization in major angiosperm phyla. *Nature, 449*(7161), 463–467.

Jiang, S., An, H., Xu, F., & Zhang, X., (2020). Chromosome-level genome assembly & annotation of the loquat (*Eriobotrya japonica*) genome. *GigaScience, 9*(3), giaa015.

Kilian, B., & Graner, A., (2012). NGS technologies for analyzing germplasm diversity in genebanks. *Briefings in Functional Genomics, 11*(1), 38–50.

Klein, H., Xiao, Y., Conklin, P. A., Govindarajulu, R., Kelly, J. A., Scanlon, M. J., Whipple, C. J., & Bartlett, M., (2018). Bulked-segregant analysis coupled to whole genome sequencing (BSA-Seq) for rapid gene cloning in maize. *G3, 8*(11), 3583–3592.

Kole, C., Muthamilarasan, M., Henry, R., Edwards, D., Sharma, R. Abberton, M., Batley, J., et al., (2015). Application of genomics-assisted breeding for generation of climate resilient crops: Progress and prospects. *Frontiers in Plant Science, 6*, 563.

Kuo, L. Y., & Li, F. W., (2019). A roadmap for fern genome sequencing. *American Fern Journal, 109*(3), 212–223.

Kuzmin, D. A., Feranchuk, S. I., Sharov, V. V., Cybin, A. N., Makolov, S. V., Putintseva, Y. A., Oreshkova, N. V., & Krutovsky, K. V., (2019). Stepwise large genome assembly approach: A case of *Siberian larch* (*Larix sibirica* Ledeb). *BMC Bioinformatics, 20*(1), 35–46.

Lam, H. Y. K., Clark, M. J., Chen, R., Chen, R., Natsoulis, G., O'Huallachain, M., Dewey, F. E., et al., (2011). Performance comparison of whole-genome sequencing platforms. *Nature Biotechnology, 30*(1), 78–82.

Li, Y., Sun, P., Lu, Z., Chen, J., Wang, Z., Du, X., Zheng, Z., et al., (2021). The *Corylus mandshurica* genome provides insights into the evolution of *Betulaceae* genomes & hazelnut breeding. *Horticulture Research, 8*(1), 1–13.

Liu, C., Feng, C., Peng, W., Hao, J., Wang, J., Pan, J., & He, Y., (2020). Chromosome-level draft genome of a diploid plum (*Prunus salicina*). *GigaScience, 9*(12), giaa130.

Liu, J., Osbourn, A., & Ma, P., (2015). MYB transcription factors as regulators of phenylpropanoid metabolism in plants. *Molecular Plant, 8*(5), 689–708.

Luo, H., Pandey, M. K., Khan, A. W., Wu, B., Guo, J., Ren, X., Zhou, X., et al., (2019). Next-generation sequencing identified genomic region and diagnostic markers for resistance to bacterial wilt on chromosome B02 in peanut (*Arachis hypogaea* L.). *Plant Biotechnology Journal, 17*(12), 2356–2369.

Ma, Q., Sun, T., Li, S., Wen, J., Zhu, L., Yin, T., Yan, K., et al., (2020). The *Acer truncatum* genome provides insights into nervonic acid biosynthesis. *The Plant Journal, 104*(3), 662–678.

Marco-Puche, G., Lois, S., Benítez, J., & Trivino, J. C., (2019). RNA-seq perspectives to improve clinical diagnosis. *Frontiers in Genetics, 10*, 1152.

Mardis, E. R., (2008). Next-generation DNA sequencing methods. *Annual Review of Genomics and Human Genetics, 9*, 387–402.

Margulies, M., Egholm, M., Altman, W. E., Attiya, S., Bader, J. S., Bemben, L. A., Berka, J., et al., (2005). Genome sequencing in microfabricated high-density picolitre reactors. *Nature, 437*(7057), 376–380.

Meyerson, M., Gabriel, S., & Getz, G., (2010). Advances in understanding cancer genomes through second-generation sequencing. *Nature Reviews Genetics, 11*(10), 685–696.

Ming, R., Hou, S., Feng, Y., Yu, Q., Dionne-Laporte, A., Saw, J. H., Senin, P., et al., (2008). The draft genome of the transgenic tropical fruit tree papaya (*Carica papaya* Linnaeus). *Nature, 452*(7190), 991–996.

Moose, S. P., & Mumm, R. H., (2008). Molecular plant breeding as the foundation for 21st century crop improvement. *Plant Physiology, 147*(3), 969–977.

Mosca, E., Cruz, F., Gómez-Garrido, J., Bianco, L., Rellstab, C., Brodbeck, S., Csilléry, K., et al., (2019). A reference genome sequence for the European silver fir (*Abies alba* Mill.): A community-generated genomic resource. *G3: Genes, Genomes, Genetics, 9*(7), 2039–2049.

Mozo, T., Dewar, K., Dunn, P., Ecker, J. R., Fischer, S., Kloska, S., Lehrach, H., et al., (1999). A complete BAC-based physical map of the *Arabidopsis thaliana* genome. *Nature Genetics, 22*(3), 271–275.

Murigneux, V., Rai, S. K., Furtado, A., Bruxner, T. J., Tian, W., Harliwong, I., Wei, H., et al., (2020). Comparison of long-read methods for sequencing & assembly of a plant genome. *GigaScience, 9*(12), g146.

Nock, C. J., Baten, A., Mauleon, R., Langdon, K. S., Topp, B., Hardner, C., Furtado, A., Henry, R. J., & King, G. J., (2020). Chromosome-scale assembly & annotation of the macadamia genome (Macadamia integrifolia HAES 741). *G3: Genes, Genomes, Genetics, 10*(10), 3497–3504.

Osuna, M. C., Rubio De, C. R., Landis, J. B., & Perfectti, F., (2021). Genomic resources for *Erysimum* spp. (*Brassicaceae*): Transcriptome & chloroplast genomes. *Frontiers in Ecology & Evolution, 9*, 206.

Pal, N. (2001) Method of sequencing DNA based on the detection of the release of pyrophosphate and enzymatic nucleotide degradation. U.S. Patent 6,258,568, issued July 10, 2001. https://patents.google.com/patent/EP0946752B1/en.

Paterson, A. H., Bowers, J. E., Bruggmann, R., Dubchak, I., Grimwood, J., Gundlach, H., Haberer, G., et al., (2009). The sorghum bicolor genome and the diversification of grasses. *Nature, 457*(7229), 551–556.

Paterson, R. R. M., & Lima, N., (2010). How will climate change affect mycotoxins in food? *Food Research International, 43*(7), 1902–1914.

Pollock, D. D., Eisen, J. A., Doggett, N. A., & Cummings, M. P., (2000). A case for evolutionary genomics and the comprehensive examination of sequence biodiversity. *Molecular Biology and Evolution, 17*(12), 1776–1788.

Rabbani, B., Tekin, M., & Mahdieh, N., (2014). The promise of whole-exome sequencing in medical genetics. *Journal of Human Genetics, 59*(1), 5–15.

Ramirez-Gonzalez, R. H., Segovia, V., Bird, N., Fenwick, P., Holdgate, S., Berry, S., Jack, P., et al., (2015). RNA-S eq bulked segregant analysis enables the identification of high-resolution genetic markers for breeding in hexaploid wheat. *Plant Biotechnology Journal, 13*(5), 613–624.

Rensing, S. A., Lang, D., Zimmer, A. D., Terry, A., Salamov, A., Shapiro, H., Nishiyama, T., et al., (2008). The physcomitrella genome reveals evolutionary insights into the conquest of land by plants. *Science, 319*(5859), 64–69.

Rimbert, H., Darrier, B., Navarro, J., Kitt, J., Choulet, F., Leveugle, M., Duarte, J., Rivière, N., Eversole, K., International Wheat Genome Sequencing Consortium, Le Gouis, J., on behalf The Breed Wheat Consortium, Davassi, A., Balfourier, F., Le Paslier, M. C., Berard, A., Brunel, D., Feuillet, C., Poncet, C., & Paux, E., (2018). High throughput SNP discovery and genotyping in hexaploid wheat. *PLOS One, 13*(1), e0186329.

Rothberg, J. M., Hinz, W., Rearick, T. M., Schultz, J., Mileski, W., Davey, M., Leamon, J. H., et al., (2011). An integrated semiconductor device enabling non-optical genome sequencing. *Nature, 475*(7356), 348–352.

Sanger, F., Nicklen, S., & Coulson, A. R., (1977). DNA sequencing with chain-terminating inhibitors. *Proceedings of the National Academy of Sciences of the United States of America, 74*(12), 5463–5467.

Schadt, E. E., Turner, S., & Kasarskis, A., (2010). A window into third-generation sequencing. *Human Molecular Genetics, 19*(R2), R227–R240.

Schmutz, J., Cannon, S. B., Schlueter, J., Ma, J., Mitros, T., Nelson, W., Hyten, D. L., et al., (2010). Genome sequence of the palaeopolyploid soybean. *Nature, 463*(7278), 178–183.

Schnable, P. S., Ware, D., Fulton, R. S., Stein, J. C., Wei, F., Pasternak, S., Liang, C., et al., (2009). The B73 maize genome: Complexity, diversity, and dynamics. *Science, 326*(5956), 1112–1115.

Shendure, J., & Ji, H., (2008). Next-generation DNA sequencing. *Nature Biotechnology, 26*(10), 1135–1145.

Strijk, J. S., Hinsinger, D. D., Roeder, M. M., Chatrou, L. W., Couvreur, T. L., Erkens, R. H., Sauquet, H., et al., (2021). Chromosome-level reference genome of the soursop (Annona muricata): A new resource for magnoliid research & tropical pomology. *Molecular Ecology Resources.*

Sturtevant, D., Lu, S., Zhou, Z. W., Shen, Y., Wang, S., Song, J. M., Zhong, J., et al., (2020). The genome of jojoba (*Simmondsia chinensis*): A taxonomically isolated species that directs wax ester accumulation in its seeds. *Science Advances, 6*(11), eaay3240.

Su, W., Jing, Y., Lin, S., Yue, Z., Yang, X., Xu, J., Wu, J., et al., (2021). Polyploidy underlies co-option & diversification of biosynthetic triterpene pathways in the apple tribe. *Proceedings of the National Academy of Sciences, 118*(20).

Thompson, J. F., & Milos, P. M., (2011). The properties and applications of single-molecule DNA sequencing. *Genome Biol., 12*, 217.

Trimarchi, M. P., Murphy, M., Frankhouser, D., Rodriguez, B. A. T., Curfman, J., Marcucci, G., Yan, P., & Bundschuh, R., (2012). Enrichment-based DNA methylation analysis using next-generation sequencing: Sample exclusion, estimating changes in global methylation, and the contribution of replicate lanes. In: Trimarchi, et al. (eds.), *BMC Genomics* (Vol. 13, No. 8, p. S6).

Turner, S. D., (2009). *Promoting Food Security in Lesotho: Issues and Options*. Priority Support Program.

Van, K., Rastogi, K., Kim, K. H., & Lee, S. H., (2013). Next-generation sequencing technology for crop improvement. *SABRAO Journal of Breeding and Genetics, 45*(1), 84–99.

Varshney, R. K., Nayak, S. N., May, G. D., & Jackson, S. A., (2009). Next-generation sequencing technologies and their implications for crop genetics and breeding. *Trends in Biotechnology, 27*(9), 522–530.

Varshney, R. K., Song, C., Saxena, R. K., Azam, S., Yu, S., Sharpe, A. G., Cannon, S., et al., (2013). Draft genome sequence of chickpea (*Cicer arietinum*) provides a resource for trait improvement. *Nature Biotechnology, 31*(3), 240–246.

Velasco, R., Zharkikh, A., Affourtit, J., Dhingra, A., Cestaro, A., Kalyanaraman, A., Fontana, P., et al., (2010). The genome of the domesticated apple (*Malus*× *domestica* Borkh.). *Nature Genetics, 42*(10), 833–839.

Voelker, J., Shepherd, M., & Mauleon, R., (2021). A high-quality draft genome for *Melaleuca alternifolia* (tea tree): A new platform for evolutionary genomics of myrtaceous terpene-rich species. *Gigabyte*, *2021*, 1–15.

Wan, T., Liu, Z. M., Li, L. F., Leitch, A. R., Leitch, I. J., Lohaus, R., Liu, Z. J., et al., (2018). A genome for gnetophytes & early evolution of seed plants. *Nature Plants, 4*(2), 82–89.

Wang, L., Ma, G., Wang, H., Cheng, C., Mu, S., Quan, W., Jiang, L., et al., (2019). A draft genome assembly of halophyte *Suaeda aralocaspica*, a plant that performs C4 photosynthesis within individual cells. *GigaScience, 8*(9), giz116.

Wang, W., Das, A., Kainer, D., Schalamun, M., Morales-Suarez, A., Schwessinger, B., & Lanfear, R., (2020). The draft nuclear genome assembly of *Eucalyptus pauciflora*: A pipeline for comparing de novo assemblies. *GigaScience, 9*(1), giz160.

Weber, J. L., & Myers, E. W., (1997). Human whole-genome shotgun sequencing. *Genome Research, 7*(5), 401–409.

Xu, W., Wu, D., Yang, T., Sun, C., Wang, Z., Han, B., Wu, S., et al., (2021). Genomic insights into the origin, domestication & genetic basis of agronomic traits of castor bean. *Genome Biology, 22*(1), 1–27.

Yan, J., Kandianis, C. B., Harjes, C. E., Bai, L., Kim, E. H., Yang, X., Skinner, D. J., et al., (2010). Rare genetic variation at Zea mays crtRB1 increases β-carotene in maize grain. *Nature Genetics, 42*(4), 322–327.

You, F. M., Huo, N., Deal, K. R., Gu, Y. Q., Luo, M. C., McGuire, P. E., Dvorak, J., & Anderson, O. D., (2011). Annotation-based genome-wide SNP discovery in the large and complex *Aegilops tauschii* genome using next-generation sequencing without a reference genome sequence. *BMC Genomics, 12*(1), 59.

Yu, J., Li, L., Wang, S., Dong, S., Chen, Z., Patel, N., Goffinet, B., et al., (2020). Draft genome of the aquatic moss *Fontinalis antipyretica* (Fontinalaceae, Bryophyta). *Gigabyte*, 1–9.

Zhao, T., Ma, W., Yang, Z., Liang, L., Chen, X., Wang, G., Ma, Q., & Wang, L., (2021). A chromosome-level reference genome of the hazelnut, *Corylus heterophylla* Fisch. *GigaScience, 10*(4), giab027.

Zhong, C., Sun, S., Li, Y., Duan, C., & Zhu, Z., (2018). Next-generation sequencing to identify candidate genes and develop diagnostic markers for a novel Phytophthora resistance gene, RpsHC18, in soybean. *TAG. Theoretical and Applied Genetics. Theoretische und Angewandte Genetik, 131*(3), 525–538.

Index

B

C

D

E

M

O

P